Current Topics in Behavioral Neurosciences

Volume 52

Current Topics in Behavioral Neurosciences provides critical and comprehensive discussions of the most significant areas of behavioral neuroscience research, written by leading international authorities. Each volume in the series represents the most informative and contemporary account of its subject available, making it an unrivalled reference source. Each volume will be made available in both print and electronic form.

With the development of new methodologies for brain imaging, genetic and genomic analyses, molecular engineering of mutant animals, novel routes for drug delivery, and sophisticated cross-species behavioral assessments, it is now possible to study behavior relevant to psychiatric and neurological diseases and disorders on the physiological level. The Behavioral Neurosciences series focuses on translational medicine and cutting-edge technologies. Preclinical and clinical trials for the development of new diagnostics and therapeutics as well as prevention efforts are covered whenever possible. Special attention is also drawn on epigenetical aspects, especially in psychiatric disorders.

CTBN series is indexed in PubMed and Scopus.

More information about this series at https://link.springer.com/bookseries/7854

Styliani Vlachou • Kevin Wickman
Editors

Behavioral Neurobiology of $GABA_B$ Receptor Function

Springer

Editors
Styliani Vlachou
School of Psychology
Faculty of Science and Health
Dublin City University
Dublin, Ireland

Kevin Wickman
Department of Pharmacology
University of Minnesota
Minneapolis, MN, USA

ISSN 1866-3370 ISSN 1866-3389 (electronic)
Current Topics in Behavioral Neurosciences
ISBN 978-3-030-91334-2 ISBN 978-3-030-91335-9 (eBook)
https://doi.org/10.1007/978-3-030-91335-9

This Springer imprint is published by the registered company Springer Nature Switzerland AG.
The registered company address is: Gewerbestrasse 11, 6330 Cham, Switzerland

Preface

It has been just over 40 years since Norman G. Bowery discovered and named the $GABA_B$ receptor. It has been 10 years since the last comprehensive book presentation focused on $GABA_B$ receptors.

The main goal of this book is to provide the field with a contemporary and comprehensive perspective on the $GABA_B$ receptor, its physiological relevance, and its therapeutic potential. The volume is organized into introductory and special interest sections presented by experts who study the $GABA_B$ receptor from structural, signaling, pharmacologic, physiological, pathophysiological, and therapeutic perspectives. More specifically, the first chapter of this volume focuses on a brief history and the significance of the $GABA_B$ receptor, followed by a detailed presentation of the structural basis of the $GABA_B$ receptor regulation and signaling in the second chapter. The third chapter offers extensive background and knowledge on the mechanisms and regulation of neuronal $GABA_B$ receptor-dependent signaling, while the chapter "$GABA_B$ Receptor Chemistry and Pharmacology: Agonists, Antagonists and Allosteric Modulators" provides an in-depth understanding of the $GABA_B$ receptor chemistry and pharmacology through the presentation of the development efforts of agonists, antagonists, and allosteric modulators over the last few decades. Starting from the fifth chapter, a particular focus on the role of $GABA_B$ receptors in neuropsychiatric disorders is presented, with the chapter "$GABA_B$ Receptors and Drug Addiction: Psychostimulants and Other Drugs of Abuse" covering the role of $GABA_B$ receptors on addiction, with a focus on psychostimulants. The sixth and seventh chapters further add to the above, by extensively presenting all the preclinical and clinical work, respectively, on the $GABA_B$ receptor as it relates to alcohol use disorder. The eighth chapter provides a detailed account of the role of $GABA_B$ receptors in pain, followed by an extensive presentation of the role of $GABA_B$ receptors on anxiety and mood disorders in the ninth chapter. Finally, the tenth chapter focuses on neurodegeneration and the role $GABA_B$ receptors can play in this setting, and the eleventh and final chapter of this book volume explores the role of $GABA_B$ receptors and cognitive processing in health and disease.

The editors would like to thank all excellent contributors for their participation in this volume. Their expertise and dedication to this research work over decades has formulated an excellent background for the potential use of $GABA_B$ receptors as therapeutic targets for a variety of psychiatric disorders and medical conditions.

Except for the essential contributions to this book volume, this work would not have been possible without the excellent support by Springer and all assistant editors and project managers, specifically Alamelu Damodharan, Gerit Rother, Coral Zhou, Susanne Dathe, and Arunkumar Kathiravan.

Most importantly, we would like to thank Mark A. Geyer, Charles A. Marsden, Bart A. Ellenbroek, and Thomas R.E. Barnes, editors of Current Topics in Behavioral Neurosciences (CTBN) without whom this book would not have taken shape.

I (Styliani Vlachou) would love to dedicate this CTBN book volume to my husband Bruno Di Micco whose incredible support and patience accompanies me in all my academic steps in the last few years and gives me more strength to continue. I cannot mention the extensive support in the form of cuddles by all of our paw babies – four dogs and a cat – in the house while working on this book volume, especially during the COVID-19 pandemic. Further, I would like to thank my co-editor, Kevin Wickman, for his extensive work on our book volume, but mainly for his expert input, especially at crucial time points, and his patience, kindness, and collaborative spirit throughout this process. Although we still have not met in person, it was an absolute pleasure working with you, Kevin. A special thanks and dedication goes out to Mark Geyer, who was never my mentor, but I always wish he was! This work would not have been possible without you.

I (Kevin Wickman) dedicate this volume to my many mentors in science, the exceptional colleagues I have had the great pleasure of collaborating with over the years, and all of the early-career scientists who have worked with me on pursuits that include the $GABA_B$ receptor. I sincerely appreciate the invitation from Styliani Vlachou to contribute as co-editor for this volume. I can sense your positive energy and passion for science across an ocean, and I recognize all of the hard work you put in to making this vision a reality.

The book is meant to appeal to a broad spectrum of biomedical and clinical scientists, to any scholars with an interest in the $GABA_B$ receptor.

The editors hope readers find this work to be thought-provoking, instructive, and informative.

Dublin, Ireland
Minneapolis, MN, USA
12 October 2021

Styliani Vlachou
Kevin Wickman

Contents

A Brief History and the Significance of the $GABA_B$ Receptor

Styliani Vlachou

Contents

Abstract γ-Aminobutyric acid (GABA) is the main inhibitory neurotransmitter in the brain. GABA type B ($GABA_B$) receptors ($GABA_BRs$) are the only metabotropic G protein-coupled receptors for GABA and can be found distributed not only in the central nervous system, but also in the periphery. This chapter introduces important, fundamental knowledge related to $GABA_BR$ function and the various potential therapeutic applications of the development of novel $GABA_BR$-active compounds, as documented through extensive studies presented in subsequent chapters of this Current Topic in Behavioral Neurosciences volume on the role of the neurobiology of $GABA_BR$ function. The compounds that have received increased attention in the last few years compared to $GABA_BR$ agonists and antagonists – the positive allosteric modulators – exhibit better pharmacological profiles and fewer side effects. As we continue to unveil the mystery of $GABA_BRs$ at the molecular and cellular levels, we further understand the significance of these receptors. Future

S. Vlachou (✉)
Neuropsychopharmacology Division, Behavioural Neuroscience Laboratory,
School of Psychology, Faculty of Science and Health, Dublin City University, Dublin, Ireland
e-mail: stella.vlachou@dcu.ie

Curr Topics Behav Neurosci (2022) 52: 1–18
https://doi.org/10.1007/7854_2021_264
Published Online: 1 October 2021

directions should aim for developing highly selective $GABA_BR$ compounds for treating neuropsychiatric disorders and their symptomatology.

Keywords Animal models · Cognition · Drug and alcohol addiction · $GABA_B$ receptors · Humans · Neurochemistry · Neurodegeneration · Neurodevelopmental disorders · Pharmacology · Pharmacotherapy · Treatment

1 Introduction

It is an absolute honor to introduce the content of this book volume on the neurobiology of the metabotropic γ-aminobutyric acid (GABA) type B ($GABA_B$) receptor ($GABA_BR$). I cannot help point out that the goal this chapter aims to achieve is challenging. How can anyone introduce and capture in a few pages the extensive investigations conducted by esteemed colleagues on the neurobiology of $GABA_B$ receptor function since its discovery and description in mammalian tissue by Norman G. Bowery over 40 years ago? This chapter will only present a summary of the overall work on the $GABA_BR$. The chapters to follow in this volume constitute excellent contributions of the state-of-the-art in the $GABA_BR$ research field, as they provide extensive details on the structure, signal transduction, pharmacology, and neurochemistry of $GABA_BRs$, but also on the role of $GABA_BRs$ in a variety of settings, such as drug and alcohol addiction, anxiety and mood disorders, neurodevelopment, neurodegeneration, and cognitive processes in health and disease.

GABA is the primary inhibitory neurotransmitter in the brain and one of the most studied neurotransmitters in the brain over the last 50 years (Smart and Stephenson 2019). The $GABA_BR$ is the only metabotropic G protein-coupled receptor (GPCR) of the three identified and characterized receptors of GABA, and it belongs to Class C of GPCRs. The history of the $GABA_BR$ begins before these receptors were identified and characterized, at the time when Norman G. Bowery was conducting many highly innovative studies on the extrasynaptic and asynaptic GABA systems in an effort to help solve the mystery of their presence in the ganglia and the peripheral nerves, and their significance [as extensively reviewed by (Brown 2018)]. This early work took place in the early 1970s and it proved preparatory to the discovery of the $GABA_BR$.

The pioneering work on the $GABA_BR$ can only but be traced back to Norman G. Bowery and his research team, when in 1979 and 1980, respectively, they described for the first time a receptor that was responsive to GABA. This receptor was activated by β-(4-chlorophenyl)-γ-aminobutyric acid (baclofen), the most extensively studied $GABA_BR$ agonist, but it was not responsive to bicuculline, a light-sensitive competitive antagonist of $GABA_A$ receptors, in mammalian tissue (Bowery et al. 1979, 1980). This response differentiated that newly identified receptor from the already identified and characterized $GABA_A$ receptor, which was responsive to bicuculline.

Many years went by without any further advances, until extensive work by Klemmens Kaupmann, Wolfgang Froestl, Bernard Bettler and colleagues, as well as the groups of Jones and White and their colleagues, described the molecular structure of the $GABA_BR$, characterized the gene sequence and identified the heteromerization of the $GABA_BR$ into subunits necessary to be assembled together in order to generate a functional $GABA_BR$ (Kaupmann et al. 1997, 1998; Jones et al. 1998; Kaupmann and Bettler 1998; White et al. 1998).

2 $GABA_B$ Receptor Structure, Function, and Distribution

All the experts participating in this volume have made major contributions to the understanding of the neurobiology of the $GABA_BR$. Details of the structure, characteristics, trafficking, and function of the $GABA_BR$ heterodimer can be found in many contributions in this volume, especially in Fritzius et al. (2020), Rose and Wickman (2020), Nieto et al. (2021).

In more recent years, there is a lot more understanding of the $GABA_BR$ at the molecular and functional level (Bettler et al. 2004; Emson 2007; Schwenk et al. 2016; Shaye et al. 2021). $GABA_BRs$ play a crucial role in mediating slow and long-lasting synaptic inhibition through indirect neuronal K^+ and Ca^{2+} channel gating and through effects on other second messenger targets like cAMP. $GABA_BRs$ consist of principal and auxiliary subunits that influence receptor properties in different ways. The principal subunits affect the surface expression and the axonal versus dendritic distribution of these receptors, whereas the auxiliary subunits determine the potency of agonists on the receptor and kinetics of the receptor response to them (Gassmann and Bettler 2012). The two main subunits of the $GABA_BRs$ are $GABA_BR1$ and $GABA_BR2$. In order for the $GABA_BR$ to be active and functional, these subunits need to interact to form a stable heterodimer. Importantly, orthosteric agonists and antagonists bind to $GABA_BR1$, while PAMs bind to the $GABA_BR2$ subunit. Most recent findings show that $GABA_BRs$ are receptor complexes consisting of primary or other subunits, but also linked with and affecting numerous factors and proteins. The homeostatic interaction between all these components plays an important role in $GABA_BR$ function.

Interestingly, $GABA_BRs$ can be found on both presynaptic and postsynaptic membranes. When presynaptic receptors are activated, they inhibit either the release of GABA (in the case of autoreceptors on GABA neuronal terminals) or the release of other neurotransmitters and peptides (in the case of heteroreceptors on other-than-GABA-neurotransmitter neuronal terminals). When postsynaptic receptors are activated, they in turn activate K^+ channels and induce slow inhibitory postsynaptic potentials.

$GABA_BRs$ are widely expressed and distributed in the mammalian central nervous system (CNS) [e.g., (Bowery et al. 1984; Boyes and Bolam 2007; Metz et al.

2011)], with the highest receptor densities occurring in the cerebral cortex, the cerebellum, the interpeduncular nucleus, and the dorsal horn of the spinal horn (Chu et al. 1990). Importantly, $GABA_BRs$ can also be found in other brain areas, such as the hippocampus, the thalamus, the amygdala and the basal ganglia, and outside the CNS (please see below the section "Other conditions outside the CNS" in this chapter). Thus, they are very widely expressed in mammalian tissues. As expected, their distribution is not equal in the different brain areas. However, lower distribution does not mean lower significance of these receptors in that particular area (Bowery 2010).

3 $GABA_B$ Receptor Ligands: Agonists, Antagonists, and Allosteric Modulators

The synthesis of the first exogenous GABA analogue (β-(4-chlorophenyl)-GABA (baclofen) in 1962, followed by the identification of the three GABA receptor types, paved the way for developing numerous $GABA_BR$ agonists, partial agonists, inverse agonists, and antagonists (Malcangio and Bowery 1995). Baclofen has been the most extensively studied compound in both preclinical and clinical studies, and the only $GABA_BR$-selective drug clinically approved for muscle relaxation in humans.

Positive allosteric modulators (PAMs) of $GABA_BR$ were described 20 years ago by Stephan Urwyler and colleagues (Urwyler et al. 2001; Urwyler 2011). Since then, numerous PAMs have been reported in the literature by different research groups and/or pharmaceutical companies. A major difference between $GABA_BR$ agonists and allosteric modulators is that the latter bind to a region of the $GABA_BR$ different from and outside of the ligand-binding (i.e., orthosteric) site. Through this action, they either increase (i.e., in the case of PAMs) or decrease (i.e., in the case of negative allosteric modulators; NAMs) the effects of GABA, without possessing intrinsic agonistic activity, but only by modulating the endogenous GABA release effects on the $GABA_BRs$ (Ong and Kerr 2005; Kniazeff 2020). Thus, they produce fewer and/or less severe side effects compared to the $GABA_BR$ agonists or antagonists, and they activate the $GABA_BR$ without inducing desensitization (Gjoni and Urwyler 2008, 2009; Froestl 2010). Although numerous PAMs of the $GABA_BR$ have been developed over the years, it is only recently that the first NAM was described (Porcu et al. 2021), leading the way toward the development of more NAMs for therapeutic purposes.

It is critical to note that through extensive efforts spanning approximately four decades, there has been only one $GABA_BR$ compound – baclofen – approved for clinical use as a muscle relaxant. The extensive and informative presentation of the history of, as well as the development and characteristics of $GABA_BR$ agonists, partial agonists, inverse agonists, antagonists and allosteric modulators can be found in Nieto et al. (2021) of this volume. This chapter, with other contributions in this volume, will help shed light on the future directions necessary for developing $GABA_B$R-active compounds as potential therapeutic targets.

4 $GABA_B$ Receptors as Potential Therapeutic Targets

4.1 Drug and Alcohol Use Disorders

$GABA_B$Rs are considered mediators or potential therapeutic targets for many neuropsychiatric disorders, including drug addiction. The story of understanding of the role of $GABA_B$Rs in drug and alcohol use disorders has two main components, one that focuses on the effects of the addictive drugs on $GABA_B$R signaling and the other that focuses on the effects of $GABA_B$R compounds on those of drugs of abuse. Extensive efforts in recent years have showcased the important role of these receptors and their ligands on the effects of drugs of abuse, such as psychostimulants, opioids and alcohol, in both preclinical and clinical studies. Alternatively, drugs of abuse can evoke plasticity of $GABA_B$R-dependent signaling in the brain, with changes marked at various molecular or cellular levels, such as RNA expression, receptor trafficking, G protein coupling, and changes in the actions of effector proteins, and affecting various neurotransmitter systems.

Although dopaminergic agents have been the primary therapeutic targets for drug dependence treatment, in recent years, GABA and glutamate, with their metabotropic $GABA_B$Rs and glutamatergic receptors, have received increased attention as potential "alternative" approaches to dopamine-targeting compounds. More specifically, considering the enhanced dopamine (DA) signaling in the mesocorticolimbic circuitry through molecular and cellular mechanisms by drugs of abuse (McCall et al. 2019), earlier efforts were mainly focused on the blocking and/or inhibitory effects of DA-active compounds on the rewarding and reinforcing effects of drugs of abuse in rodents. Furthermore, nowadays, the focus is not only on a plethora of overlapping but distinct neurotransmitter pathways, including GABA and glutamate systems, but also on further exploring and highlighting key $GABA_B$R-related sex differences, which can potentially affect susceptibility to drugs of abuse (DePoy et al. 2016; DeBaker et al. 2021). The presence of $GABA_B$Rs in different areas of the brain reward circuitry and how drugs of abuse induce alterations in $GABA_B$R signaling is well-presented in Li and Slesinger (2021) of this volume.

Numerous early studies at the preclinical level have shown that baclofen and other $GABA_B$R agonists or mediators, such as GABA metabolism or reuptake inhibitors, reduced the rewarding effects of cocaine, nicotine, morphine, and alcohol in animal models mainly using rats and mice at different phases of drug dependence [e.g., (Brebner et al. 1999; Xi and Stein 1999; Corrigall et al. 2000; Fattore et al. 2002; Paterson et al. 2004, 2005); for thorough reviews, please see Vlachou and Markou (2010); Phillips and Reed (2014); Jacobson et al. (2018), as well as Holtyn and Weerts (2020), Logge et al. (2020), Li and Slesinger (2021) of this volume].

While these effects appeared promising, the side effects of the $GABA_BR$ agonists also presented in some of these studies (e.g., muscle relaxation, sedation, cognitive deficits, fatigue, tolerance development, seizures) indicated that the development of more selective agents with less side effects was necessary for many of these efforts to translate to clinical testing and possible approval for therapeutic purposes.

In more recent years, studies have focused on the effects of PAMs and NAMs of $GABA_BRs$ on drugs of abuse [for thorough reviews, please see Agabio and Colombo (2015); Filip et al. (2015); Maccioni and Colombo (2019)], as these compounds have yielded more promising findings than those of $GABA_BR$ agonists, due to their mechanism of action and thus, fewer side effects. Extensive studies have been conducted on the effects of $GABA_BR$ compounds on the behavioral, cognitive, locomotor, and rewarding effects of psychostimulants in particular; however, studies on other drugs of abuse are still in their infancy.

Importantly, a limited number of studies with $GABA_BR$ knock-out (KO) mice lacking either the $GABA_BR1$ or the $GABA_BR2$ subunit (Vigot et al. 2006) have been conducted in relation to drug addiction. The vast majority of them have focused on the role the two subunits in the different phases of nicotine dependence (e.g., acute administration or withdrawal phase) and some explore their relation to reward and stress [e.g., (Varani et al. 2012, 2014, 2015, 2018; Jacobson et al. 2016)].

At clinical level, after so many years of research, still in 2021 only baclofen, a $GABA_BR$ agonist, is approved for use in humans as a muscle relaxant. Importantly, it was also approved in France for treating alcohol use disorder in 2018, although with controversy (Braillon et al. 2020). Numerous other studies have examined the effects of baclofen on drug dependence in humans over many years and they have not led to clinical approval. With regard to PAMs and their potential use in humans, a study on only one compound, the PAM ADX71441 (developed by Addex Therapeutics), was recently funded by the NIH for clinical studies on cocaine-dependent individuals (Kalinichev et al. 2017). We can only but hope that in the next few years more compounds will proceed to clinical trials and will be approved as safe therapies for substance use disorders.

4.2 Pain and Analgesia

There is extensive evidence that $GABA_BRs$ are involved in the processing of pain signals and the induction of analgesia for chronic pain conditions (Pan et al. 2008; Malcangio 2018). The contribution of $GABA_BRs$ to the generation and modulation of pain signals, their involvement in chronic pain states, and their potential use as targets for developing novel analgesic compounds are discussed in detail by Dietmar Benke in Benke (2020) of this volume.

Briefly, numerous studies at preclinical level have shown that $GABA_BRs$ are present in nociceptors (Enna and McCarson 2006; Hanack et al. 2015), which are sensory neurons that respond to potentially harmful heat, mechanical, or chemical stimuli. Nociceptors can transmit slow or fast signals, depending on whether they are unmyelinated or myelinated, respectively. Nociceptor synapses can be found in the dorsal horn of the spinal cord where $GABA_BRs$ have also been identified [e.g., (Zhou et al. 2017)]. More specifically, $GABA_BRs$ or their subunits have been identified in both central and peripheral axon endings of nociceptors, as well as in dorsal horn interneurons and projection neurons, in supraspinal areas and in descending pain control pathways. Considering the extensive presence of $GABA_BRs$ in all the loci and pathways mentioned above, researchers have investigated the potential therapeutic effects of $GABA_BR$ compounds in different kinds of chronic pathological pain, such as inflammatory or neuropathic pain [e.g., (Barr et al. 2013; Meuwissen et al. 2020)], and the overall results have suggested that the enhancement of $GABA_BR$ activity can relieve or alleviate pain, although the exact mechanisms with which the nociceptive processes are activated are complex.

4.3 *Anxiety and Mood Disorders*

Many studies over the last few years have shown that alteration of the function of the $GABA_BR$ can contribute to stress- and anxiety-related conditions and in affective disorders. At the preclinical level, numerous studies have shown that $GABA_BR$ agonists reduce anxiety-like and depressive-like symptomatology in animal models, although contradictory findings also exist (Mombereau et al. 2004; Cryan and Kaupmann 2005; Cryan and Slattery 2010; Jacobson et al. 2018).

As noted above, the side-effect profile of $GABA_BR$ agonists such as baclofen (Agabio et al. 2013) is pronounced, reducing hope for the use of these types of compounds as therapeutics for a number of conditions, including mood disorders. Clinical studies of $GABA_BR$ agonists and their effects in the setting of anxiety disorders are more sparse. However, overall, preclinical and clinical findings, presented in detail in Felice et al. (2020) of this volume suggest a possible therapeutic role of $GABA_BR$ compounds in anxiety and depression, especially when the focus switches to novel PAMs and NAMs, and proteins that affect $GABA_BR$ activity, which have been tested more extensively in recent years.

Importantly, research advancements in diagnostic and therapeutic procedures such as the use of transcranial magnetic stimulation in major and treatment-resistant depression patients (Fatih et al. 2021; Kinjo et al. 2021) have recently shown deficits in $GABA_BR$ neurophysiology (i.e., $GABA_BR$-mediated cortical inhibition) (Premoli et al. 2014a, b, 2017; Veronezi et al. 2016; Lissemore et al. 2018) that is affected by anti-depressant medication. Further advances in the use of innovative techniques are expected to help elucidate the role of $GABA_BR$ in anxiety and affective disorders.

4.4 Neurodegeneration

A considerable role of the $GABA_BR$ has been identified in neurodegeneration in relation to both its neuropathology and symptomatology, and the potential use of $GABA_BR$ compounds as therapeutics. The GABAergic system has been linked with numerous neurodevelopmental and neurodegenerative conditions (Kleppner and Tobin 2001; Smart and Stephenson 2019; Murrell et al. 2020), although the initial focus was on the $GABA_A$ receptors, and their agonists and antagonists [for reviews see, (Solomon et al. 2019; Bhagat et al. 2021; Castellano et al. 2021)].

Several studies have implicated $GABA_BRs$ and their subunits in the pathophysiology of Alzheimer's disease, with changes in the density of expression of the $GABA_BRs$ identified [e.g., (Iwakiri et al. 2005; Pilipenko et al. 2018, 2019; Dinamarca et al. 2019; Martín-Belmonte et al. 2020)]. Further, although the main neuropathological features are diverse among neurodegenerative disorders, $GABA_BRs$ seem to be strongly implicated in the pathophysiology of Parkinson's disease [e.g., (de Groote et al. 1999; Enna et al. 2006; Yang et al. 2021); for a recent review, see Roberts et al. (2021)] and Huntington's disease [e.g., (Allen et al. 2009; Rekik et al. 2011; Kim and Seo 2014; Holley et al. 2019; Barry et al. 2020)]. Extensive work has focused on the role of $GABA_BRs$ on epilepsy, specifically on temporal lobe epilepsy (TLE) (Princivalle et al. 2002, 2003; Chandler et al. 2003; Straessle et al. 2003; Dugladze et al. 2013; Rocha et al. 2015; Sheilabi et al. 2018)] and absence epilepsy [e.g., (Sperk et al. 2004; Han et al. 2013; Jafarian et al. 2021; Pagès et al. 2021)]. $GABA_BR$ antagonists may be of most therapeutic benefit for the treatment of seizures in epilepsy. However, the difficulty lies not only in the selectivity of these agents, which would minimize side effects, but mainly in the route of administration (central or systemic) and the overall pharmacokinetics of $GABA_BR$ compounds.

An extensive presentation of the most recent developments on the role of GABA and $GABA_BRs$ in neurodegeneration, with a particular focus on Alzheimer's and Parkinson's disease and epilepsy is presented in Princivalle (2021) of this volume.

4.5 Cognitive Processes in Neurodevelopment, Health and Disease

The role of $GABA_BRs$ in cognitive processes has been extensively studied from different perspectives. The vast majority of the existing literature focuses on animal models of learning and memory to assess the role of $GABA_BRs$ in these processes. Examples of the animal models used are active and passive avoidance paradigms, the Morris water maze and radial arm maze, many working memory tasks (e.g., T and Y mazes), and the novel-object recognition, and (dis) location tasks (Bowery 2006; Serrats et al. 2017). However, a lot of work has also been conducted on how cognition functions in neurodevelopment and neurodegeneration, and what may be the potential beneficial or therapeutic role that the $GABA_BR$ can play in cognitive development, decline, or enhancement (Heaney and Kinney 2016).

Examples of neurodevelopmental conditions more extensively studied under the scope of the potential therapeutic role of the $GABA_BR$ include autism spectrum disorders, fragile X syndrome, and Down syndrome; examples of neurodegenerative conditions are Alzheimer's disease, epilepsy, and autoimmune anti-$GABA_B$ encephalitis. Many studies have specifically examined the role of $GABA_BRs$ in cognition, learning, and memory processes in these conditions. Although there are controversial findings in the literature, altogether, results from studies in this area, presented in detail in the last chapter of this volume on the role of the $GABA_B$ receptors and cognitive processing in health and disease (Vlachou 2021), indicate a potential therapeutic role of $GABA_BR$ compounds for treating cognitive dysfunction and learning/memory impairments for some of these conditions, especially in neurodegenerative disorders. However, the ongoing effort to develop more selective $GABA_BR$ compounds with fewer side effects will help further elucidate how alteration of $GABA_BR$ function can help battle cognitive processes symptomatology in health, neurodevelopment, and neurodegeneration.

4.6 Other Conditions Outside the CNS

Considering that the $GABA_BR$ is present not only in the CNS, but also in the periphery and even outside the nervous system, the potential therapeutic use of $GABA_BR$ compounds for symptoms and conditions outside the CNS has been examined over a number of years. Although studies in this direction are not the focus of this volume, the $GABA_BR$ has been investigated as a potential target for the treatment of gastroesophageal reflux disease, chronic abdominal pain, and overactive bladder (Alstermark et al. 2008; Lehmann 2009; Hyland and Cryan 2010; Brozmanová et al. 2013; Scarpellini et al. 2015). The focus on these conditions was based on the findings that the $GABA_BR$ is present in the gastrointestinal tract and plays a role in the regulation of a number of processes including intestinal motility, gastric acid secretion and gastric emptying, and esophageal sphincter relaxation. AZD3355 (Lesogaberan) was a selective peripheral $GABA_BR$ agonist, which was tested for the treatment of gastroesophageal reflux disease in earlier years in animal models or in human patients (Boeckxstaens et al. 2011; Canning et al. 2012; Shaheen et al. 2013), but more recent studies with the PAMs ADX71441 and ADX71943 have shown analgesic effects in a rat model of bladder pain and in the acetic acid-induced writhing test in mice and formalin tests in mice and rats, respectively (Kalinichev et al. 2014; Kannampalli et al. 2017). The efforts in this direction are focused on developing compounds that are peripherally selective and thus show less side effects than centrally active $GABA_BR$ agonists. Thus, peripherally selective allosteric modulators may be the drugs of choice in this direction.

Other efforts try to shed light on the role of $GABA_BRs$ in the treatment of different types of gastrointestinal system-related cancers, such as colorectal or pancreatic cancer [e.g., (Al-Wadei et al. 2011; Banerjee et al. 2014; Schuller 2017; An et al. 2021; Wang et al. 2021)]. Importantly, in the case of colorectal cancer, it was recently shown that $GABA_BR1$ expression was significantly lower in tumor tissues than those in non-tumor normal tissues, and that those colorectal cancer patients with high $GABA_BR1$ expression lived longer (Wang et al. 2021). In the case of pancreatic cancer, preclinical findings suggest that GABA itself, but not baclofen, may be promising for the treatment of pancreatic cancer in tobacco smokers, as low doses of nicotine within the range of nicotine replacement therapy induce gemcitabine resistance in pancreatic cancer and GABA significantly reverses this effect (Banerjee et al. 2014).

Most interesting findings derive from the recent profiling of $GABA_A$ and $GABA_BR$ expression in the myometrium of the human uterus (Söderhielm et al. 2018) and the significant role of the $GABA_{B1}R$ subunit in embryo implantation and uterine decidualization in preclinical studies in mice (Chen et al. 2021). Studies in the near future will shed light on the role of $GABA_BR$ expression in the female reproductive organs, pregnancy, and embryonic development.

5 Conclusion

Altogether, the most important conclusion deriving from all research conducted over the last four decades is that the clinical promise of the $GABA_BR$ agonists, antagonists, PAMs, and NAMs is extensive (Enna and Bowery 2004). Most of the efforts in that direction focus on PAMs and NAMs in the last few years, due to their improved and more selective pharmacological profile exhibiting more direct effects and less side effects in the various neuropsychiatric and other disorders. To fully understand the significant role of the $GABA_BR$ research findings, one must delve into the following chapters of this volume, which I hope you will find informative, enjoyable and thought provoking.

Ultimately, I want to point out that the expanded "$GABA_BR$ family" consists of esteemed colleagues from around the world, all of whom have significantly contributed to the evolving history of the $GABA_BR$, and many of whom have kindly participated and offered excellent contributions to this volume. In addition to the recent loss of Norman G. Bowery (1944–2016), the "$GABA_BR$ family" has lost two more esteemed members in the last few years – including Wolfgang Froestl (1946–2015), who was a leading scientist for the discovery of most of the early $GABA_BR$ agonists and antagonists, and Athina Markou (1961–2016), who offered major contributions to the role of $GABA_BR$ on drug dependence, specifically on nicotine dependence, reward mechanisms, and depressive-like behavior. The author of the first and last chapter and editor of this volume worked closely with both. I want to dedicate this introductory chapter of the Current Topics in Behavioral Neurosciences volume on the neurobiology of $GABA_BRs$ to Athina Markou, my postdoctoral mentor, who introduced me to the $GABA_BR$ research and "family" and made me love it as much as she did, and to Wolfgang Froestl, who was not only an excellent chemist and neuroscientist, but also one of the most gentle people I have met in our field.

References

Agabio R, Colombo G (2015) GABAB receptor as therapeutic target for drug addiction: from baclofen to positive allosteric modulators. Psychiatr Pol 49(2):215–223. https://doi.org/10.12740/PP/33911
Agabio R, Preti A, Gessa GL (2013) Efficacy and tolerability of baclofen in substance use disorders: a systematic review. Eur Addict Res 19(6):325–345. https://doi.org/10.1159/000347055
Allen KL et al (2009) Cannabinoid (CB1), GABAA and GABAB receptor subunit changes in the globus pallidus in Huntington's disease. J Chem Neuroanat 37(4):266–281. https://doi.org/10.1016/J.JCHEMNEU.2009.02.001
Alstermark C et al (2008) Synthesis and pharmacological evaluation of novel γ-aminobutyric acid type B (GABAB) receptor agonists as gastroesophageal reflux inhibitors. J Med Chem 51 (14):4315–4320
Al-Wadei HAN, Ullah MF, Al-Wadei M (2011) GABA (γ-aminobutyric acid), a non-protein amino acid counters the β-adrenergic cascade-activated oncogenic signaling in pancreatic cancer: a review of experimental evidence. Mol Nutr Food Res 55(12):1745–1758. https://doi.org/10.1002/MNFR.201100229
An JJ, Seok H, Ha EM (2021) GABA-producing Lactobacillus plantarum inhibits metastatic properties and induces apoptosis of 5-FU-resistant colorectal cancer cells via GABAB receptor signaling. J Microbiol 59(2):202–216. https://doi.org/10.1007/S12275-021-0562-5
Banerjee J et al (2014) Differential modulation of nicotine-induced gemcitabine resistance by GABA receptor agonists in pancreatic cancer cell xenografts and in vitro. BMC Cancer 14(1). https://doi.org/10.1186/1471-2407-14-725
Barr M et al (2013) Measuring GABAergic inhibitory activity with TMS-EEG and its potential clinical application for chronic pain. J Neuroimmune Pharmacol 8(3):535–546. https://doi.org/10.1007/S11481-012-9383-Y
Barry J et al (2020) Mechanisms underlying the enhancement of γ-aminobutyric acid responses in the external globus pallidus of R6/2 Huntington's disease model mice. J Neurosci Res 98 (11):2349–2356. https://doi.org/10.1002/JNR.24710
Benke D (2020) GABAB receptors and pain. Curr Top Behav Neurosci. https://doi.org/10.1007/7854_2020_130
Bettler B et al (2004) Molecular structure and physiological functions of GABA(B) receptors. Physiol Rev 84(3):835–867. https://doi.org/10.1152/physrev.00036.2003
Bhagat K et al (2021) Rational approaches for the design of various GABA modulators and their clinical progression. Mol Divers 25(1):551–601. https://doi.org/10.1007/S11030-020-10068-4
Boeckxstaens GE et al (2011) Translational gastrointestinal pharmacology in the 21st century: "the lesogaberan story". Curr Opin Pharmacol 11(6):630–633. https://doi.org/10.1016/J.COPH.2011.10.011
Bowery NG (2006) GABAB receptor: a site of therapeutic benefit. Curr Opin Pharmacol 6(1 Spec Iss):37–43. https://doi.org/10.1016/j.coph.2005.10.002
Bowery NG (2010) Historical perspective and emergence of the GABAB receptor. Adv Pharmacol 58:1–18. https://doi.org/10.1016/s1054-3589(10)58001-3
Bowery N et al (1979) Baclofen: a selective agonist for a novel type of GABA receptor proceedings. Br J Pharmacol 67:444P–445P
Bowery N et al (1980) (−)Baclofen decreases neurotransmitter release in the mammalian CNS by an action at a novel GABA receptor. Nature 283:92–94

Bowery N et al (1984) GABA receptor multiplicity. Visualization of different receptor types in the mammalian CNS. Neuropharmacology 23(2B):219–231. https://doi.org/10.1016/0028-3908(84)90063-7

Boyes J, Bolam J (2007) Localization of GABA receptors in the basal ganglia. Prog Brain Res 160:229–243. https://doi.org/10.1016/S0079-6123(06)60013-7

Braillon A et al (2020) Baclofen and alcohol use disorders: breakthrough or great white elephant? Alcohol Alcohol 55(1):49–50. https://doi.org/10.1093/ALCALC/AGZ083

Brebner K et al (1999) The GABA(B) agonist CGP 44532 decreases cocaine self-administration in rats: demonstration using a progressive ratio and a discrete trials procedure. Neuropharmacology 38(11):1797–1804. https://doi.org/10.1016/S0028-3908(99)00094-5

Brown DA (2018) Norman Bowery's discoveries about extrasynaptic and asynaptic GABA systems and their significance. Neuropharmacology 136(Pt A):3–9. https://doi.org/10.1016/j.neuropharm.2017.11.006

Brozmanová M et al (2013) Evaluation of the effect of GABAB agonists on the vagal nodose C-fibers in the esophagus. Physiol Res 62(3):285–295. https://doi.org/10.33549/PHYSIOLRES.932429

Canning BJ, Mori N, Lehmann A (2012) Antitussive effects of the peripherally restricted GABAB receptor agonist lesogaberan in guinea pigs: comparison to baclofen and other GABAB receptor-selective agonists. Cough 8(1). https://doi.org/10.1186/1745-9974-8-7

Castellano D, Shepard R, Lu W (2021) Looking for novelty in an "old" receptor: recent advances toward our understanding of GABA A Rs and their implications in receptor pharmacology. Front Neurosci 14. https://doi.org/10.3389/FNINS.2020.616298

Chandler KE et al (2003) Plasticity of GABAB receptor-mediated heterosynaptic interactions at mossy fibers after status epilepticus. J Neurosci 23(36):11382–11391. https://doi.org/10.1523/JNEUROSCI.23-36-11382.2003

Chen W et al (2021) Unique and independent role of the GABAB1 subunit in embryo implantation and uterine decidualization in mice. Genes Dis 8(1):79–86. https://doi.org/10.1016/J.GENDIS.2019.06.005

Chu D et al (1990) Distribution and kinetics of GABAB binding sites in rat central nervous system: a quantitative autoradiographic study. Neuroscience 34(2):341–357. https://doi.org/10.1016/0306-4522(90)90144-S

Corrigall WA et al (2000) Response of nicotine self-administration in the rat to manipulations of mu-opioid and gamma-aminobutyric acid receptors in the ventral tegmental area. Psychopharmacology 149(2):107–114. http://www.ncbi.nlm.nih.gov/pubmed/10805604

Cryan JF, Kaupmann K (2005) Don't worry "B" happy!: a role for GABA(B) receptors in anxiety and depression. Trends Pharmacol Sci 26(1):36–43. https://doi.org/10.1016/j.tips.2004.11.004

Cryan JF, Slattery DA (2010) GABAB receptors and depression. Current status. Adv Pharmacol (S Diego, CA) 58(C):427–451. https://doi.org/10.1016/S1054-3589(10)58016-5

de Groote C et al (1999) Functional characterization and expression of thalamic GABA (B) receptors in a rodent model of Parkinson's disease. Neuropharmacology 38 (11):1683–1689. https://doi.org/10.1016/S0028-3908(99)00125-2

DeBaker M et al (2021) Differential impact of inhibitory G-protein signaling pathways in ventral tegmental area dopamine neurons on behavioral sensitivity to cocaine and morphine. eNeuro 8 (2). https://doi.org/10.1523/ENEURO.0081-21.2021

DePoy LM, Allen AG, Gourley SL (2016) Adolescent cocaine self-administration induces habit behavior in adulthood: sex differences and structural consequences. Transl Psychiatry 6(8): e875. https://doi.org/10.1038/tp.2016.150

Dinamarca MC et al (2019) Complex formation of APP with GABA B receptors links axonal trafficking to amyloidogenic processing. Nat Commun 10(1). https://doi.org/10.1038/S41467-019-09164-3

Dugladze T et al (2013) GABA(B) autoreceptor-mediated cell type-specific reduction of inhibition in epileptic mice. Proc Natl Acad Sci U S A 110(37):15073–15078. https://doi.org/10.1073/PNAS.1313505110

Emson PC (2007) GABA(B) receptors: structure and function. Prog Brain Res 160:43–57. https://doi.org/10.1016/S0079-6123(06)60004-6
Enna SJ, Bowery NG (2004) GABA B receptor alterations as indicators of physiological and pharmacological function. Biochem Pharmacol 68(8):1541–1548. https://doi.org/10.1016/j.bcp.2004.06.037
Enna S, McCarson K (2006) The role of GABA in the mediation and perception of pain. Adv Pharmacol 54:1–27. https://doi.org/10.1016/S1054-3589(06)54001-3
Enna S, Reisman S, Stanford J (2006) CGP 56999A, a GABA(B) receptor antagonist, enhances expression of brain-derived neurotrophic factor and attenuates dopamine depletion in the rat corpus striatum following a 6-hydroxydopamine lesion of the nigrostriatal pathway. Neurosci Lett 406(1–2):102–106. https://doi.org/10.1016/J.NEULET.2006.07.004
Fatih P et al (2021) A systematic review of long-interval intracortical inhibition as a biomarker in neuropsychiatric disorders. Front Psych 12. https://doi.org/10.3389/FPSYT.2021.678088
Fattore L et al (2002) Baclofen antagonizes intravenous self-administration of nicotine in mice and rats. Alcohol Alcohol 37(5):495–498. https://doi.org/10.1093/alcalc/37.5.495
Felice D, Cryan J, O'Leary O (2020) GABA B receptors: anxiety and mood disorders. Curr Top Behav Neurosci. https://doi.org/10.1007/7854_2020_171
Filip M et al (2015) GABAB receptors as a therapeutic strategy in substance use disorders: focus on positive allosteric modulators. Neuropharmacology 88:36–47. https://doi.org/10.1016/J.NEUROPHARM.2014.06.016
Fritzius T et al (2020) Structural basis of GABAB receptor regulation and signaling. Curr Top Behav Neurosci. https://doi.org/10.1007/7854_2020_147
Froestl W (2010) Chemistry and pharmacology of GABAB receptor ligands. Adv Pharmacol (S Diego, CA). https://doi.org/10.1016/S1054-3589(10)58002-5
Gassmann M, Bettler B (2012) Regulation of neuronal GABA(B) receptor functions by subunit composition. Nat Rev Neurosci 13(6):380–394. https://doi.org/10.1038/NRN3249
Gjoni T, Urwyler S (2008) Receptor activation involving positive allosteric modulation, unlike full agonism, does not result in GABAB receptor desensitization. Neuropharmacology 55(8):1293–1299. https://doi.org/10.1016/J.NEUROPHARM.2008.08.008
Gjoni T, Urwyler S (2009) Changes in the properties of allosteric and orthosteric GABAB receptor ligands after a continuous, desensitizing agonist pretreatment. Eur J Pharmacol 603(1–3):37–41. https://doi.org/10.1016/J.EJPHAR.2008.12.014
Han HA, Cortez MA, Snead OC (2013) GABAB receptor and absence epilepsy. In: Jasper's basic mechanisms of the epilepsies. Oxford University Press, pp 242–256. https://doi.org/10.1093/med/9780199746545.003.0019
Hanack C et al (2015) GABA blocks pathological but not acute TRPV1 pain signals. Cell 160(4):759–770. https://doi.org/10.1016/J.CELL.2015.01.022
Heaney CF, Kinney JW (2016) Role of GABAB receptors in learning and memory and neurological disorders. Neurosci Biobehav Rev 63:1–28. https://doi.org/10.1016/j.neubiorev.2016.01.007
Holley SM et al (2019) Major contribution of somatostatin-expressing interneurons and cannabinoid receptors to increased GABA synaptic activity in the striatum of Huntington's disease mice. Front Synaptic Neurosci 11(MAY):14. https://doi.org/10.3389/fnsyn.2019.00014
Holtyn AF, Weerts EM (2020) GABAB receptors and alcohol use disorders: preclinical studies. Curr Top Behav Neurosci. https://doi.org/10.1007/7854_2020_178
Hyland NP, Cryan JF (2010) A gut feeling about GABA: focus on GABAB receptors. Front Pharmacol. https://doi.org/10.3389/FPHAR.2010.00124
Iwakiri M et al (2005) Changes in hippocampal GABABR1 subunit expression in Alzheimer's patients: association with Braak staging. Acta Neuropathol 109(5):467–474. https://doi.org/10.1007/s00401-005-0985-9
Jacobson LH et al (2016) Differential roles of GABAB1 subunit isoforms on locomotor responses to acute and repeated administration of cocaine. Behav Brain Res 298:12–16. https://doi.org/10.1016/J.BBR.2015.10.039

Jacobson LH et al (2018) The gamma-aminobutyric acid B receptor in depression and reward. Biol Psychiatry 83(11):963–976. https://doi.org/10.1016/j.biopsych.2018.02.006
Jafarian M et al (2021) The effect of GABAergic neurotransmission on the seizure-related activity of the laterodorsal thalamic nuclei and the somatosensory cortex in a genetic model of absence epilepsy. Brain Res 1757:147304. https://doi.org/10.1016/J.BRAINRES.2021.147304
Jones KA et al (1998) GABA(B) receptors function as a heteromeric assembly of the subunits GABA(B)R1 and GABA(B)R2. Nature 396(6712):674–679. https://doi.org/10.1038/25348
Kalinichev M et al (2014) Evaluation of peripheral versus central effects of GABAB receptor activation using a novel, positive allosteric modulator of the GABAB receptor ADX71943, a pharmacological tool compound with a fully peripheral activity profile. Br J Pharmacol 171 (21):4941–4954. https://doi.org/10.1111/BPH.12812
Kalinichev M et al (2017) The drug candidate, ADX71441, is a novel, potent and selective positive allosteric modulator of the GABAB receptor with a potential for treatment of anxiety, pain and spasticity. Neuropharmacology 114:34–47. https://doi.org/10.1016/j.neuropharm.2016.11.016
Kannampalli P et al (2017) Analgesic effect of ADX71441, a positive allosteric modulator (PAM) of GABAB receptor in a rat model of bladder pain. Neuropharmacology 126:1–11. https://doi.org/10.1016/J.NEUROPHARM.2017.08.023
Kaupmann K, Bettler B (1998) Heteromerization of GABA B receptors: a new principle for G protein-coupled receptors. Satellite symposium to the 28 th annual meeting of the Society for Neuroscience Los Angeles, CA, November 5-7, 1998. CNS Drug Rev 4(4):376–379. https://doi.org/10.1111/J.1527-3458.1998.TB00077.X
Kaupmann K et al (1997) Expression cloning of GABA(B) receptors uncovers similarity to metabotropic glutamate receptors. Nature 386(6622):239–246. https://doi.org/10.1038/386239A0
Kaupmann K et al (1998) GABA(B)-receptor subtypes assemble into functional heteromeric complexes. Nature 396(6712):683–687. https://doi.org/10.1038/25360
Kim W, Seo H (2014) Baclofen, a GABAB receptor agonist, enhances ubiquitin-proteasome system functioning and neuronal survival in Huntington's disease model mice. Biochem Biophys Res Commun 443(2):706–711. https://doi.org/10.1016/J.BBRC.2013.12.034
Kinjo M et al (2021) Transcranial magnetic stimulation neurophysiology of patients with major depressive disorder: a systematic review and meta-analysis. Psychol Med 51(1):1–10. https://doi.org/10.1017/S0033291720004729
Kleppner S, Tobin A (2001) GABA signalling: therapeutic targets for epilepsy, Parkinson's disease and Huntington's disease. Expert Opin Ther Targets 5(2):219–239. https://doi.org/10.1517/14728222.5.2.219
Kniazeff J (2020) The different aspects of the GABAB receptor allosteric modulation. Adv Pharmacol 88:83–113. https://doi.org/10.1016/BS.APHA.2020.02.003
Lehmann A (2009) GABAB receptors as drug targets to treat gastroesophageal reflux disease. Pharmacol Ther 122(3):239–245. https://doi.org/10.1016/J.PHARMTHERA.2009.02.008
Li X, Slesinger PA (2021) GABAB receptors and drug addiction: psychostimulants and other drugs of abuse. Curr Topics Behav Neurosci. https://doi.org/10.1007/7854_2020_187
Lissemore JI et al (2018) Reduced GABAergic cortical inhibition in aging and depression. Neuropsychopharmacology 43(11):2277–2284. https://doi.org/10.1038/S41386-018-0093-X
Logge WB, Morley KC, Haber PS (2021) GABAB receptors and alcohol use disorders: clinical studies. Curr Top Behav Neurosci. https://doi.org/10.1007/7854_2020_182
Maccioni P, Colombo G (2019) Potential of GABA B receptor positive allosteric modulators in the treatment of alcohol use disorder. CNS Drugs 33(2):107–123. https://doi.org/10.1007/S40263-018-0596-3
Malcangio M (2018) GABAB receptors and pain. Neuropharmacology 136(Pt A):102–105. https://doi.org/10.1016/j.neuropharm.2017.05.012
Malcangio M, Bowery NG (1995) Possible therapeutic application of GABA(B) receptor agonists and antagonists. Clin Neuropharmacol 18:285–305. https://doi.org/10.1097/00002826-199508000-00001

Martín-Belmonte A et al (2020) Density of GABAB receptors is reduced in granule cells of the hippocampus in a mouse model of Alzheimer's disease. Int J Mol Sci 21(7). https://doi.org/10.3390/ijms21072459

McCall NM, De Velasco EM, Wickman K (2019) GIRK channel activity in dopamine neurons of the ventral tegmental area bidirectionally regulates behavioral sensitivity to cocaine. J Neurosci 39(19):3600–3610. https://doi.org/10.1523/JNEUROSCI.3101-18.2019

Metz M et al (2011) Distribution of the auxiliary GABAB receptor subunits KCTD8, 12, 12b, and 16 in the mouse brain. J Comp Neurol 519(8):1435–1454. https://doi.org/10.1002/CNE.22610

Meuwissen K et al (2020) Burst and tonic spinal cord stimulation both activate spinal GABAergic mechanisms to attenuate pain in a rat model of chronic neuropathic pain. Pain Pract 20 (1):75–87. https://doi.org/10.1111/PAPR.12831

Mombereau C et al (2004) Genetic and pharmacological evidence of a role for GABA(B) receptors in the modulation of anxiety- and antidepressant-like behavior. Neuropsychopharmacology 29 (6):1050–1062. https://doi.org/10.1038/sj.npp.1300413

Murrell E et al (2020) Classics in neuroimaging: development of positron emission tomography tracers for imaging the GABAergic pathway. ACS Chem Neurosci 11(14):2039–2044. https://doi.org/10.1021/ACSCHEMNEURO.0C00343

Nieto A et al (2021) GABAB receptor chemistry and pharmacology: agonists, antagonists, and allosteric modulators. Curr Top Behav Neurosci. https://doi.org/10.1007/7854_2021_232

Ong J, Kerr DIB (2005) Clinical potential of GABA B receptor modulators. CNS Drug Rev 11 (3):317–334

Pagès N et al (2021) Anticonvulsive profile of two GABA B receptor antagonists on acute seizure mice models. Epilepsy Res 174. https://doi.org/10.1016/J.EPLEPSYRES.2021.106644

Pan H et al (2008) Modulation of pain transmission by G-protein-coupled receptors. Pharmacol Ther 117(1):141–161. https://doi.org/10.1016/J.PHARMTHERA.2007.09.003

Paterson NE, Froestl W, Markou A (2004) The GABAB receptor agonists baclofen and CGP44532 decreased nicotine self-administration in the rat. Psychopharmacology 172(2):179–186. https://doi.org/10.1007/s00213-003-1637-1

Paterson NE, Froestl W, Markou A (2005) Repeated administration of the GABAB receptor agonist CGP44532 decreased nicotine self-administration, and acute administration decreased cue-induced reinstatement of nicotine-seeking in rats. Neuropsychopharmacology 30 (1):119–128. https://doi.org/10.1038/sj.npp.1300524

Phillips TJ, Reed C (2014) Targeting $GABA_B$ receptors for anti-abuse drug discovery. Expert Opin Drug Discovery 9(11):1307–1317. https://doi.org/10.1517/17460441.2014.956076

Pilipenko V et al (2018) Very low doses of muscimol and baclofen ameliorate cognitive deficits and regulate protein expression in the brain of a rat model of streptozocin-induced Alzheimer's disease. Eur J Pharmacol 818:381–399. https://doi.org/10.1016/j.ejphar.2017.11.012

Pilipenko V et al (2019) GABA-containing compound gammapyrone protects against brain impairments in Alzheimer's disease model male rats and prevents mitochondrial dysfunction in cell culture. J Neurosci Res 97(6):708–726. https://doi.org/10.1002/jnr.24396

Porcu A et al (2021) COR758, a negative allosteric modulator of GABA B receptors. Neuropharmacology 189. https://doi.org/10.1016/J.NEUROPHARM.2021.108537

Premoli I, Castellanos N et al (2014a) TMS-EEG signatures of GABAergic neurotransmission in the human cortex. J Neurosci 34(16):5603–5612. https://doi.org/10.1523/JNEUROSCI.5089-13.2014

Premoli I, Rivolta D et al (2014b) Characterization of GABAB-receptor mediated neurotransmission in the human cortex by paired-pulse TMS-EEG. NeuroImage 103:152–162. https://doi.org/10.1016/J.NEUROIMAGE.2014.09.028

Premoli I et al (2017) The impact of GABAergic drugs on TMS-induced brain oscillations in human motor cortex. NeuroImage 163:1–12. https://doi.org/10.1016/J.NEUROIMAGE.2017.09.023

Princivalle AP (2021) GABAB Receptors in Neurodegeneration. Curr Top Behav Neurosci. https://doi.org/10.1007/7854_2021_222

Princivalle A et al (2002) Studies of GABA(B) receptors labelled with [(3)H]-CGP62349 in hippocampus resected from patients with temporal lobe epilepsy. Br J Pharmacol 136 (8):1099–1106. https://doi.org/10.1038/SJ.BJP.0704812

Princivalle AP et al (2003) Modification of GABAB1 and GABAB2 receptor subunits in the somatosensory cerebral cortex and thalamus of rats with absence seizures (GAERS). Epilepsy Res 55(1–2):39–51

Rekik L et al (2011) γ-Aminobutyric acid type B receptor changes in the rat striatum and substantia nigra following intrastriatal quinolinic acid lesions. J Neurosci Res 89(4):524–535. https://doi.org/10.1002/JNR.22574

Roberts B, Lopes E, Cragg S (2021) Axonal modulation of striatal dopamine release by local γ-aminobutyric acid (GABA) signalling. Cell 10(3). https://doi.org/10.3390/CELLS10030709

Rocha L et al (2015) GABAergic alterations in neocortex of patients with pharmacoresistant temporal lobe epilepsy can explain the comorbidity of anxiety and depression: the potential impact of clinical factors. Front Cell Neurosci 8(Jan). https://doi.org/10.3389/FNCEL.2014.00442

Rose TR, Wickman K (2020) Mechanisms and regulation of neuronal GABAB receptor-dependent signaling. Curr Top Behav Neurosci. https://doi.org/10.1007/7854_2020_129

Scarpellini E et al (2015) Effect of baclofen on the acid pocket at the gastroesophageal junction. Dis Esophagus 28(5):488–495. https://doi.org/10.1111/DOTE.12224

Schuller HM (2017) Regulatory role of G protein-coupled receptors in pancreatic cancer development and progression. Curr Med Chem 25(22):2566–2575. https://doi.org/10.2174/0929867324666170303121708

Schwenk J et al (2016) Modular composition and dynamics of native GABAB receptors identified by high-resolution proteomics. Nat Neurosci 19(2):233–242. https://doi.org/10.1038/NN.4198

Serrats J, Cunningham MO, Davies CH (2017) GABAB receptor modulation — to B or not to be B a pro-cognitive medicine? Curr Opin Pharmacol 35:125–132. https://doi.org/10.1016/j.coph.2017.09.012

Shaheen NJ et al (2013) Efficacy and safety of lesogaberan in gastro-oesophageal reflux disease: a randomised controlled trial. Gut 62(9):1248–1255. https://doi.org/10.1136/GUTJNL-2012-302737

Shaye H et al (2021) Molecular mechanisms of metabotropic GABAB receptor function. Sci Adv 7 (22). https://doi.org/10.1126/SCIADV.ABG3362

Sheilabi M et al (2018) Quantitative expression and localization of GABA B receptor protein subunits in hippocampi from patients with refractory temporal lobe epilepsy. Neuropharmacology 136(Pt A):117–128. https://doi.org/10.1016/J.NEUROPHARM.2017.08.001

Smart T, Stephenson F (2019) A half century of γ-aminobutyric acid. Brain Neurosci Adv 3:239821281985824. https://doi.org/10.1177/2398212819858249

Söderhielm PC et al (2018) Profiling of GABAA and GABAB receptor expression in the myometrium of the human uterus. Life Sci 214:145–152. https://doi.org/10.1016/J.LFS.2018.10.033

Solomon VR et al (2019) GABA allosteric modulators: An overview of recent developments in non-benzodiazepine modulators. Eur J Med Chem 171:434–461. https://doi.org/10.1016/J.EJMECH.2019.03.043

Sperk G et al (2004) GABA and its receptors in epilepsy. Adv Exp Med Biol 548:92–103. https://doi.org/10.1007/978-1-4757-6376-8_7

Straessle A et al (2003) Rapid and long-term alterations of hippocampal GABAB receptors in a mouse model of temporal lobe epilepsy. Eur J Neurosci 18(8):2213–2226. https://doi.org/10.1046/J.1460-9568.2003.02964.X

Urwyler S (2011) Allosteric modulation of family C G-protein-coupled receptors: from molecular insights to therapeutic perspectives. Pharmacol Rev 63(1):59–126

Urwyler S et al (2001) Positive allosteric modulation of native and recombinant γ-aminobutyric acidb receptors by 2,6-Di-tert-butyl-4-(3-hydroxy-2,2-dimethyl-propyl)-phenol (CGP7930) and its aldehyde analog CGP13501. Mol Pharmacol 60(5):963–971

Vlachou S (2021) GABA-B receptors and cognitive processing in health and disease. Curr Top Behav Neurosci. https://doi.org/10.1007/7854_2021_231
Varani AP et al (2012) Acute behavioural responses to nicotine and nicotine withdrawal syndrome are modified in GABAB1 knockout mice. Neuropharmacology 63(5):863–872. https://doi.org/10.1016/J.NEUROPHARM.2012.06.006
Varani AP et al (2014) Involvement of GABAB receptors in biochemical alterations induced by anxiety-related responses to nicotine in mice: genetic and pharmacological approaches. Neuropharmacology 81:31–41. https://doi.org/10.1016/J.NEUROPHARM.2014.01.030
Varani AP et al (2015) Lack of GABAB receptors modifies behavioural and biochemical alterations induced by precipitated nicotine withdrawal. Neuropharmacology 90:90–101. https://doi.org/10.1016/J.NEUROPHARM.2014.11.013
Varani AP et al (2018) Nicotine-induced molecular alterations are modulated by GABA B receptor activity. Addict Biol 23(1):230–246. https://doi.org/10.1111/ADB.12506
Veronezi B et al (2016) Evidence for increased motor cortical facilitation and decreased inhibition in atypical depression. Acta Psychiatr Scand 134(2):172–182. https://doi.org/10.1111/ACPS.12565
Vigot R et al (2006) Differential compartmentalization and distinct functions of GABAB receptor variants. Neuron 50(4):589–601. https://doi.org/10.1016/j.neuron.2006.04.014
Vlachou S, Markou A (2010) $GABA_B$ receptors in reward processes. Adv Pharmacol. https://doi.org/10.1016/S1054-3589(10)58013-X
Wang H et al (2021) Gabab receptor inhibits tumor progression and epithelial-mesenchymal transition via the regulation of hippo/yap1 pathway in colorectal cancer. Int J Biol Sci 17(8):1953–1962. https://doi.org/10.7150/IJBS.58135
White JH et al (1998) Heterodimerization is required for the formation of a functional GABA(B) receptor. Nature 396(6712):679–682
Xi ZX, Stein EA (1999) Baclofen inhibits heroin self-administration behavior and mesolimbic dopamine release. J Pharmacol Exp Ther 290(3):1369–1374
Yang Y et al (2021) Blockade of pre-synaptic and post-synaptic GABA B receptors in the lateral habenula produces different effects on anxiety-like behaviors in 6-hydroxydopamine hemiparkinsonian rats. Neuropharmacology 196:108705. https://doi.org/10.1016/J.NEUROPHARM.2021.108705
Zhou Y et al (2017) The role of spinal GABAB receptors in cancer-induced bone pain in rats. J Pain 18(8):933–946. https://doi.org/10.1016/J.JPAIN.2017.02.438

Structural Basis of $GABA_B$ Receptor Regulation and Signaling

Thorsten Fritzius, Michal Stawarski, Shin Isogai, and Bernhard Bettler

Contents

Abstract $GABA_B$ receptors (GBRs), the G protein-coupled receptors for the inhibitory neurotransmitter γ-aminobutyric acid (GABA), activate Go/i-type G proteins that regulate adenylyl cyclase, Ca^{2+} channels, and K^{+} channels. GBR signaling to enzymes and ion channels influences neuronal activity, plasticity processes, and network activity throughout the brain. GBRs are obligatory heterodimers composed of GB1a or GB1b subunits with a GB2 subunit. Heterodimeric GB1a/2 and GB1b/2 receptors represent functional units that associate in a modular fashion with regulatory, trafficking, and effector proteins to generate receptors with distinct physiological functions. This review summarizes current knowledge on the structure, organization, and functions of multi-protein GBR complexes.

T. Fritzius, M. Stawarski, and B. Bettler (✉)
Department of Biomedicine, Institute of Physiology, Pharmazentrum, University of Basel, Basel, Switzerland
e-mail: bernhard.bettler@unibas.ch

S. Isogai (✉)
Biozentrum, Focal Area Structural Biology and Biophysics, University of Basel, Basel, Switzerland

Microbial Downstream Process Development, Lonza AG, Visp, Switzerland
e-mail: shin.isogai@lonza.com

Curr Topics Behav Neurosci (2022) 52: 19–38
https://doi.org/10.1007/7854_2020_147
Published Online: 19 August 2020

Keywords AJAP-1 · Amyloid precursor protein · APP · GABA-B · HCN · KCTD · PIANP · TRPV1

1 Heterodimers Are the Minimal Functional Receptor Units

The GB1a and GB1b subunit isoforms were cloned in 1997 using a radioligand binding approach (Kaupmann et al. 1997). This showed that GBRs belong to class C GPCRs, which include metabotropic glutamate receptors, Ca^{2+}-sensing receptors, and taste receptors (Kaupmann et al. 1997). As for other class C GPCRs, GB1 subunits contain a large extracellular venus fly trap domain (VFTD), the typical heptahelical transmembrane domain (7TMD), and an intracellular C-terminal domain (Fig. 1). When expressed in heterologous cells, GB1 subunits exhibited ~tenfold lower affinity for GABA than native GBRs (Kaupmann et al. 1997, 1998; White et al. 1998). In addition, GB1 subunits failed to exit the endoplasmic reticulum (Couve et al. 1998) and to efficiently inhibit adenylyl cyclase or activate Kir3 channels (Kaupmann et al. 1997). This showed that GB1 subunits do not form functional receptors by themselves. Soon after cloning of GB1 subunits, cDNA homology searches and yeast-two-hybrid screens identified the sequence-related GB2 subunit (Kaupmann et al. 1998; Kuner et al. 1999; Ng et al. 1999; White et al. 1998). GB2 was again non-functional when expressed in heterologous cells (Galvez et al. 2001). In situ hybridization studies showed that neurons generally co-express GB1 and GB2 transcripts, which indicated that GB1 and GB2 subunits might function together in a heterodimeric receptor. Co-expression of GB1 with GB2 subunits indeed generated receptors with tenfold higher affinity for GABA that also efficiently signaled to neuronal effectors, including adenylyl cyclase, Kir3 channels, and P/Q/N-type Ca^{2+} channels (Kaupmann et al. 1998; Kuner et al. 1999; Ng et al. 1999; White et al. 1998). This finding represented the earliest demonstration of an obligatory heterodimeric G protein-coupled receptor (Marshall et al. 1999). The GB1a and GB1b subunit isoforms derive from the same gene by differential promoter usage and exhibit distinct expression patterns in the central nervous system (Bischoff et al. 1999). Structurally, GB1a differs from GB1b by the presence of two sushi domains in the N-terminal domain (SD1, SD2) (Blein et al. 2004) (Fig. 1). The N-terminal SD1 has an intrinsically disordered structure, while SD2 is more compactly folded. The SDs function as axonal trafficking signals (Biermann et al. 2010) and stabilize the receptor at the cell surface (Hannan et al. 2012). Accordingly, axons predominantly express GB1a/2 receptors, while the somatodendritic compartment expresses GB1b/2 receptors. However, GB1a/2 receptors are also present in the dendrites but excluded from the spine heads, in contrast to GB1b/2 receptors (Biermann et al. 2010; Dinamarca et al. 2019; Vigot et al. 2006).

GBRs have evolved quality control signals that prevent unfolded or unassembled subunits from exiting the endoplasmic reticulum (ER) and the Golgi apparatus. The

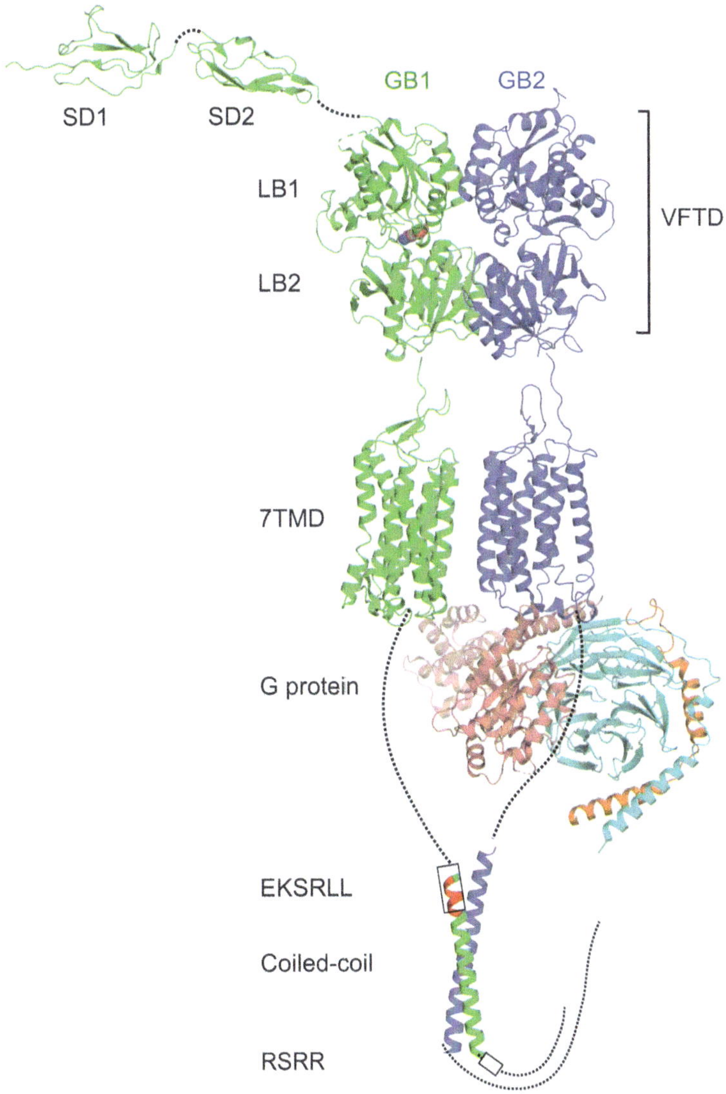

Fig. 1 Structural model of the GBR heterodimer. The model is based on the published structures of SD1 [PDB ID: 6HKC (Rice et al. 2019)], SD2 [1SRZ (Blein et al. 2004)], baclofen-bound VFTDs [4MS4 (Geng et al. 2013)], active and inactive mGlu5 [6N51 and 6N52 (Koehl et al. 2019)], the coiled-coil domain of GBRs [4PAS (Burmakina et al. 2014)] and the heterotrimeric G protein complex [3SN6 (Rasmussen et al. 2011)]. GB1 and GB2 are colored in green and slate, respectively. The active and inactive conformations of the $7TMD_{mGlu5}$ were used in GB1 and GB2, respectively. The boxes in the C-terminal domain of GB1 indicate the retention motifs RSRR and EKSRLL that control heterodimer assembly during biosynthesis

intracellular C-terminal domain of GB1 subunits encodes the ER retention signal RSRR, which is located distal to a coiled-coil heterodimerization domain (Margeta-Mitrovic et al. 2000; Pagano et al. 2001) (Fig. 1). Prenylated rab acceptor family 2 (PRAF2) protein, an ER-resident molecule, binds to the ER retention signal and prevents exit of the GB1 subunit from the ER (Doly et al. 2016). Coiled-coil heteromerization of GB1 with GB2 subunits competitively displaces PRAF2 from its binding motif and enables forward trafficking of the GB1/2 heterodimer in the biosynthetic pathway. The coat protein complex I (COPI) also binds to the ER retention signal and shuttles unassembled GB1 subunits from the cis-Golgi back to the ER (Brock et al. 2005). An additional signal within the coiled-coil domain of GB1, the di-leucine motif EKSRLL (Fig. 1), controls release of receptors from the trans-Golgi network (Restituito et al. 2005). Msec7-1, a guanine-exchange factor protein of the ARF family of GTPases, binds to this di-leucine motif and prevents exit of unassembled GB1 subunits from the Golgi apparatus. Structural data show that heterodimerization of GB1 with GB2 subunits occludes the di-leucine signal and prevents Msec7-1 from binding (Burmakina et al. 2014).

At high cell surface density, GB1/2 heterodimers assemble by random collision into higher-order oligomers of two or more heterodimers (Calebiro et al. 2013; Comps-Agrar et al. 2011; Maurel et al. 2008; Schwenk et al. 2010; Stewart et al. 2018). It appears that in higher-order oligomers GB1 subunits arrange in a line via the opposite sides of their 7TMDs, while GB2 subunits are on the side (Xue et al. 2019). Mutation of either E380 + L382, T410 + E412, or E413 in the VFTD of rat GB1a ($VFTD_{GB1a}$) disrupts the formation of higher-order oligomers (Comps-Agrar et al. 2011; Stewart et al. 2018). Molecular modeling indicates that two G proteins can couple to one GBR tetramer (Xue et al. 2019). Interestingly, however, higher-order oligomerization limits the capacity of GBRs to activate G proteins, presumably because only one of the agonist binding sites in the two neighboring GB1 subunits of a GBR tetramer can be occupied (Stewart et al. 2018). It is unknown whether suppression of G protein signaling in higher-order oligomers is of regulatory significance or not.

2 Signal Transduction in the Receptor Heterodimer

GB1 and GB2 subunits fulfill distinct functions in the receptor heterodimer. Only GB1 contains a GABA binding site (Galvez et al. 1999), whereas GB2 couples to the G protein (Duthey et al. 2002; Havlickova et al. 2002). GB2 additionally allosterically increases GABA affinity at GB1 (Kaupmann et al. 1998; White et al. 1998). After binding of GABA at GB1, multiple allosteric interactions between subunit domains are necessary to activate the G protein at GB2. X-ray structures of the extracellular dimerization interface are now available (Geng et al. 2012, 2013). The $VFTD_{GB1}$/$VFTD_{GB2}$ dimer structure shows that GB1 and GB2 subunits interact sideways, facing opposite directions (Geng et al. 2013) (Fig. 2). Each VFTD has a bi-lobed structure, where lobe 1 (LB1) is positioned more distant from the plasma

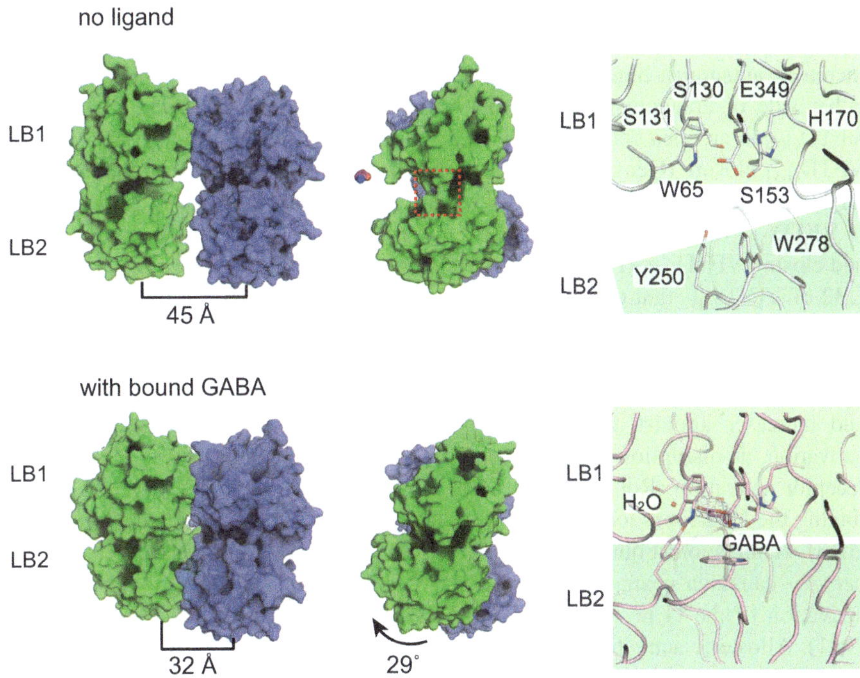

Fig. 2 Conformational changes in VFTDs during receptor activation. Structures of the VFTDs in the absence of a ligand [PDB ID: 4MQE, top] and with GABA bound to VFTD$_{GB1}$ [4MS3, bottom (Geng et al. 2013)] are shown. Upon GABA binding, LB2 of VFTD$_{GB1}$ (green) rotates by 29° and moves toward VFTD$_{GB2}$ (slate) by ~10 Å. In contrast, LB2 of VFTD$_{GB2}$ rotates only by 9° upon activation. Amino acid residues in VFTD$_{GB1}$ involved in the binding of GABA are shown on the right, with numbering of residues according to human GB1b (Geng et al. 2013). Y250, W278, and S131 (water mediated contact) on LB2 interact with GABA and subsequently close the interface between the LB domains. The 2F(o)-F(c) electron density map of GABA is shown as mesh at $\sigma = 1.5$ (right bottom)

membrane than LB2. A peptide hinge connects LB1 and LB2, enabling LB2 to move in relation to LB1. In the heterodimer, VFTD$_{GB1}$ and VFTD$_{GB2}$ bind to each other via LB1. The LB1/LB1 interaction is stabilized by multiple hydrophobic contacts (Y113/Y117 in GB1b; Y118/W149 in GB2, amino acid numbering refers to human sequences in the remainder of the manuscript), salt bridges (R141 in GB1b, D109 in GB2), hydrogen bonds (E138 in GB1b, N110 in GB2), and multiple van der Waals contacts (Geng et al. 2013). The hydrophobic patch is located at the center of the interface and flanked by sites forming hydrogen bonds and water-mediated contacts.

Mutagenesis and X-ray crystallography studies show that agonists and antagonists bind to a large crevice between LB1 and LB2 in VFTD$_{GB1}$ (Galvez et al. 1999; Geng et al. 2013). Agonists and antagonists form multiple interactions with residues in LB1, including hydrogen bonds with S130, S153, H170, and E349, van der Waals contacts with W65, and water-mediated contacts with S131 (GB1b numbering)

(Geng et al. 2013) (Fig. 2). Agonists additionally bind to Y250 and W278 in LB2. The antagonists CGP54626 and SCH50911 also bind to W278 in LB2, which increases antagonist-binding affinity (Geng et al. 2013). The large substituents at either side of antagonists physically prevent $VFTD_{GB1}$ closure, which stabilizes $VFTD_{GB1}$ in an open conformation (Geng et al. 2013). Conversely, agonists induce $VFTD_{GB1}$ closure (Geng et al. 2012; Kniazeff et al. 2004). Mutations that stabilize $VFTD_{GB1}$ closure therefore lead to constitutive activity. Upon agonist binding, LB2 in $VFTD_{GB1}$ rotates 29° about a nearly horizontal axis, bringing LB2 close to LB1 and closing $VFTD_{GB1}$ (Fig. 2). This rotation additionally moves $VFTD_{GB1}$ closer to LB2 of $VFTD_{GB2}$ that remains in a constitutively open conformation stabilized by hydrogen bonds between LB1 and LB2 (Geng et al. 2012). LB2 of $VFTD_{GB2}$ twists by ~9° about a nearly vertical axis, moving it toward LB2 of $VFTD_{GB1}$. As a result, a new LB2/LB2 interface forms that stabilizes $VFTD_{GB1}$ in the closed conformation and increases agonist affinity. The LB2/LB2 interface is essential for receptor activation, as disruption of the interface by insertion of a glycan wedge precludes receptor activation (Rondard et al. 2008). Conversely, a covalent disulfide bridge linking the LB2 lobes locks the receptor in a constitutively active state (Geng et al. 2013). While agonist binding promotes $VFTD_{GB1}$/$VFTD_{GB2}$ interaction, it simultaneously causes a spatial reorientation of $7TMD_{GB1}$ and $7TMD_{GB2}$ that enables activation of the G protein at $7TMD_{GB2}$ (Matsushita et al. 2010; Monnier et al. 2011). Allosteric activation of $7TMD_{GB2}$ occurs in *cis* and in *trans* via $VFTD_{GB2}$ and $7TMD_{GB1}$, respectively (Monnier et al. 2011). Receptor activation disrupts an ionic lock at the intracellular side of $7TMD_{GB2}$ (Binet et al. 2007). The ionic lock is formed by a salt bridge between D688 in TM6 and K574 in TM3, which prevents outward movement of TM6 and stabilizes the inactive closed conformation of the $7TMD_{GB2}$. Disruption of the ionic lock by mutation allosterically increases agonist affinity. Recent studies show that crosslinking of the TM6-TM6 interaction between GB1 and GB2 is sufficient for receptor activation and leads to constitutive activity (Xue et al. 2019). Interestingly, GBRs also exhibit constitutive activity in the absence of agonists (Galvez et al. 2001; Grunewald et al. 2002). This suggests that GBRs exhibit high intrinsic conformational flexibility and spontaneously oscillate between inactive and active states, similar as shown for the isolated VFTDs of metabotropic glutamate receptor 2 (Olofsson et al. 2014).

Almost all known allosteric modulators of GBRs bind to $7TMD_{GB2}$ (Binet et al. 2004; Dupuis et al. 2006; Sun et al. 2016). The only exception is Ca^{2+}, which binds to S269 in LB1 of GB1a (S153 in GB1b) and thereby increases affinity for GABA (Galvez et al. 2000; Wise et al. 1999) (Fig. 2). Positive allosteric modulators (PAMs) at GBRs generally increase agonist potency and efficacy (Urwyler et al. 2005). Some PAMs also have agonistic properties and activate the receptor in the absence of orthosteric agonists, presumably by stabilizing the active conformation of $7TMD_{GB2}$. The negative allosteric modulator (NAM) CLH304 has inverse agonist properties, suppressing basal activity as well as agonist-induced receptor activation, likely by preventing $7TMD_{GB2}$ from reaching the active conformation (Chen et al. 2014; Sun et al. 2016). The binding sites of allosteric modulators in $7TMD_{GB2}$ are unknown. Crystal structures of other class C G protein-coupled receptors suggest that allosteric modulators enter a cavity located between transmembrane domains 3, 5, 6, and 7 (Christopher et al. 2015; Dore et al. 2014; Gregory et al. 2011; Wu et al. 2014). A GB2 homology model predicts a hydrophobic binding pocket in $7TMD_{GB2}$ and identified potential amino acid residues involved in binding of allosteric modulators (Freyd et al. 2017).

3 Auxiliary KCTD Subunits

Several observations pointed to native GBRs being composed of more than just a GB1 and a GB2 subunit. For example, GBR complexes isolated from brain tissue had molecular masses of 0.6–1.1 MDa, while the heterodimer only accounts for 240 kDa (Schwenk et al. 2010). Moreover, the kinetic properties of native GBR responses varied and differed from those of GB1/2 heterodimers expressed in heterologous cells (Turecek et al. 2014). Quantitative proteomic approaches identified approximately 30 proteins that interact with GB1 or GB2 in the brain (Dinamarca et al. 2019; Schwenk et al. 2010, 2016; Turecek et al. 2014). These proteins provide a molecular basis to explain the functional diversity of native GBRs. Known interactions between components of GBR complexes have been summarized recently (Fritzius and Bettler 2020).

Abundant GBR-interacting proteins are the K^+ channel tetramerization domain (KCTD) proteins KCTD8, KCTD12, KCTD12b, and KCTD16 (herein collectively referred to as the KCTDs) (Schwenk et al. 2010). The KCTDs are part of a larger family of KCTD proteins comprising 26 members with sequence similarity to the cytoplasmic tetramerization (T1, also known as BTB or POZ) domain of voltage-gated K^+ channels (Correale et al. 2013; Zheng et al. 2019). The KCTDs are composed of the N-terminal T1 domain and a H1 domain, with both isolated domains capable of forming oligomers (Correale et al. 2013; Fritzius et al. 2017). KCTD8 and KCTD16 additionally encode a C-terminal H2 domain that scaffolds effector channels and other receptor-associated proteins (see below). Structural studies demonstrate that $T1_{KCTD12}$ and $T1_{KCTD16}$ form homopentamers (Pinkas et al. 2017; Smaldone et al. 2016; Zheng et al. 2019; Zuo et al. 2019) (Figs. 3a and 5). The $T1_{KCTD16}$ pentamer is open, with a gap of 8–16 Å at its narrowest and widest points. Since one $T1_{KCTD16}$ monomer in the pentamer occupies 25 Å, the gap in the pentamer is too small to accommodate a sixth $T1_{KCTD16}$ monomer. Multiple electrostatic interactions and nonpolar associations stabilize adjacent $T1_{KCTD16}$ domains. Most of the conserved amino acid residues are involved in $T1_{KCTD16}$ interactions, supporting that all four KCTDs assemble as pentamers. Co-crystallization of $T1_{KCTD16}$ with a C-terminal GB2 peptide shows that the $T1_{KCTD16}$ pentamer wraps around the peptide (Fig. 3a, b). The GB2 peptide loops inside the central opening of the pentamer, entering and leaving the pentamer at its N-terminal surface. The apex of the GB2 peptide loop forms a short helix that contains the Y903 residue critical for KCTD binding (Correale et al. 2013; Schwenk et al. 2010; Zheng et al. 2019) (Fig. 3a). X-ray crystallography reveals that Y903 is

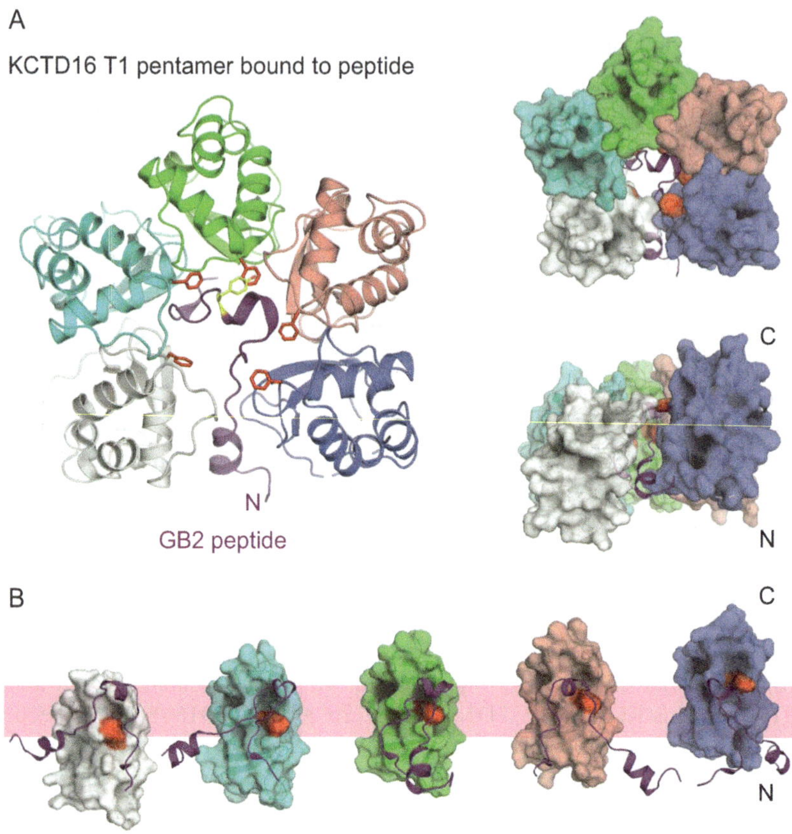

Fig. 3 Binding of the GB2 C-terminus to the $T1_{KCTD16}$ pentamer. (**a**) Structure of a GB2 C-terminal peptide bound to the $T1_{KCTD16}$ pentamer [PDB ID: 6M8R (Zheng et al. 2019)]. The F80 (red) and Y903 (yellow) residues in $T1_{KCTD16}$ and GB2, respectively, are highlighted. $T1_{KCTD16}$ domains form an open pentamer with C6 symmetry. A twist of the ring prevents the sixth subunit from being inserted. Upon binding to the GB2 C-terminal peptide, the open pentamer contracts by roughly 4–5 Å and creates a tight channel for the peptide (right). The orientation of the complex is indicated by the N- and C-terminus of KCTD16. (**b**) The twisted ring structure enables each $T1_{KCTD16}$ subunit to form a distinct binding interface with the peptide. A cross-section of the pentamer shows the interaction of each of the five $T1_{KCTD16}$ domains with the GB2 peptide. The F80 residues (red) and the $T1_{KCTD16}$ domains are aligned vertically according to the position of GB2 peptide

located in the middle of an extensive interaction interface (Zuo et al. 2019). The interaction takes place off center in the central pore of the pentamer, opposite of the gap. In each $T1_{KCTD16}$ domain of the pentamer, the F80 residue protrudes into the central pore. A slight offset due to a tilt of each $T1_{KCTD16}$ monomer forms a spiraling ladder of F80 residues in the inner wall of the pentamer (Fig. 3b). This arrangement allows F80 residues to bind many side chains in a GB2 peptide of 25 amino acid residues (Zheng et al. 2019). Consistent with the X-ray data, the F80A mutation completely abrogates KCTD16 binding to GBRs (Zuo et al. 2019). Interestingly, binding of GB2 results in a compaction of the $T1_{KCTD16}$ pentamer (Fig. 3a).

Reverse affinity purification experiments using KCTD-specific antibodies revealed that KCTD8, KCTD12, KCTD12b, and KCTD16 not only bind to GB2 but also to the Gβγ subunits of the G protein (Turecek et al. 2014; Zheng et al. 2019). Co-crystallization studies show that each $H1_{KCTD12}$ pentamer binds five Gβγ subunits in a near perfect C5 rotational symmetry (Zheng et al. 2019) (Fig. 4). The five Gβγ molecules form a tightly packed outer ring in which every Gβ subunit directly contacts neighboring Gβ subunits as well as two adjacent H1 domains of the pentamer. $H1_{KCTD12}$ folds into a β sheet made up by five antiparallel β strands (β1-5) interspersed with two short α helices (α1 and α2). The amino acids at and around the loops between β1/β2 as well as β3/α2 bind to an acidic patch at the top of the Gβ propeller (interface I) and a groove between the N-terminal helix and the β propeller of Gβ (interface II) (Fig. 4). In $H1_{KCTD12}$, R232 (contacting interface I on Gβ) and R257 (contacting interface II on Gβ) are particularly important for the interaction, as mutation of either residue completely abolishes Gβγ binding and modulation of G protein signaling by KCTD12. The Gγ subunit is located peripherally and does not interact with the H1 domain. However, Gγ allows anchoring of the complex at the plasma membrane (Figs. 4 and 5). When incubating $KCTD12_{H1}$ with a substoichiometric amount of Gβγ, only full 5/5 complexes and free KCTD12 were observed, with no evidence of partial oligomers (Zheng et al. 2019). This suggests that binding of $H1_{KCTD12}$ to Gβγ is highly cooperative. Supported by 3D reconstructions of electron microscopy images of the full-length KCTD12 protein in complex with Gβγ (Zheng et al. 2019), the picture of a large multi-protein complex emerges, in which KCTDs simultaneously bind via their T1 and H1 domains GB2 and Gβγ subunits, respectively (Fig. 5). Of note, Gβγ binding to the $H1_{KCTD12}$ pentamer partially occludes the Gα binding-site on the surface of Gβγ, indicating that the trimeric G protein does not assemble with KCTD12 into a pentameric complex. This contrasts earlier biochemical findings that support that GBRs and KCTDs form a complex with the heterotrimeric G protein (Turecek et al. 2014). Co-crystallization of the H1 domain with Gβγ may therefore favor a structure that differs from the structure of the full-length KCTD protein assembled with receptor and the heterotrimeric G protein.

Dual binding of the KCTDs to the receptor and the G protein enables KCTDs to regulate the kinetics of receptor signaling (Fritzius et al. 2017; Schwenk et al. 2010; Seddik et al. 2012; Turecek et al. 2014). Pre-assembly of the G protein via the KCTDs at the receptor significantly accelerates G protein signaling, most likely by overcoming slow diffusion-limited association of the G protein with the receptor (Turecek et al. 2014). When studying GBR-mediated K^+ current responses, KCTDs shorten both the rise time and the delay between agonist binding and the onset of K^+ currents. The KCTDs are therefore responsible for the fast kinetics observed with GBR-induced current responses in neurons (Schwenk et al. 2010; Turecek et al. 2014). While all KCTDs accelerate GBR signaling, selectively KCTD12 and KCTD12b induce a rapid desensitization of GBR-mediated K^+ currents (Schwenk et al. 2010; Seddik et al. 2012) through an activity-dependent uncoupling of Gβγ

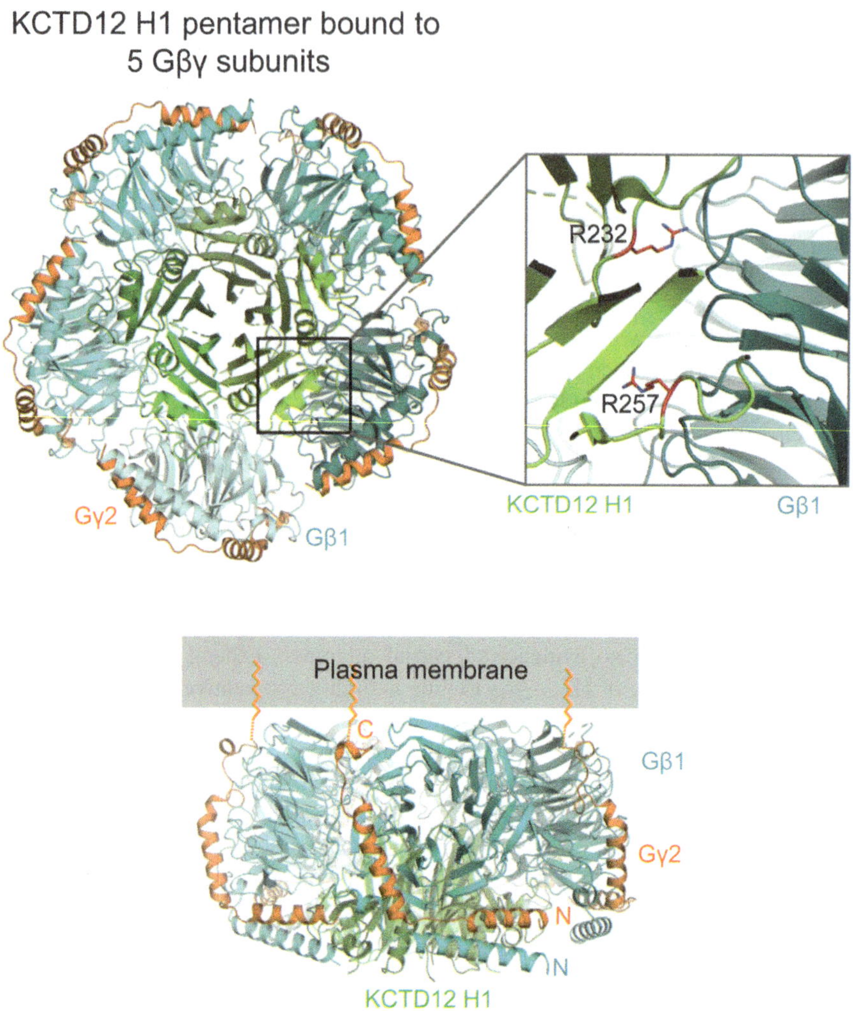

Fig. 4 The $H1_{KCTD12}$/Gβγ complex. Top: $H1_{KCTD12}$ pentamers and five Gβ1γ2 subunits form together a complex with C5 symmetry [PDB ID: 6M8S, (Zheng et al. 2019)]. Each of the five Gβ1γ2 subunits binds two $H1_{KCTD12}$ subunits. The cutout shows R232 (interface I) and R257 (interface II) that are crucial for Gβ1 recognition. Bottom: Due to lipidation of Gβγ subunits, the complex is expected to be tethered to the plasma membrane

from effector channels (Turecek et al. 2014; Zheng et al. 2019). Some neuronal populations simultaneously express multiple KCTDs, raising the possibility that the KCTDs form hetero-oligomers. Indeed, in heterologous cells, KCTD8, KCTD12, and KCTD16 form hetero-oligomers in all possible combinations (Balasco et al. 2019; Fritzius et al. 2017). Association of KCTD12/16 hetero-oligomers with GBRs in hippocampal pyramidal cells confers unique kinetic properties to GBR-induced K^+ currents, showing that hetero-oligomers increase the kinetic repertoire of GBR

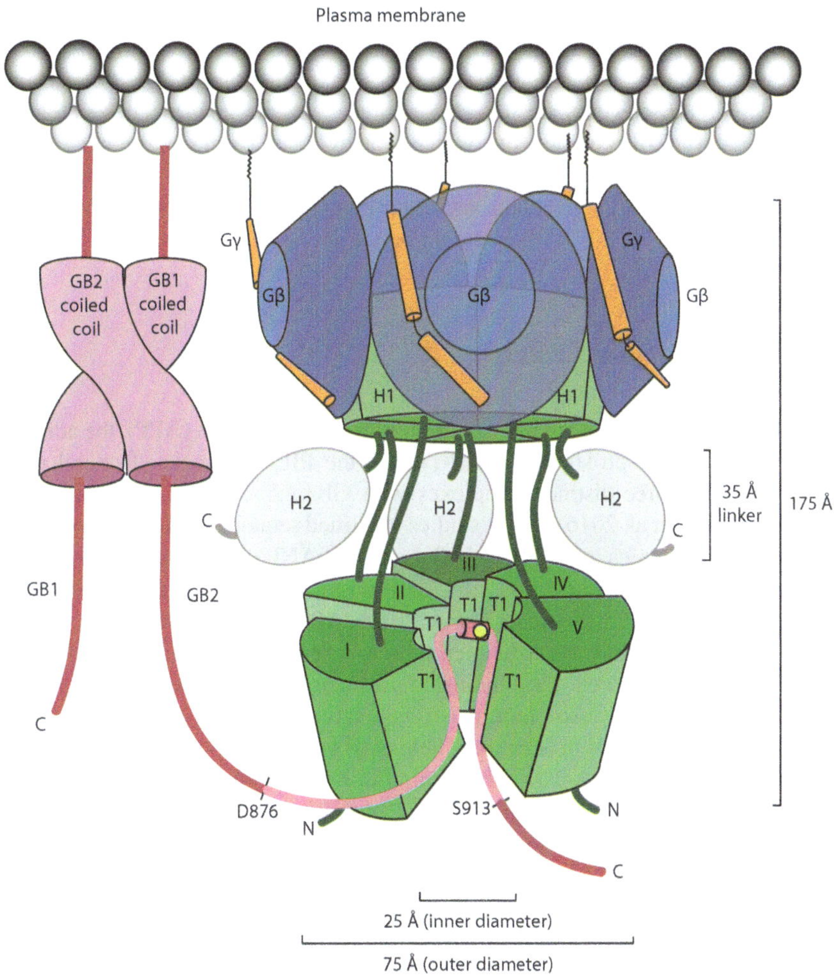

Fig. 5 Scheme of the multi-protein GBR/KCTD12/G protein signaling complex. The intracellular part of the GBR (dark pink), a pentamer of KCTD12 proteins (green), Gβ (blue), and Gγ (orange) subunits are depicted. The N-terminal T1 domain of the KCTDs forms an open pentamer that interacts with the cytoplasmic tail of GB2 (the GB2 peptide, containing the amino acids D888 to S913, co-crystallized with the $T1_{KCTD16}$ is highlighted in bright pink). This part of GB2 loops inside the central opening of the T1 pentamer, entering and leaving it at its N-terminal surface. The amino acid Y903 (yellow circle) at the apex of the GB2 loop is critical for KCTD binding. A slight offset due to a tilt of each $T1_{KCTD16}$ monomer allows the pentamer to bind a large number of amino acid side chains within the cytoplasmic tail of GB2. A short linker (35 Å) connects the N-terminal T1 domain with the C-terminal H1 domain of KCTDs, which binds to the Gβγ heterodimer of the G protein. The scheme depicts the closed $H1_{KCTD12}$ pentamer bound to five copies of Gβγ. Anchoring of the Gγ subunit to the phospholipid bilayer tethers KCTDs to the plasma membrane. The expected location of the C-terminal H2 domain present in KCTD16 and KCTD8 is indicated for three of the KCTDs in the pentamer. Distances are derived from three-dimensional negative-stain electron microscopy reconstructions (Zheng et al. 2019)

signaling (Fritzius et al. 2017). Of note, the KCTDs exert little influence on allosteric and orthosteric binding sites of GBRs (Rajalu et al. 2015).

Reverse-affinity purification experiments support that the KCTDs do not bind to other GPCRs (Schwenk et al. 2016; Turecek et al. 2014). The KCTDs are non-obligatory GBR components, which, however, are expressed by most neurons (and some glial cells) in the vertebrate brain (Metz et al. 2011). Since the KCTDs stably associate with the receptor and control receptor kinetics and surface expression (Ivankova et al. 2013), they should be viewed as auxiliary receptor subunits.

4 SD-Interacting Proteins

Proteomic studies showed that the β-amyloid precursor protein (APP), the adherence junction-associated protein 1 (AJAP-1), and the PILRα-associated neural protein (PIANP) form three distinct complexes with GB1a/2 receptors (Dinamarca et al. 2019; Schwenk et al. 2016). NMR studies identified sequence-related epitopes in the extracellular domains of APP, AJAP-1, and PIANP that bind with nanomolar affinities to the N-terminal SD1 of GB1a, with a rank order of affinities AJAP-1 > PIANP >> APP (Dinamarca et al. 2019). APP is best known as the source of β-amyloid (Aβ) peptides in Alzheimer's disease. Electrophysiological and biochemical experiments showed that binding of APP to GB1a is necessary for vesicular trafficking of GBRs to axon terminals (Dinamarca et al. 2019), consistent with the proposed role of the SDs in axonal trafficking (Biermann et al. 2010). Proteomic data show that APP associates with calsyntenins and c-Jun N-terminal kinase-interacting proteins (JIPs) that link the APP/GBR complex in cargo vesicles to the axonal kinesin-1 motor. Of potential relevance for Alzheimer's disease, complex formation with GBRs stabilizes APP at the cell surface and reduces proteolysis of APP to Aβ (Dinamarca et al. 2019). A related study showed that binding of the soluble form of APP (sAPP) to the SD1 of GB1a inhibits neurotransmitter release, synaptic transmission and spontaneous neuronal activity (Rice et al. 2019). The fact that a GBR antagonist disinhibits sAPP-inhibited neurotransmitter release supports that sAPP acts as a GBR agonist or positive allosteric modulator. However, it was also reported that sAPP has no functional effects on GBR signaling in heterologous cells (Dinamarca et al. 2019). Therefore, additional studies need to confirm sAPP effects on GBR signaling. AJAP-1 and PIANP, the two other proteins binding to the SD of GB1a, do not play a role in axonal trafficking of GBRs (Dinamarca et al. 2019). These proteins localize to adherens junctions that stabilize cell-cell interactions (Winkler et al. 2019; Yamada and Nelson 2007) and may be important for anchoring GB1a/2 receptors at presynaptic sites, either in *cis* or through trans-synaptic interactions. In support of this hypothesis, PIANP knock-out mice exhibit a deficit in GBR-mediated inhibition of glutamate release in the hippocampus (Winkler et al. 2019).

Amyloid-like protein 2 (APLP2) and integral membrane protein 2B (ITM2B) and ITM2C are additional transmembrane proteins that selectively co-purify with GB1a/

2 receptors (Dinamarca et al. 2019; Schwenk et al. 2016). These proteins associate with APP and are secondary interactors of GBRs. GBRs can therefore assemble with multi-protein APP complexes into super-complexes (complexes of complexes).

5 Effector Channels

The best-studied GBR functions in the central nervous system are the gating of voltage-sensitive Ca^{2+} (Ca_v) channels and inwardly rectifying Kir3-type K^+ channels by the Gβγ subunits of the activated G protein (Gassmann and Bettler 2012). GBRs inhibit N- and P/Q-type Ca_v channels, which suppress neurotransmitter release at most synapses in the brain. GBR activation of Kir3 channels hyperpolarizes the membrane, shunts postsynaptic currents in the dendrites, and inhibits neuronal firing. The α1B, α2, δ1, and δ2 subunits of N-type Ca_V channels co-purify with the GB1, GB2, and KCTD16 subunits, supporting that these channels bind to GBRs via KCTD16 (Schwenk et al. 2016). Association of GBR with N-type Ca_V channels directly links the receptor to the presynaptic release machinery. Proteomic work did not support a physical association of native GBRs with Kir3 channels (Schwenk et al. 2016), in contrast to earlier studies in heterologous expression systems (Ciruela et al. 2010; David et al. 2006; Fowler et al. 2007). It is possible that proteomic approaches miss weak interactions of Kir3 channels with GBRs. Alternatively, overexpression of two membrane proteins in heterologous cells may lead to artificial aggregates detected in BRET and immunoprecipitation experiments. Proteomic work additionally identified novel effector channels of GBRs, such as the transient receptor potential vanilloid 1 (TRPV1) (Hanack et al. 2015) and HCN2 channels (Schwenk et al. 2016). Sensitization of TRPV1 channels is central to the initiation of pathological forms of pain. TRPV1 assembles in a complex with GB1 (Hanack et al. 2015). Since agonist activity at GB1 reverts the sensitized state of TRPV1 channels, it may be possible to exploit the TRPV1/GB1 complex for anti-pain therapy. HCN1 and HCN2, like N-type Ca_v channels, appear to associate with GBRs through KCTD16 (Schwenk et al. 2016). HCN channels are widely expressed in the heart and the central nervous system, where they are involved in the generation of rhythmic activity (Biel et al. 2009). GBRs activate HCN currents in dopaminergic neurons of the ventral tegmental area and thereby shorten the duration of inhibitory postsynaptic potentials (Schwenk et al. 2016). The mechanism of GBR-induced HCN channel activation is unknown but may include (1) membrane hyperpolarization via Kir3 channels, (2) allosteric gating of HCN channels by conformational changes in the receptor, and/or (3) dynamic interactions of HCN channels with G protein subunits or second messengers.

6 Additional Receptor-Associated Proteins

Additional components of native GBR signaling complexes are calnexin, reticulocalbin-2, inactive dipeptidyl-peptidases 6/10, 14-3-3 proteins, synaptotagmin-11, and neuroligin-3 (Schwenk et al. 2016). The anatomically and temporally restricted expression of these proteins in the brain limits the set of available receptor constituents in individual cells and further supports a modular GBR architecture. For some of these receptor components binding sites on GB1, GB2, or the KCTDs have been identified (Fritzius and Bettler 2020). Yeast-two-hybrid screens identified several additional proteins that potentially interact with GB1 or GB2 (Pin and Bettler 2016). These proteins may represent low-abundance or transiently interacting GBR components that escaped detection in proteomic approaches.

7 Concluding Remarks

During the past decade, numerous structural and biophysical studies have greatly improved our understanding of the sequence of allosteric events involved in the activation of heterodimeric GBRs. However, the structures of the full-length heterodimeric GBR at atomic resolution in its active and inactive state, with and without bound G protein or allosteric modulators, are still missing. Cryo-electron microscopy appears to be a promising approach to obtain such high-resolution structural information, which is necessary to validate and extend current concepts. The functional relevance of higher-order GBR complexes is still unclear and needs to be addressed in native tissue. The recognition that GBR heterodimers interact with an inventory of ~30 proteins to form a variety of multi-protein complexes with distinct kinetic properties, localizations, and functions represents a departure from earlier concepts based on receptor protomers working in isolation. For some GBR interacting proteins (KCTDs, APP, HCN channels), we have identified functional effects and/or obtained high-resolution structures in association with the receptor. However, we still lack functional and structural information for most of the receptor components identified in proteomic approaches. Furthermore, much effort needs to be devoted to the study of the structural dynamics in GBR complexes during physiological processes. Understanding the structure and function of identified GBR complexes in the brain hopefully will help to identify promising molecular targets for therapeutic intervention.

Acknowledgments We thank M. Gassmann for helpful discussions. This work was supported by grants of the Swiss Science Foundation (31003A-172881) and the National Center for Competences in Research (NCCR) "Synapsy, Synaptic Basis of Mental Health Disease" (to B.B).

References

Balasco N, Smaldone G, Vitagliano L (2019) The structural versatility of the BTB domains of KCTD proteins and their recognition of the $GABA_B$ receptor. Biomol Ther 9(8). pii: E323

Biel M, Wahl-Schott C, Michalakis S, Zong X (2009) Hyperpolarization-activated cation channels: from genes to function. Physiol Rev 89:847–885

Biermann B, Ivankova-Susankova K, Bradaia A, Abdel Aziz S, Besseyrias V, Kapfhammer JP, Missler M, Gassmann M, Bettler B (2010) The sushi domains of $GABA_B$ receptors function as axonal targeting signals. J Neurosci 30:1385–1394

Binet V, Brajon C, Le Corre L, Acher F, Pin JP, Prezeau L (2004) The heptahelical domain of $GABA_{B2}$ is activated directly by CGP7930, a positive allosteric modulator of the $GABA_B$ receptor. J Biol Chem 279:29085–29091

Binet V, Duthey B, Lecaillon J, Vol C, Quoyer J, Labesse G, Pin JP, Prezeau L (2007) Common structural requirements for heptahelical domain function in class A and class C G protein-coupled receptors. J Biol Chem 282:12154–12163

Bischoff S, Leonhard S, Reymann N, Schuler V, Shigemoto R, Kaupmann K, Bettler B (1999) Spatial distribution of $GABA_BR1$ receptor mRNA and binding sites in the rat brain. J Comp Neurol 412:1–16

Blein S, Ginham R, Uhrin D, Smith BO, Soares DC, Veltel S, McIlhinney RA, White JH, Barlow PN (2004) Structural analysis of the complement control protein (CCP) modules of $GABA_B$ receptor 1a: only one of the two CCP modules is compactly folded. J Biol Chem 279:48292–48306

Brock C, Boudier L, Maurel D, Blahos J, Pin JP (2005) Assembly-dependent surface targeting of the heterodimeric $GABA_B$ receptor is controlled by COPI but not 14-3-3. Mol Biol Cell 16:5572–5578

Burmakina S, Geng Y, Chen Y, Fan QR (2014) Heterodimeric coiled-coil interactions of human $GABA_B$ receptor. Proc Natl Acad Sci U S A 111:6958–6963

Calebiro D, Rieken F, Wagner J, Sungkaworn T, Zabel U, Borzi A, Cocucci E, Zurn A, Lohse MJ (2013) Single-molecule analysis of fluorescently labeled G-protein-coupled receptors reveals complexes with distinct dynamics and organization. Proc Natl Acad Sci U S A 110:743–748

Chen LH, Sun B, Zhang Y, Xu TJ, Xia ZX, Liu JF, Nan FJ (2014) Discovery of a negative allosteric modulator of $GABA_B$ receptors. ACS Med Chem Lett 5:742–747

Christopher JA, Aves SJ, Bennett KA, Dore AS, Errey JC, Jazayeri A, Marshall FH, Okrasa K, Serrano-Vega MJ, Tehan BG et al (2015) Fragment and structure-based drug discovery for a class C GPCR: discovery of the mGlu5 negative allosteric modulator HTL14242 (3-chloro-5-[6-(5-fluoropyridin-2-yl)pyrimidin-4-yl]benzonitrile). J Med Chem 58:6653–6664

Ciruela F, Fernandez-Duenas V, Sahlholm K, Fernandez-Alacid L, Nicolau JC, Watanabe M, Lujan R (2010) Evidence for oligomerization between GABAB receptors and GIRK channels containing the GIRK1 and GIRK3 subunits. Eur J Neurosci 32:1265–1277

Comps-Agrar L, Kniazeff J, Norskov-Lauritsen L, Maurel D, Gassmann M, Gregor N, Prezeau L, Bettler B, Durroux T, Trinquet E, Pin JP (2011) The oligomeric state sets $GABA_B$ receptor signalling efficacy. EMBO J 30:2336–2349

Correale S, Esposito C, Pirone L, Vitagliano L, Di Gaetano S, Pedone E (2013) A biophysical characterization of the folded domains of KCTD12: insights into interaction with the $GABA_{B2}$ receptor. J Mol Recognit 26:488–495

Couve A, Filippov AK, Connolly CN, Bettler B, Brown DA, Moss SJ (1998) Intracellular retention of recombinant $GABA_B$ receptors. J Biol Chem 273:26361–26367

David M, Richer M, Mamarbachi AM, Villeneuve LR, Dupre DJ, Hebert TE (2006) Interactions between GABA-B1 receptors and Kir 3 inwardly rectifying potassium channels. Cell Signal 18:2172–2181

Dinamarca MC, Raveh A, Schneider A, Fritzius T, Fruh S, Rem PD, Stawarski M, Lalanne T, Turecek R, Choo M et al (2019) Complex formation of APP with $GABA_B$ receptors links axonal trafficking to amyloidogenic processing. Nat Commun 10:1331

Doly S, Shirvani H, Gata G, Meye FJ, Emerit MB, Enslen H, Achour L, Pardo-Lopez L, Yang SK, Armand V et al (2016) $GABA_B$ receptor cell-surface export is controlled by an endoplasmic reticulum gatekeeper. Mol Psychiatry 21:480–490

Dore AS, Okrasa K, Patel JC, Serrano-Vega M, Bennett K, Cooke RM, Errey JC, Jazayeri A, Khan S, Tehan B et al (2014) Structure of class C GPCR metabotropic glutamate receptor 5 transmembrane domain. Nature 511:557–562

Dupuis DS, Relkovic D, Lhuillier L, Mosbacher J, Kaupmann K (2006) Point mutations in the transmembrane region of $GABA_{B2}$ facilitate activation by the positive modulator N,N-′-dicyclopentyl-2-methylsulfanyl-5-nitro-pyrimidine-4,6-diamine (GS39783) in the absence of the $GABA_{B1}$ subunit. Mol Pharmacol 70:2027–2036

Duthey B, Caudron S, Perroy J, Bettler B, Fagni L, Pin JP, Prezeau L (2002) A single subunit (GB2) is required for G-protein activation by the heterodimeric $GABA_B$ receptor. J Biol Chem 277:3236–3241

Fowler CE, Aryal P, Suen KF, Slesinger PA (2007) Evidence for association of GABA(B) receptors with Kir3 channels and regulators of G protein signalling (RGS4) proteins. J Physiol 580:51–65

Freyd T, Warszycki D, Mordalski S, Bojarski AJ, Sylte I, Gabrielsen M (2017) Ligand-guided homology modelling of the $GABA_{B2}$ subunit of the $GABA_B$ receptor. PLoS One 12:e0173889

Fritzius T, Bettler B (2020) The organizing principle of $GABA_B$ receptor complexes: physiological and pharmacological implications. Basic Clin Pharmacol Toxicol 126(Suppl 6):25–34

Fritzius T, Turecek R, Seddik R, Kobayashi H, Tiao J, Rem PD, Metz M, Kralikova M, Bouvier M, Gassmann M, Bettler B (2017) KCTD hetero-oligomers confer unique kinetic properties on hippocampal $GABA_B$ receptor-induced K^+ currents. J Neurosci 37:1162–1175

Galvez T, Parmentier ML, Joly C, Malitschek B, Kaupmann K, Kuhn R, Bittiger H, Froestl W, Bettler B, Pin JP (1999) Mutagenesis and modeling of the $GABA_B$ receptor extracellular domain support a venus flytrap mechanism for ligand binding. J Biol Chem 274:13362–13369

Galvez T, Urwyler S, Prezeau L, Mosbacher J, Joly C, Malitschek B, Heid J, Brabet I, Froestl W, Bettler B et al (2000) Ca^{2+} requirement for high-affinity g-aminobutyric acid (GABA) binding at $GABA_B$ receptors: involvement of serine 269 of the $GABA_BR1$ subunit. Mol Pharmacol 57:419–426

Galvez T, Duthey B, Kniazeff J, Blahos J, Rovelli G, Bettler B, Prezeau L, Pin JP (2001) Allosteric interactions between GB1 and GB2 subunits are required for optimal $GABA_B$ receptor function. EMBO J 20:2152–2159

Gassmann M, Bettler B (2012) Regulation of neuronal $GABA_B$ receptor functions by subunit composition. Nat Rev Neurosci 13:380–394

Geng Y, Xiong D, Mosyak L, Malito DL, Kniazeff J, Chen Y, Burmakina S, Quick M, Bush M, Javitch JA et al (2012) Structure and functional interaction of the extracellular domain of human $GABA_B$ receptor GBR2. Nat Neurosci 15:970–978

Geng Y, Bush M, Mosyak L, Wang F, Fan QR (2013) Structural mechanism of ligand activation in human $GABA_B$ receptor. Nature 504:254–259

Gregory KJ, Dong EN, Meiler J, Conn PJ (2011) Allosteric modulation of metabotropic glutamate receptors: structural insights and therapeutic potential. Neuropharmacology 60:66–81

Grunewald S, Schupp BJ, Ikeda SR, Kuner R, Steigerwald F, Kornau HC, Kohr G (2002) Importance of the g-aminobutyric $acid_B$ receptor C-termini for G-protein coupling. Mol Pharmacol 61:1070–1080

Hanack C, Moroni M, Lima WC, Wende H, Kirchner M, Adelfinger L, Schrenk-Siemens K, Tappe-Theodor A, Wetzel C, Kuich PH et al (2015) GABA blocks pathological but not acute TRPV1 pain signals. Cell 160:759–770

Hannan S, Wilkins ME, Smart TG (2012) Sushi domains confer distinct trafficking profiles on $GABA_B$ receptors. Proc Natl Acad Sci U S A 109:12171–12176

Havlickova M, Prezeau L, Duthey B, Bettler B, Pin JP, Blahos J (2002) The intracellular loops of the GB2 subunit are crucial for G-protein coupling of the heteromeric g-aminobutyrate B receptor. Mol Pharmacol 62:343–350

Ivankova K, Turecek R, Fritzius T, Seddik R, Prezeau L, Comps-Agrar L, Pin JP, Fakler B, Besseyrias V, Gassmann M, Bettler B (2013) Up-regulation of $GABA_B$ receptor signaling by constitutive assembly with the K^+ channel tetramerization domain-containing protein 12 (KCTD12). J Biol Chem 288:24848–24856

Kaupmann K, Huggel K, Heid J, Flor PJ, Bischoff S, Mickel SJ, McMaster G, Angst C, Bittiger H, Froestl W, Bettler B (1997) Expression cloning of $GABA_B$ receptors uncovers similarity to metabotropic glutamate receptors. Nature 386:239–246

Kaupmann K, Malitschek B, Schuler V, Heid J, Froestl W, Beck P, Mosbacher J, Bischoff S, Kulik A, Shigemoto R et al (1998) $GABA_B$-receptor subtypes assemble into functional heteromeric complexes. Nature 396:683–687

Kniazeff J, Saintot PP, Goudet C, Liu J, Charnet A, Guillon G, Pin JP (2004) Locking the dimeric $GABA_B$ G-protein-coupled receptor in its active state. J Neurosci 24:370–377

Koehl A, Hu H, Feng D, Sun B, Zhang Y, Robertson MJ, Chu M, Kobilka TS, Laeremans T, Steyaert J et al (2019) Structural insights into the activation of metabotropic glutamate receptors. Nature 566:79–84

Kuner R, Kohr G, Grunewald S, Eisenhardt G, Bach A, Kornau HC (1999) Role of heteromer formation in $GABA_B$ receptor function. Science 283:74–77

Margeta-Mitrovic M, Jan YN, Jan LY (2000) A trafficking checkpoint controls $GABA_B$ receptor heterodimerization. Neuron 27:97–106

Marshall FH, Jones KA, Kaupmann K, Bettler B (1999) $GABA_B$ receptors – the first 7TM heterodimers. Trends Pharmacol Sci 20:396–399

Matsushita S, Nakata H, Kubo Y, Tateyama M (2010) Ligand-induced rearrangements of the $GABA_B$ receptor revealed by fluorescence resonance energy transfer. J Biol Chem 285:10291–10299

Maurel D, Comps-Agrar L, Brock C, Rives ML, Bourrier E, Ayoub MA, Bazin H, Tinel N, Durroux T, Prezeau L et al (2008) Cell-surface protein-protein interaction analysis with time-resolved FRET and snap-tag technologies: application to GPCR oligomerization. Nat Methods 5:561–567

Metz M, Gassmann M, Fakler B, Schaeren-Wiemers N, Bettler B (2011) Distribution of the auxiliary $GABA_B$ receptor subunits KCTD8, 12, 12b, and 16 in the mouse brain. J Comp Neurol 519:1435–1454

Monnier C, Tu H, Bourrier E, Vol C, Lamarque L, Trinquet E, Pin JP, Rondard P (2011) Trans-activation between 7TM domains: implication in heterodimeric $GABA_B$ receptor activation. EMBO J 30:32–42

Ng GY, Clark J, Coulombe N, Ethier N, Hebert TE, Sullivan R, Kargman S, Chateauneuf A, Tsukamoto N, McDonald T et al (1999) Identification of a $GABA_B$ receptor subunit, gb2, required for functional $GABA_B$ receptor activity. J Biol Chem 274:7607–7610

Olofsson L, Felekyan S, Doumazane E, Scholler P, Fabre L, Zwier JM, Rondard P, Seidel CA, Pin JP, Margeat E (2014) Fine tuning of sub-millisecond conformational dynamics controls metabotropic glutamate receptors agonist efficacy. Nat Commun 5:5206

Pagano A, Rovelli G, Mosbacher J, Lohmann T, Duthey B, Stauffer D, Ristig D, Schuler V, Meigel I, Lampert C et al (2001) C-terminal interaction is essential for surface trafficking but not for heteromeric assembly of $GABA_B$ receptors. J Neurosci 21:1189–1202

Pin JP, Bettler B (2016) Organization and functions of mGlu and $GABA_B$ receptor complexes. Nature 540:60–68

Pinkas DM, Sanvitale CE, Bufton JC, Sorrell FJ, Solcan N, Chalk R, Doutch J, Bullock AN (2017) Structural complexity in the KCTD family of Cullin3-dependent E3 ubiquitin ligases. Biochem J 474:3747–3761

Rajalu M, Fritzius T, Adelfinger L, Jacquier V, Besseyrias V, Gassmann M, Bettler B (2015) Pharmacological characterization of $GABA_B$ receptor subtypes assembled with auxiliary KCTD subunits. Neuropharmacology 88:145–154

Rasmussen SG, DeVree BT, Zou Y, Kruse AC, Chung KY, Kobilka TS, Thian FS, Chae PS, Pardon E, Calinski D et al (2011) Crystal structure of the beta2 adrenergic receptor-Gs protein complex. Nature 477:549–555

Restituito S, Couve A, Bawagan H, Jourdain S, Pangalos MN, Calver AR, Freeman KB, Moss SJ (2005) Multiple motifs regulate the trafficking of $GABA_B$ receptors at distinct checkpoints within the secretory pathway. Mol Cell Neurosci 28:747–756

Rice HC, de Malmazet D, Schreurs A, Frere S, Van Molle I, Volkov AN, Creemers E, Vertkin I, Nys J, Ranaivoson FM et al (2019) Secreted amyloid-b precursor protein functions as a $GABA_BR1a$ ligand to modulate synaptic transmission. Science 363(6423). pii: eaao4827

Rondard P, Huang S, Monnier C, Tu H, Blanchard B, Oueslati N, Malhaire F, Li Y, Trinquet E, Labesse G et al (2008) Functioning of the dimeric $GABA_B$ receptor extracellular domain revealed by glycan wedge scanning. EMBO J 27:1321–1332

Schwenk J, Metz M, Zolles G, Turecek R, Fritzius T, Bildl W, Tarusawa E, Kulik A, Unger A, Ivankova K et al (2010) Native $GABA_B$ receptors are heteromultimers with a family of auxiliary subunits. Nature 465:231–235

Schwenk J, Perez-Garci E, Schneider A, Kollewe A, Gauthier-Kemper A, Fritzius T, Raveh A, Dinamarca MC, Hanuschkin A, Bildl W et al (2016) Modular composition and dynamics of native $GABA_B$ receptors identified by high-resolution proteomics. Nat Neurosci 19:233–242

Seddik R, Jungblut SP, Silander OK, Rajalu M, Fritzius T, Besseyrias V, Jacquier V, Fakler B, Gassmann M, Bettler B (2012) Opposite effects of KCTD subunit domains on $GABA_B$ receptor-mediated desensitization. J Biol Chem 287:39869–39877

Smaldone G, Pirone L, Pedone E, Marlovits T, Vitagliano L, Ciccarelli L (2016) The BTB domains of the potassium channel tetramerization domain proteins prevalently assume pentameric states. FEBS Lett 590:1663–1671

Stewart GD, Comps-Agrar L, Norskov-Lauritsen LB, Pin JP, Kniazeff J (2018) Allosteric interactions between $GABA_{B1}$ subunits control orthosteric binding sites occupancy within $GABA_B$ oligomers. Neuropharmacology 136:92–101

Sun B, Chen L, Liu L, Xia Z, Pin JP, Nan F, Liu J (2016) A negative allosteric modulator modulates $GABA_B$-receptor signalling through GB2 subunits. Biochem J 473:779–787

Turecek R, Schwenk J, Fritzius T, Ivankova K, Zolles G, Adelfinger L, Jacquier V, Besseyrias V, Gassmann M, Schulte U et al (2014) Auxiliary $GABA_B$ receptor subunits uncouple G protein bg subunits from effector channels to induce desensitization. Neuron 82:1032–1044

Urwyler S, Gjoni T, Koljatic J, Dupuis DS (2005) Mechanisms of allosteric modulation at $GABA_B$ receptors by CGP7930 and GS39783: effects on affinities and efficacies of orthosteric ligands with distinct intrinsic properties. Neuropharmacology 48:343–353

Vigot R, Barbieri S, Brauner-Osborne H, Turecek R, Shigemoto R, Zhang YP, Lujan R, Jacobson LH, Biermann B, Fritschy JM et al (2006) Differential compartmentalization and distinct functions of $GABA_B$ receptor variants. Neuron 50:589–601

White JH, Wise A, Main MJ, Green A, Fraser NJ, Disney GH, Barnes AA, Emson P, Foord SM, Marshall FH (1998) Heterodimerization is required for the formation of a functional $GABA_B$ receptor. Nature 396:679–682

Winkler M, Biswas S, Berger SM, Küchler M, Enkel T, Preisendörfer L, Choo M, Früh S, Rem PD, Enkel T et al (2019) Pianp deficiency links $GABA_B$ receptor signaling and hippocampal and cerebellar neuronal cell composition to autism-like behavior. Mol Psychiatry. https://doi.org/10.1038/s41380-019-0519-9

Wise A, Green A, Main MJ, Wilson R, Fraser N, Marshall FH (1999) Calcium sensing properties of the $GABA_B$ receptor. Neuropharmacology 38:1647–1656

Wu H, Wang C, Gregory KJ, Han GW, Cho HP, Xia Y, Niswender CM, Katritch V, Meiler J, Cherezov V et al (2014) Structure of a class C GPCR metabotropic glutamate receptor 1 bound to an allosteric modulator. Science 344:58–64

Xue L, Sun Q, Zhao H, Rovira X, Gai S, He Q, Pin JP, Liu J, Rondard P (2019) Rearrangement of the transmembrane domain interfaces associated with the activation of a GPCR hetero-oligomer. Nat Commun 10:2765

Yamada S, Nelson WJ (2007) Synapses: sites of cell recognition, adhesion, and functional specification. Annu Rev Biochem 76:267–294

Zheng S, Abreu N, Levitz J, Kruse AC (2019) Structural basis for KCTD-mediated rapid desensitization of $GABA_B$ signalling. Nature 567:127–131

Zuo H, Glaaser I, Zhao Y, Kurinov I, Mosyak L, Wang H, Liu J, Park J, Frangaj A, Sturchler E et al (2019) Structural basis for auxiliary subunit KCTD16 regulation of the $GABA_B$ receptor. Proc Natl Acad Sci U S A 116:8370–8379

Mechanisms and Regulation of Neuronal $GABA_B$ Receptor-Dependent Signaling

Timothy R. Rose and Kevin Wickman

Contents

Abstract γ-Aminobutyric acid B receptors ($GABA_BRs$) are broadly expressed throughout the central nervous system where they play an important role in regulating neuronal excitability and synaptic transmission. $GABA_BRs$ are G protein-coupled receptors that mediate slow and sustained inhibitory actions via modulation of several downstream effector enzymes and ion channels. $GABA_BRs$ are obligate heterodimers that associate with diverse arrays of proteins to form modular

T. R. Rose and K. Wickman (✉)
Department of Pharmacology, University of Minnesota, Minneapolis, MN, USA
e-mail: wickm002@umn.edu

Curr Topics Behav Neurosci (2022) 52: 39–80
https://doi.org/10.1007/7854_2020_129
Published Online: 18 August 2020

complexes that carry out distinct physiological functions. $GABA_BR$-dependent signaling is fine-tuned and regulated through a multitude of mechanisms that are relevant to physiological and pathophysiological states. This review summarizes the current knowledge on $GABA_BR$ signal transduction and discusses key factors that influence the strength and sensitivity of $GABA_BR$-dependent signaling in neurons.

Keywords Adenylyl cyclase · GABA · $GABA_B$ receptor · GIRK channel · Phosphorylation · Plasticity · RGS · Voltage-gated Ca^{2+} channel

1 Introduction

γ-Aminobutyric acid (GABA) is the main inhibitory neurotransmitter in the adult mammalian brain that crucially regulates the balance of excitation and inhibition that is necessary for proper brain function (McCormick 1989, Wu and Sun 2015). GABA acts on either the ionotropic $GABA_A$ or $GABA_C$ receptors ($GABA_ARs$ or $GABA_CRs$) or metabotropic $GABA_B$ receptors ($GABA_BRs$). $GABA_ARs$ and $GABA_CRs$ are ligand-gated chloride channels that mediate fast synaptic inhibition (Kittler et al. 2002), while $GABA_BRs$ are G protein-coupled receptors (GPCRs) that mediate slow synaptic inhibition throughout the central nervous system (CNS). $GABA_BRs$ were first identified by Dr. Norman Bowery in 1979 and became better understood due to the development of selective agonists (e.g., baclofen) and antagonists, as well as the cloning of the receptor in 1997 (Bowery and Hudson 1979, Bowery and Brown 1997, Kaupmann et al. 1997). Since then, our understanding of $GABA_BR$ structure and function has evolved dramatically. Discovery of the existence and functional relevance of $GABA_BR$ subunit isoforms, interacting proteins, macromolecular complexes, and regulatory mechanisms has revealed the diverse and dynamic nature of $GABA_BR$-dependent signaling throughout the brain. At the same time, dysregulation of $GABA_BR$ activity has become increasingly recognized as a driver of neurological and neuropsychiatric disorders (Gassmann and Bettler 2012, Fritzius and Bettler 2019). The goal of this review is to highlight some of the key proteins involved in mediating and regulating $GABA_BR$-dependent signaling in the brain, with an emphasis on mechanisms that fine-tune $GABA_BR$ activity that are relevant to physiology and disease.

2 $GABA_BR$ Structure

Evidence from biochemical, electrophysiological, and behavioral studies revealed that the predominant native $GABA_BR$ is an obligatory heterodimer composed of the $GABA_{B1}$ (GB1) and $GABA_{B2}$ (GB2) subunits (Fan et al. 2017, Møller et al. 2017).

The fact that most neurons co-express GB1 and GB2 and that both subunits show subcellular colocalization strongly suggested the formation of a complex (Kaupmann et al. 1998a, Charles et al. 2001, Gonchar et al. 2001, Kulik et al. 2003). Functional studies in heterologous systems and subunit-specific knockout mice ($GB1^{-/-}$ & $GB2^{-/-}$) revealed that expression of both subunits is required for electrophysiological and biochemical receptor responses (Schuler et al. 2001, Gassmann et al. 2004, Gassmann and Bettler 2012). Further support for the existence and relevance of the $GABA_BR$ heterodimer stemmed from the striking overlap in behavioral phenotypes between $GB1^{-/-}$ and $GB2^{-/-}$ mice, including spontaneous seizures, hyperalgesia, hyperlocomotion, increased anxiety, a reduced threshold for fear responses, and cognitive impairments (Gassmann et al. 2004, Gassmann and Bettler 2012). Immunoprecipitation experiments provided some of the first direct evidence of complex formation between GB1 and GB2, and later studies confirmed these results using other biochemical assays and ultrastructural techniques (Schwenk et al. 2016, Fan et al. 2017, Frangaj and Fan 2018). Crystal structures of $GABA_BR$ subunit domains have revealed that both GB1 and GB2 contain an extracellular Venus flytrap (VFT) domain, a seven-transmembrane (7TM) domain, and an intracellular carboxyl (C)-terminal domain (Frangaj and Fan 2018).

2.1 *Ligand Binding*

The VFT domain of GB1 contains the orthosteric binding site for GABA, as well as other agonists and antagonists. GABA interacts with key residues in the GB1 VFT to induce closure of the domain and stabilize its active conformation (Frangaj and Fan 2018). The GABA-binding site of GB1 is well conserved across species, unlike the VFT domain of GB2 (Freyd et al. 2017). Consistent with a lack of genetic conservation, the GB2 VFT cannot bind ligands, and the receptor still functions upon deletion of the domain (Monnier et al. 2011, Møller et al. 2017). However, the GB2 VFT does enhance the agonist affinity for the GB1 VFT through direct interactions that stabilize the agonist-bound state (Galvez et al. 2001, Liu et al. 2004).

2.2 *Coupling to G Proteins*

The 7TM domain of GB2 facilitates coupling between the receptor and G proteins. The intracellular loops of the GB2 7TM domain are required for functional coupling, as mutations in either the second or third intracellular loops prevent G protein activation (Robbins et al. 2001, Duthey et al. 2002, Havlickova et al. 2002, Binet et al. 2004). While GB1 is not required for G protein coupling, the GB1 7TM domain enhances coupling efficiency (Fan et al. 2017).

2.3 Cell Surface Trafficking

The C-terminal domains of both subunits form a coiled-coil structure that facilitates heterodimerization and surface expression of $GABA_BRs$ (Jiang et al. 2012, Burmakina et al. 2014). When expressed alone in heterologous cells or native neurons, GB1 did not reach the cell surface and was retained in the endoplasmic reticulum (ER) (Couve et al. 1998, Gassmann et al. 2004). Although GB2 did reach the cell surface when expressed alone, it remained non-functional: unable to respond to GABA or the $GABA_BR$ agonist baclofen (Schuler et al. 2001). When co-expressed, GB2 traffics GB1 to the cell surface to generate a receptor capable of high ligand affinity, G protein binding, and effector activation (Galvez et al. 2001).

Two sequences on the C-terminal tail of GB1 prevent exit from the ER (Restituito et al. 2005, Fan et al. 2017). The first is the arginine-based ER retention/retrieval (RSRR) motif, and the second is an upstream di-leucine (LL) motif. Mutations in the RSRR motif enable GB1 to exit the ER and reach the cell surface in the absence of GB2 (Margeta-Mitrovic et al. 2000, Calver et al. 2001, Pagano et al. 2001). Combining RSRR mutations with LL mutations enhanced the exit of GB1 from the ER (Margeta-Mitrovic et al. 2000, Restituito et al. 2005). These findings, alongside the crystal structure of the GB1/GB2 coiled-coil complex (Burmakina et al. 2014), revealed that interactions between the C-termini of GB1 and GB2 mask the ER retention motif to enable trafficking of the receptor to the cell surface. Subsequent investigations would discover that prenylated Rab acceptor 1 domain family, member 2 (PRAF2) sequesters GB1 in the ER to prevent its progression to the Golgi apparatus (Doly et al. 2016). PRAF2 directly interacts with the RSRR and LL motifs of GB1 to prevent ER exit and is competitively displaced from GB1 by GB2. The stoichiometry of PRAF2, GB1, and GB2 concentrations delicately controls surface density of the $GABA_BR$ (Doly et al. 2016). Beyond PRAF2, the C-terminus of GB1 associates with a number of intracellular proteins involved in receptor trafficking and heterodimerization (Table 1).

2.4 Alternative Splicing

The expression of multiple GB1 isoforms contributes to the diverse functions of $GABA_BRs$. There are 14 known isoforms of GB1 (GB1a–n), which can be generated by differential transcription or splicing of the mRNA (Bettler et al. 2004, Jiang et al. 2012, Xu et al. 2014). GB1a and GB1b are the most abundant isoforms in the brain and are the only isoforms that are highly conserved across vertebrate species (Kaupmann et al. 1997, 1998b, Benke et al. 1999, Bettler et al. 2004). Both isoforms have some differences in their spatial and temporal expression patterns in the rodent brain, as well as distinct subcellular localizations (Bettler et al. 2004, Jiang et al. 2012, Kasten et al. 2015, Castelli and Gessa 2016). In general, GB1a is expressed

Table 1 Potential $GABA_B$R-interacting proteins

	Interaction site(s)	*Impact(s) on signaling*	*References*
Core components			
GB1a	GB2	Agonist binding	Gassmann and Bettler (2012), Frangaj and Fan (2018)
GB1b	GB2	Agonist binding	Gassmann and Bettler (2012), Frangaj and Fan (2018)
GB2	GB1a, GB1b, G protein	G protein coupling	Gassmann and Bettler (2012), Frangaj and Fan (2018)
G proteins ($G\alpha_{i/o}$)	GB2 (intracellular loops), KCTD proteins	Mediate $GABA_B$R signaling through downstream effectors	Schwenk et al. (2016), Fritzius and Bettler (2019)
KCTD proteins	GB2 (C-terminus), G protein ($G\beta\gamma$)	Accelerate G protein signaling, mediate fast desensitization (KCTD 12/12b), increase agonist potency (KCTD 12/16)	Turecek et al. (2014), Fritzius and Bettler (2019)
Peripheral components			
RGS proteins	G proteins ($G\alpha_{i/o}$), $GABA_B$Rs (GB1/GB2)	Negatively modulate G protein ($G\alpha_{i/o}$) signaling	Benians et al. (2005), Fowler et al. (2007)
$G\alpha$ inhibitory interacting protein (GINIP)	G protein ($G\alpha_i$)	Promotes G protein signaling	Gaillard et al. (2014)
Ubiquitin-specific protease 14 (USP14)	GB1 (2nd intracellular loop)	Regulates post-endocytotic deubiquitination and degradation of $GABA_B$Rs	Lahaie et al. (2016)
G protein-coupled receptor interacting scaffolding protein (GISP)	GB1 (C-terminus)	Promotes $GABA_B$R surface expression	Kantamneni et al. (2007)
Shroom 4 (shrm4)	GB1 (C-terminus)	Facilitates $GABA_B$R trafficking to cell surface	Zapata et al. (2017)
Ras-associated protein 1 (Rap1)	GB1 (C-terminus)	Promotes $GABA_B$R surface expression by stimulating receptor recycling	Zhang et al. (2015)
Marlin-1	GB1 (C-terminus)	Regulates GB2 protein synthesis or vesicular trafficking?	Couve et al. (2004)
NEM-sensitive fusion (NSF) protein	GB1 (C-terminus), GB2 (C-terminus)	Facilitates PKC-mediated $GABA_B$R desensitization	Pontier et al. (2006)

(continued)

Table 1 (continued)

	Interaction site(s)	*Impact(s) on signaling*	*References*
Syntaxin-1A	GB1 (C-terminus), N-type channels, G proteins (Gβγ)	Facilitates Gβγ-mediated inhibition of N-type channels, regulates probability of synaptic vesicle fusion	Jarvis et al. (2000), Vertkin et al. (2015)
Rat homologue of B2-1/cytohesin-1 (msec-7-1)	GB1 (C-terminus, LL motif)	Promotes $GABA_BR$ surface expression	Restituito et al. (2005)
Prenylated Rab acceptor 1 domain family, member 2 (PRAF2)	GB1 (C-terminus, LL/RSRR motifs)	Intracellular retention of GB1 in ER	Doly et al. (2016)
Coat protein I complex (COPI)	GB1 (C-terminus, RSRR motif)	Intracellular retention of GB1 (facilitate retrograde transport of GB1 from Golgi to ER?)	McMahon and Mills (2004), Brock et al. (2005), Bettler and Tiao (2006)
14-3-3 proteins	GB1 (C-terminus, RSRR motif), KCTD8/16	Disrupt $GABA_BR$ heterodimerization, decouple $GABA_BRs$ from GIRK channels	Couve et al. (2001), Brock et al. (2005), Laffray et al. (2012), Workman et al. (2015), Schwenk et al. (2016)
Tenascin/HNK-1	GB1 (N-terminus)	Inhibit of postsynaptic $GABA_BR$-dependent signaling	Saghatelyan et al. (2003)
CCAAT/enhancer-binding protein homologous protein (CHOP)	GB1a (N-terminus), GB2 (C-terminus)	Intracellular retention of $GABA_BRs$ in ER	Steiger et al. (2004), Sauter et al. (2005)
Fibulin-2	GB1a (SD1)	Receptor anchoring?	Blein et al. (2004)
β-Amyloid precursor protein (APP)	GB1a (SD1)	Facilitates GB1a/GB2 receptor axonal trafficking	Schwenk et al. (2016), Dinamarca et al. (2019)
Secreted β-amyloid precursor protein (sAPP)	GB1a (SD1)	Acts as an agonist/positive allosteric modulator on GB1a/GB2 receptors?	Dinamarca et al. (2019), Rice et al. (2019)
PILR-associating neural protein (PIANP)	GB1a (SD1)	Anchors GB1a/GB2 receptors at presynaptic terminals?	Schwenk et al. (2016), Dinamarca et al. (2019)
Adherence junction-associated protein 1 (AJAP1)	GB1a (SD1)	Anchors GB1a/GB2 receptors at presynaptic terminals?	Schwenk et al. (2016), Dinamarca et al. (2019)
G protein-coupled receptor kinases 4 and 5 (GRK4/5)	GB2	Promote agonist-induced desensitization	Perroy et al. (2003), Kanaide et al. (2007)
Multi-PDZ domain protein 1 (Mupp1)	GB2 (C-terminus)	Promotes $GABA_BR$ surface stability and signaling	Balasubramanian et al. (2006)

(continued)

Table 1 (continued)

	Interaction site(s)	*Impact(s) on signaling*	*References*
β-Filamin	GB2 (C-terminus)	Anchors GABA$_B$Rs to the cytoskeleton	Bettler et al. (2004)
Plakophilin-related *armadillo* repeat protein-interacting PDZ protein (PAPIN)	GB2 (C-terminus)	Promotes GABA$_B$R surface stability and signaling	Balasubramanian et al. (2006)
Tamalin	GB2 (C-terminus)	[a]	Kitano et al. (2002)
Inactive dipeptidyl-peptidases 6 and 10 (DPP-6/10)	KCTD12	[a]	Schwenk et al. (2016)
Calsyntenin-3	APP	Link APP/GABA$_B$R to kinesin-1 motor to facilitate axonal trafficking	Schwenk et al. (2016), Dinamarca et al. (2019)
c-Jun N-terminal kinase-interacting protein 3 (JIP-3)	APP	Link APP/GABA$_B$R to kinesin-1 motor to facilitate axonal trafficking	Schwenk et al. (2016), Dinamarca et al. (2019)
Amyloid-like protein 2 (APLP2)	APP	[a]	Schwenk et al. (2016)
Integral membrane proteins 2B and 2C (ITM2B/C)	APP	[a]	Schwenk et al. (2016)
Potassium chloride cotransporter (KCC2)	[a]	GABA$_B$R agonism triggers internalization of KCC2, altering driving force for Cl^--permeable GABA$_A$Rs	Wright et al. (2017)
Neuroligin-3	[a]	[a]	Schwenk et al. (2016)
Reticulocalbin-2	[a]	[a]	Schwenk et al. (2016)
Synaptotagmin-11	[a]	[a]	Schwenk et al. (2016)
CLC-10	[a]	[a]	Schwenk et al. (2016)
Neuron-specific gene family members 1 and 2 (NSG1/2)	[a]	[a]	Schwenk et al. (2016)
Calnexin	[a]	[a]	Schwenk et al. (2016)

[a]Not determined
"?" represents presumed interaction sites or impacts on signaling

presynaptically in axon terminals, while GB1b is expressed postsynaptically in dendritic spines (Kasten et al. 2015). However, GB1a is also expressed postsynaptically in dendritic branches, although it is mostly excluded from dendritic spines (Vigot et al. 2006, Kasten et al. 2015). In line with their distinct subcellular distribution patterns, studies in hippocampal neurons from GB1a$^{-/-}$ and GB1b$^{-/-}$ mice revealed that presynaptic (GB1a/GB2) GABA$_B$Rs mediate inhibition of neurotransmitter release, while postsynaptic (predominantly GB1b/GB2) $GABA_BRs$ generate slow inhibitory postsynaptic currents (Vigot et al. 2006).

GB1a is longer than GB1b (961 vs 841aa), as it contains two N-terminal protein interaction motifs, known as sushi domains (Lee et al. 2010). Sushi domains are highly conserved among species, are present in several GPCRs, and mediate protein interactions in a wide array of adhesion proteins (Grace et al. 2004, Lehtinen et al. 2004). Interestingly, the two GB1a sushi domains (SD1 and SD2) are structurally distinct (Blein et al. 2004), which may help to explain the different protein interactions observed between domains (Blein et al. 2004, Biermann et al. 2010, Hannan et al. 2012, Schwenk et al. 2016, Dinamarca et al. 2019, Rice et al. 2019). In addition to stabilizing GB1a/GB2 receptors at the cell surface (Hannan et al. 2012, 2016), the GB1a sushi domains are necessary and sufficient for axonal transport. Sushi domain mutations prevent GB1a from reaching axon terminals, and fusing the sushi domains to metabotropic glutamate receptor 1 (mGluR1) enables the somatodendritic protein to traffic down axons (Biermann et al. 2010). Thus, sushi domains act as axonal targeting signals, interacting with proteins to facilitate presynaptic transport. Several studies have identified a variety of proteins that directly or indirectly interact with the GB1a sushi domains (Table 1), with known or unknown functional influence. For example, experiments combining affinity purifications with mass spectrometry identified a number of proteins in the rodent brain that associate with the GB1a sushi domains. Of those identified, the β-amyloid precursor protein (APP), adherence junction-associated protein 1 (AJAP-1), and PILRα-associated neural protein (PIANP) directly bind, in a mutually exclusive manner, to GB1a SD1 (Schwenk et al. 2016, Dinamarca et al. 2019). Binding of APP to SD1 is necessary for vesicular trafficking of $GABA_BRs$ to axon terminals and presynaptic receptor function (Dinamarca et al. 2019). APP binds kinesin-1 adaptors of the c-Jun N-terminal kinase-interacting protein (JIP) and calsyntenin (CSTN) protein families to link APP/GB1a complexes to kinesin-1 motors that drive axonal transport (Valdés et al. 2012, Schwenk et al. 2016, Dinamarca et al. 2019, Fritzius and Bettler 2019). APP/GB1a complexes also stabilize APP at the cell surface to limit endosomal processing of APP to amyloid-beta (Aβ), a major component of plaques in Alzheimer's disease patients (Dinamarca et al. 2019). Taken together, APP/GB1a complex formation links $GABA_BR$ axonal transport to Aβ formation and supports the notion that dysregulated axonal trafficking (Kins et al. 2006, Thinakaran and Koo 2008) and reduced $GABA_BR$ expression (Chu et al. 1990, Iwakiri et al. 2005, Puthiyedth et al. 2016) observed in Alzheimer's disease may promote Aβ production (Dinamarca et al. 2019). The secreted form of APP (sAPP) has also been reported to bind GB1a SD1 and activate presynaptic $GABA_BRs$ in hippocampal neurons to inhibit neurotransmitter release (Rice et al. 2019). The sAPP-induced inhibition of neurotransmitter release was prevented by a $GABA_BR$ antagonist, suggesting that sAPP may function as an agonist or positive allosteric modulator of the $GABA_BR$ (Rice et al. 2019). In contrast to this report however, another study reported that neither APP nor sAPP influenced $GABA_BR$-dependent G protein signaling in a heterologous system (Dinamarca et al. 2019). Thus, further investigation is warranted to clarify the functional relevance of sAPP-$GABA_BR$ interactions.

Beyond GB1a and GB1b, many of the GB1c-n isoforms have only been detected at the mRNA level, and some are not conserved among species (Bettler et al. 2004, Jiang et al. 2012, Xu et al. 2014). While the physiological relevance of these isoforms remains unclear, there is a body of research to suggest that certain splice variants may be capable of fine-tuning endogenous $GABA_BR$ signaling in diverse ways – from forming functional receptors to inhibiting receptor association or function. For example, GB1c contains a single sushi domain (SD2) but shares a similar expression pattern to GB1a in the human brain (Pfaff et al. 1999, Jiang et al. 2012). When co-expressed with GB2, GB1c can form functional receptors in HEK-293 cells, suggesting that it may form a functional receptor in the rat brain, where it is expressed at the protein level (Pfaff et al. 1999). While both sushi domains are necessary for GB1a/GB2 stability, it is interesting to note that either GB1a sushi domain is sufficient for axonal transport (Biermann et al. 2010, Hannan et al. 2012). Thus, putative GB1c/GB2 receptors could theoretically function as presynaptic receptors.

GB1e/g/h/i/j/l/m/n may exist as secreted proteins, as they lack cytoplasmic domains and most, if not all, transmembrane domains (Jiang et al. 2012). GB1e is expressed at a relatively low level in the CNS of both humans and rodents but is a primary isoform in a variety of peripheral tissues. When expressed in heterologous systems, GB1e is both secreted and membrane-associated. While GB1e cannot form functional receptors when expressed with GB2, the strong interaction between the subunits is sufficient to disrupt the normal association between GB1a and GB2, but not sufficient to disrupt GB1a/GB2 receptor-mediated signaling through downstream effectors (Schwarz et al. 2000). GB1j is comprised of the two sushi domains plus 72 amino acids and is secreted when expressed in HEK-293 cells (Tiao et al. 2008). Purified sushi domains of GB1j (lacking the 72aa) impaired the inhibitory effect of $GABA_BRs$ on evoked and spontaneous glutamate release, but did not disrupt postsynaptic $GABA_BR$ activity in hippocampal neurons (Tiao et al. 2008). Although the entire protein was not studied, it was proposed that the sushi domains may scavenge an extracellular binding partner of GB1a that retains GB1a/GB2 receptors in presynaptic terminals. Given the recent identification of protein assemblies that interact with GB1a sushi domains to facilitate axonal transport (Table 1), it is theoretically possible that GB1 isoforms that contain sushi domains (GB1c/e/f/g/h/i/j) may function as dominant negative inhibitors of GB1a axonal transport and subsequent presynaptic $GABA_BR$ activity. When expressed in *Xenopus* oocytes, human-cloned GB1l and GB1m, but not GB1k, inhibited GB1a/GB2 receptor-mediated K^+ currents (Lee et al. 2010).

2.5 Oligomerization

At the cell surface, $GABA_BRs$ can exist in an equilibrium between heterodimers, tetramers, and higher-order oligomers in both heterologous systems and native neurons (Maurel et al. 2008, Schwenk et al. 2010, Comps-Agrar et al. 2011,

Gassmann and Bettler 2012, Calebiro et al. 2013). $GABA_BR$ heterodimers assemble by random collision into higher-order oligomers through weak and transient GB1-GB1 interactions (Comps-Agrar et al. 2011, Calebiro et al. 2013, Stewart et al. 2018, Xue et al. 2019). Destabilizing oligomers using competitors of the GB1-GB1 interaction, or a GB1 mutant, revealed different G protein coupling efficiencies depending on the oligomeric state of the $GABA_BR$ – suggesting a negative functional cooperativity among heterodimers within larger oligomers (Comps-Agrar et al. 2011). In addition to the reduced G protein coupling efficiency, recent reports also suggest that adjacent GABA-binding sites within $GABA_BR$ oligomers are not simultaneously occupied (Stewart et al. 2018).

2.6 $GABA_BR$ Signalosome

There is general consensus that $GABA_BRs$ function within macromolecular signaling complexes and that diverse protein interactions within these complexes can influence receptor activity, pharmacology, and localization. Proteomic approaches have identified an array of proteins that make up the $GABA_BR$ interactome and contribute to the functional diversity of native $GABA_BRs$ (Table 1) (Schwenk et al. 2010, 2016, Lujan and Ciruela 2012, Turecek et al. 2014, Fritzius and Bettler 2019). Many of the $GABA_BR$-interacting proteins show spatially and temporally restricted expression patterns throughout the brain, supporting the existence of dynamic and modular $GABA_BR$ complexes (Schwenk et al. 2016). Proteins within $GABA_BR$ complexes can be arranged in a hierarchy – from core components to peripheral components.

The receptor "core" is comprised of GB1, GB2, the heterotrimeric G protein, and K^+ channel tetramerization domain (KCTD) proteins. Obligate receptor components include GB1, GB2, and the heterotrimeric G protein, which represent the minimal components required for receptor signaling (Fritzius and Bettler 2019). Heterotrimeric G proteins critically link $GABA_BR$ heterodimers to primary downstream effectors (Gassmann and Bettler 2012, Fritzius and Bettler 2019). While $GABA_BRs$ can function without KCTD proteins, KCTD proteins are primary interactors that stably associate with $GABA_BRs$ in the brain (Schwenk et al. 2010, Turecek et al. 2014, Fritzius et al. 2017). Indeed, native neuronal KCTD proteins and $GABA_BR$ subunits robustly co-immunopurify with one another under stringent solubilization conditions (Schwenk et al. 2016).

Components of the receptor core anchor a multitude of "peripheral" components to form larger macromolecular complexes. Peripheral components include a wide variety of proteins that interact with GB1, GB2, the heterotrimeric G protein, or KCTD proteins (Tables 1 and 2). Some peripheral proteins influence receptor activity, pharmacology, or localization, while the functional relevance of others remains unknown (Schwenk et al. 2016, Fritzius and Bettler 2019).

Table 2 Potential $GABA_BR$ effectors

Effectors	*Interaction site(s)*	*Function(s)*	*References*
Adenylyl cyclase	$G\alpha_{i/o}$ (AC-I, III, V, VI, VIII, IX)	Inhibit cAMP production	Sadana and Dessauer (2009), Halls and Cooper (2017), Terunuma (2018)
	$G\beta\gamma$ (AC-I)	Inhibit cAMP production	
	$G\beta\gamma$ (AC-II, IV)	Stimulate cAMP production	
G protein-gated inwardly rectifying K^+ (GIRK) channel	$G\beta\gamma$, GB1/GB2?	Generate inhibitory postsynaptic currents, shunt excitatory input, inhibit action potential backpropagation, suppress dendritic Ca^{2+} spikes	David et al. (2006), Fowler et al. (2007), Ciruela et al. (2010), Lüscher and Slesinger (2010), Gassmann and Bettler (2012)
Voltage-gated Ca^{2+} channel	$G\beta\gamma$, KCTD16 (N-type channels)	Inhibit Ca^{2+} influx, suppress neurotransmitter release, suppress dendritic Ca^{2+} spikes	Gassmann and Bettler (2012), Schwenk et al. (2016)
		Facilitate Ca^{2+} influx, enhance neurotransmitter release (R-type and L-type channels)	Workman et al. (2013), Karls and Mynlieff (2015), Zhang et al. (2016)
		Inhibition of N-type channels can suppress BK channels	Garaycochea and Slaughter (2016)
Transient receptor potential vanilloid 1 (TRPV1) channel	GB1a	GB1 reverts TRPV1 sensitization	Hanack et al. (2015)
Hyperpolarization-activated cyclic nucleotide-gated 2 (HCN2) channel	KCTD16	Shorten the duration of inhibitory postsynaptic potentials	Schwenk et al. (2016)
TREK-2 channel	AKAP (tethers PKA to TREK-2 channels)	Reduced PKA-mediated tonic inhibition of TREK-2 channels triggers enhanced TREK-2 channel activity and postsynaptic hyperpolarization	Deng et al. (2009)
Metabotropic glutamate receptor 1 (mGluR1)	[a]	Extracellular Ca2+ interacts with $GABA_BRs$ to increase glutamate sensitivity of mGluR1	Hirono et al. (2001), Tabata et al. (2004)
		$GABA_BRs$ potentiate mGluR1/transient receptor potential canonical 3 (TRPC3) channel-mediated slow excitatory postsynaptic currents	Tian and Zhu (2018)

(continued)

Table 2 (continued)

Effectors	*Interaction site(s)*	*Function(s)*	*References*
M_2 muscarinic receptor (M_2R)	GB2 (C-terminus)	Facilitates M_2R signaling by preventing agonist-induced M_2R/GIRK channel co-internalization	Boyer et al. (2009)
Ca^{2+}-sensing receptor (CaR)	GB1, GB2	$GABA_BRs$ modulate CaR cell surface expression and signaling	Chang et al. (2007)
Transient receptor potential melastatin 3 (TRPM3) channel	Gβγ	$GABA_BRs$ inhibit TRPM3 channels	Badheka et al. (2017), Quallo et al. (2017)
$GABA_AR$ γ2s subunit	GB1	Increases GB1 cell surface expression, enhances $GABA_BR$ agonist-induced internalization	Balasubramanian et al. (2004)
ATF4/CREB2 and ATFx	GB1 (C-terminus)	$GABA_BR$-mediated transcriptional regulation	Nehring et al. (2000), White et al. (2000), Vernon et al. (2001)

[a]Not determined
"?" represents presumed interaction sites or impacts on signaling

3 $GABA_BR$-Dependent Signaling in Neurons

$GABA_BRs$ are expressed throughout the brain and are positioned within neurons at both postsynaptic (dendritic spines and shafts) and presynaptic (axon terminals) sites (Hammond and Mott 2015). In general, $GABA_BR$ activation inhibits neurons through G protein-dependent modulation of enzymes and ion channels. For example, activation of postsynaptic $GABA_BRs$ evokes a slow hyperpolarization of the postsynaptic membrane via activation of G protein-gated inwardly rectifying K^+ (Kir3/GIRK) channels (Lüscher and Slesinger 2010, Gassmann and Bettler 2012). Activation of presynaptic $GABA_BRs$ suppresses neurotransmitter release primarily through inhibition of voltage-gated Ca^{2+} channels (VGCCs) and reduced Ca^{2+} influx (Gassmann and Bettler 2012). Presynaptic $GABA_BRs$ function as either autoreceptors on GABAergic terminals or heteroreceptors on terminals releasing other neurotransmitters. Thus, presynaptic $GABA_BRs$ may be activated by GABA released from GABAergic terminals, or spillover of GABA from neighboring terminals, to suppress neurotransmitter release (Harrison et al. 1988, Wu and Saggau 1995, Boyes and Bolam 2003, Gassmann and Bettler 2012). By blocking the release of different types of neurotransmitters, $GABA_BRs$ can have excitatory or inhibitory influences at the circuit level.

3.1 GABA$_B$R Coupling to G Proteins

Heterotrimeric G proteins mediate signaling by coupling receptors to enzymes, ion channels, and other effector proteins. The heterotrimeric G protein is comprised of three distinct subunits (α, β, γ); 35 subunits (16 Gα, 5 Gβ, 14 Gγ) have been identified in humans (Milligan and Kostenis 2006, Hillenbrand et al. 2015). Inactive heterotrimeric G proteins (Gαβγ) associate with GABA$_B$Rs through direct interactions with GB2 and KCTD proteins (Schwenk et al. 2016, Frangaj and Fan 2018, Fritzius and Bettler 2019). Selective coupling of heterotrimeric G proteins to GABA$_B$Rs is primarily determined by the Gα subunit. Studies using *N*-ethylmaleimide (NEM), antisense knockdown, and G protein toxins helped reveal that prototypical GABA$_B$Rs couple to pertussis toxin (PTX)-sensitive G proteins, including most members of the Gα_i and Gα_o (G$\alpha_{i/o}$) families (Morishita et al. 1990, Knott et al. 1993, Odagaki et al. 2000, Odagaki and Koyama 2001, Milligan and Kostenis 2006).

Gα subunits may guide functional coupling of GABA$_B$Rs to different effectors. In general, adenylyl cyclase is predominately regulated by Gα_i, while GIRK channels and VGCCs are largely regulated by Gα_o. In heterologous systems, constitutively active mutants of Gα_{i1-3} proteins inhibited adenylyl cyclase, while Gα_o mutants did not (Wong et al. 1992). In a reconstituted system using G proteins extracted from bovine cerebral cortex, GABA$_B$R coupling to adenylyl cyclase was shown to involve Gα_i (Gα_{i1} or Gα_{i2}), but not Gα_o (Nishikawa et al. 1997). In transfected HEK-293 cells, functional coupling of GABA$_B$Rs to GIRK channels preferentially involved Gα_o and Gα_{i2} (Leaney and Tinker 2000). Interestingly, GB1a/GB2 receptors predominantly signaled through Gα_o, while GB1b/GB2 receptors signaled equally through Gα_o and Gα_{i2} (Leaney and Tinker 2000). Slice electrophysiological studies using myristoylated G protein peptide inhibitors in ventrolateral periaqueductal gray neurons revealed that GABA$_B$R-GIRK currents are mediated by Gα_{o1}, but not Gα_{i1-3} (Mcpherson et al. 2018). Gα_o, but not Gα_i, was also reported to couple GABA$_B$Rs to VGCCs in dorsal root ganglion neurons (Campbell et al. 1993, Menon-Johansson et al. 1993). While less is known regarding the contribution of specific Gβ and Gγ subunits to GABA$_B$R-effector coupling, Gβ2γ3 was identified as a mediator of GABA$_B$R-GIRK signaling (Schwindinger et al. 2012). In addition, Gβ1, Gβ2, and Gγ2 co-immunopurified with native neuronal GABA$_B$Rs, suggesting their potential involvement in GABA$_B$R signal transduction (Schwenk et al. 2016).

There is some evidence that GABA$_B$Rs may also couple to PTX-insensitive G proteins. Intracellular recordings in neurons from rat hippocampal slices showed that exposure to PTX did not fully prevent the baclofen-induced suppression of excitatory postsynaptic potentials (EPSPs), suggesting that presynaptic GABA$_B$R-dependent inhibition may be mediated through PTX-sensitive and PTX-insensitive G proteins (Potier and Dutar 1993). Interestingly, a recent proteomics study revealed that native GABA$_B$Rs co-immunopurify with the PTX-insensitive member of the G$\alpha_{i/o}$ family, Gα_z (Simonds 1999, Schwenk et al.

2016). While no functional evidence currently exists to support GABA$_B$R-Gα_z coupling, studies in heterologous systems show that Gα_z can regulate adenylyl cyclase, GIRK channels, and VGCCs through several G$\alpha_{i/o}$-coupled GPCRs (Jeong and Ikeda 1998, Wettschureck and Offermanns 2005). Activation of GABA$_B$Rs during a narrow window of development has also been reported to increase intracellular Ca^{2+} in a subset of neonatal hippocampal neurons through Gα_q signaling and protein kinase C (PKC) α (PKCα) activation (Carter and Mynlieff 2004, Bray and Mynlieff 2009, Karls and Mynlieff 2015). Since GABA$_B$Rs do not functionally couple to PTX-insensitive Gα_q to activate phospholipase C (PLC) in heterologous systems, it remains unclear how functional coupling may occur in neurons (Franek et al. 1999).

3.2 GABA$_B$R Regulation of Effectors

GABA$_B$Rs regulate the activity of a variety of effectors through direct or indirect interactions (Table 2). Here, we will review the functional relevance of GABA$_B$R-mediated regulation of three prototypical effectors – GIRK channels, VGCCs, and adenylyl cyclase.

3.2.1 GIRK Channels

GIRK channels are homo- or heterotetramers formed by four subunits (GIRK1–4); GIRK1–3 show broad and overlapping expression throughout the CNS, while GIRK4 is primarily found in the heart (Yang et al. 1995, Karschin et al. 1996, Luján and Aguado 2015). While multiple GIRK channel subtypes are present throughout the rodent brain, the GIRK1/2 heterotetramer is generally considered the prototypical neuronal GIRK channel (Luján et al. 2014). GIRK channels are predominantly distributed within the somatodendritic compartment, at both perisynaptic and extrasynaptic sites (Luján and Aguado 2015). Here, they mediate the postsynaptic inhibitory effect of multiple neurotransmitters through G$\alpha_{i/o}$-coupled GPCRs, including GABA$_B$Rs (Andrade et al. 1986, Misgeld et al. 1995, Lüscher et al. 1997, Lüscher and Slesinger 2010, Luján et al. 2014). Interestingly, some ultrastructural and functional data suggest that GABA$_B$Rs and GIRK channels colocalize on axon terminals and that GABA$_B$R-GIRK signaling inhibits neurotransmission (Ladera et al. 2008, Fernández-Alacid et al. 2009, Michaeli and Yaka 2010, Luján et al. 2014). However, others did not find evidence that presynaptic GABA$_B$Rs activate GIRK channels to inhibit neurotransmission (Lüscher et al. 1997, Takahashi et al. 1998).

GABA$_B$R-GIRK signaling has been detected in many neuron types throughout the brain (Lüscher and Slesinger 2010, Luján et al. 2014). GABA$_B$Rs activate GIRK channels through Gβγ dimers (Reuveny et al. 1994, Wickman et al. 1994, Lüscher and Slesinger 2010, Whorton and MacKinnon 2013), whereby direct binding of Gβγ to GIRK channels enhances gating by stabilizing an interaction between the channel and phosphatidylinositol-4,5-bisphosphate (PIP_2), a cofactor required for channel gating (Logothetis et al. 1987, Huang et al. 1998, Luján et al. 2014). Activation of GIRK channels evokes a slow hyperpolarizing conductance via K^+ efflux that can shunt excitatory input (Nicoll 2004), inhibit backpropagation of action potentials, and block the generation of dendritic Ca^{2+} spikes (Leung and Peloquin 2006). The critical role of GIRK channels in tempering cellular excitability is evident in $GIRK2^{-/-}$ mice, which are hyperactive and susceptible to spontaneous seizures (Signorini et al. 1997, Blednov et al. 2001). These behavioral phenotypes are similarly observed in $GB1^{-/-}$ and $GB2^{-/-}$ mice, underlining the importance of both forms of inhibitory signaling throughout the brain (Gassmann et al. 2004).

$GABA_B$R-GIRK coupling efficiency can differ in individual neurons based on GIRK channel subunit composition (Luján et al. 2014). For example, dopamine neurons of the ventral tegmental area (VTA) have a lower $GABA_B$R-GIRK coupling efficiency than adjacent GABA neurons. Thus, lower concentrations of baclofen (or GABA) inhibit GABA neurons, while higher concentrations are required to directly inhibit dopamine neurons. The absence of GIRK1 and presence of GIRK2 and GIRK3 in VTA dopamine neurons underlies this effect (Cruz et al. 2004).

There is evidence that $GABA_B$Rs, G proteins, and GIRK channels form macromolecular signaling complexes that enable specific and rapid signaling upon receptor activation (Lüscher and Slesinger 2010, Luján et al. 2014). $GABA_B$Rs, $G\alpha_{i/o}$-type G proteins, and GIRK channels all associate with lipid rafts, suggesting that they may interact together (Oh and Schnitzer 2001, Koyrakh et al. 2005, Becher et al. 2008). $GABA_B$Rs and GIRK channels also co-cluster in the dendrites of rodent hippocampal neurons (Kulik et al. 2006, Booker et al. 2013, Fajardo-Serrano et al. 2013) and cerebellar neurons (Ciruela et al. 2010, Luján et al. 2017). Immunoprecipitation experiments revealed $GABA_B$R/GIRK and $G\alpha_o$/GIRK co-assemblies in heterologous systems (Clancy et al. 2005, David et al. 2006, Ciruela et al. 2010) and $GABA_B$R/GIRK co-assemblies in the mouse cerebellum (Ciruela et al. 2010, Luján et al. 2017). Evidence in support of direct protein interactions largely comes from biochemical assays in heterologous systems. BRET/FRET experiments revealed close interactions (<100 Å) between $GABA_B$Rs and GIRK2 homotetramers, GIRK1/4 or GIRK1/3 heterotetramers, and $G\alpha_o$ proteins (David et al. 2006, Fowler et al. 2007, Ciruela et al. 2010).

Some functional data also support the possibility of a pre-coupling of components in the form of a macromolecular complex. GIRK channels expressed in heterologous systems or native neurons can signal even in the absence of receptor activation, supporting the possibility that some signaling components are pre-coupled (Kahanovitch et al. 2017). Increasing the surface expression of $GABA_B$Rs in *Xenopus* oocytes reduced this basal GIRK channel activation, likely via downregulation of GIRK channel surface expression during constitutive $GABA_B$R internalization, as has been reported elsewhere (Padgett et al. 2012, Hearing et al. 2013, Lecca et al. 2016). This could indicate physical interactions among $GABA_B$Rs and GIRK channels and perhaps a pre-coupling between components of this signaling cascade.

There is also evidence against the existence of pre-coupled macromolecular complexes and in favor of a collision-coupling mode of $GABA_B$R-GIRK signaling. For example, increasing the surface expression of $GABA_B$Rs in *Xenopus* oocytes accelerated GIRK channel activation, suggesting that $GABA_B$Rs or G proteins can diffuse freely in the membrane to activate GIRK channels (Kahanovitch et al. 2017). Furthermore, unlike the direct interactions reported in heterologous systems, native neuronal $GABA_B$Rs and GIRK channels did not co-immunopurify with one another in a high-resolution proteomics study (Schwenk et al. 2016). Thus, the mode of coupling between components of this signalosome remains unclear. Taken together, these results suggest that a putative $GABA_B$R-G protein-GIRK complex may be dynamic, allowing for dissociation and reassociation of components (Kahanovitch et al. 2017). The formation of dynamic complexes, with low-affinity and/or transient interactions, could explain why $GABA_B$Rs and GIRK channels did not associate in vivo (Schwenk et al. 2016).

3.2.2 Voltage-Gated Ca^{2+} Channels

VGCCs are regulated by many $G\alpha_{i/o}$-coupled GPCRs, including $GABA_B$Rs. VGCCs are typically closed at resting membrane potentials but are opened by membrane depolarization, leading to Ca^{2+} influx. Ca^{2+} influx depolarizes the cellular membrane, facilitates synaptic vesicle release, and, as a secondary messenger, regulates diverse physiological properties (Proft and Weiss 2015). VGCCs are composed of pore-forming subunits encoded by ten mammalian genes. Seven genes encode the high-voltage-activated Ca^{2+} channel subfamily including L-type (Ca_V1.1 to 1.4), P/Q-type (Ca_V2.1), N-type (Ca_V2.2), and R-type (Ca_V2.3) channels, while three genes encode low-voltage-activated T-type (Ca_V3.1–3.3) channels (Proft and Weiss 2015). In general, $GABA_B$Rs inhibit N- and P/Q-type channels in most neurons and L-, T-, and R-type channels in select neuron populations (Maguire et al. 1989, Chalifoux and Carter 2011, Proft and Weiss 2015).

$GABA_B$R activation inhibits N- and P/Q-type channels in presynaptic terminals of both glutamatergic and GABAergic neurons, as well as R-type channels in some glutamatergic neurons (Wu and Saggau 1995, Proft and Weiss 2015, Burke and Bender 2019). Inhibition of presynaptic VGCCs reduces Ca^{2+} influx and decreases the probability of neurotransmitter release (Burke and Bender 2019). $GABA_B$Rs inhibit VGCCs through direct interactions between $G\beta\gamma$ and the channel (Herlitze et al. 1996, Waard et al. 1997, Burke and Bender 2019). Mechanistically, $G\beta\gamma$ binding to VGCCs slows channel activation kinetics and induces a positive shift in the voltage dependence to inhibit Ca^{2+} influx (Bean 1989, Colecraft et al. 2000). $G\beta\gamma$-mediated inhibition can be relieved by strong depolarization or eventual dissociation of $G\beta\gamma$ from the channel (Colecraft et al. 2000, 2001, Proft and Weiss 2015). Postsynaptic $GABA_B$Rs also inhibit several VGCC subtypes in dendrites and spines (Pérez-Garci et al. 2006, Chalifoux and Carter 2011, Booker et al. 2018). Postsynaptic $GABA_B$R-VGCC signaling prevents dendritic Ca^{2+} spikes to reduce cellular excitability and limit the actions of Ca^{2+} as a secondary messenger (Malenka 1991, Pérez-Garci et al. 2006, Chalifoux and Carter 2011, Brini et al. 2014, Booker et al. 2018).

Similar to GIRK channels, $GABA_B$Rs and VGCCs have been proposed to form signaling complexes that facilitate tight functional coupling through membrane-delimited Gβγ interactions. FRET experiments revealed that GB1a/GB2 receptors associate with Gβγ and N-type channels in hippocampal pyramidal neuron boutons, suggesting the formation of signaling complexes that facilitate $GABA_B$R/VGCC-mediated presynaptic inhibition (Laviv et al. 2011). In line with this, a high-resolution proteomics approach showed that native neuronal N-type channels assemble with GB1a/GB2 receptors (Schwenk et al. 2016). Electrophysiological, biochemical, and ultrastructural evidence also support the existence of postsynaptic signaling complexes. $GABA_B$Rs co-assemble and co-cluster with P/Q-type channels in dendritic shafts of cerebellar Purkinje neurons (Luján et al. 2017) and co-cluster with L-type channels in dendrites of hippocampal somatostatin interneurons to inhibit postsynaptic Ca^{2+} influx (Booker et al. 2018).

While $GABA_B$R-VGCC signaling has well-documented inhibitory influences on neurons, under certain conditions it can also exert excitatory influences. For example, presynaptic VGCC inhibition often suppresses the release of inhibitory neurotransmitters (e.g., GABA, glycine) to disinhibit downstream neurons. $GABA_B$R-mediated inhibition of N-type channels in rat retinal neurons also suppressed big conductance Ca^{2+}-activated K^+ (BK) channels, which led to a net increase in neuronal excitability (Garaycochea and Slaughter 2016). As mentioned earlier, $GABA_B$Rs activated L-type channels via $G\alpha_q$ signaling and PKCα activation in neonatal hippocampal neurons (Carter and Mynlieff 2004, Bray and Mynlieff 2009, Karls and Mynlieff 2015). A similar activation of L-type channels through $GABA_B$Rs has been reported in response to N-methyl-D-aspartate receptor (NMDAR) blockade (Workman et al. 2013). Lastly, $GABA_B$Rs activated R-type channels on medial habenula neurons to facilitate Ca^{2+} influx and trigger neurotransmitter release into the interpeduncular nucleus (Zhang et al. 2016, Koppensteiner et al. 2017).

3.2.3 Adenylyl Cyclase

Adenylyl cyclase catalyzes the synthesis of cyclic AMP (cAMP), a key second messenger that regulates diverse cellular processes (Smit and Iyengar 1998, Halls and Cooper 2017). Ten adenylyl cyclase isoforms are expressed throughout the mammalian brain – nine transmembrane isoforms (AC-I–IX) and one soluble isoform (AC-X) (Sadana and Dessauer 2009, Halls and Cooper 2017). While all transmembrane isoforms can be stimulated by direct interactions with $G\alpha_s$, $G\alpha_{i/o}$ proteins directly inhibit AC-I, AC-III, AC-V, AC-VI, AC-VIII, and AC-IX. The Gβγ dimers also inhibit AC-I but can stimulate AC-II and AC-IV (Sadana and Dessauer 2009, Halls and Cooper 2017, Terunuma 2018). Thus, $GABA_B$Rs can bidirectionally regulate adenylyl cyclase activity through typical $G\alpha_{i/o}$- or $G\beta\gamma$-mediated inhibition or atypical $G\beta\gamma$-mediated stimulation.

Several early studies had shown that $GABA_BR$ agonists inhibit basal or forskolin-stimulated adenylyl cyclase activity in neurons via PTX-sensitive G proteins (Wojcik and Neff 1984, Knight and Bowery 1996, Simonds 1999, Bettler et al. 2004). Others found that $GABA_BRs$ can stimulate adenylyl cyclase-induced cAMP production during co-activation of $G\alpha_s$-coupled receptors by norepinephrine, isoprenaline, histamine, or vasoactive intestinal polypeptide (Bettler et al. 2004). This atypical $G\beta\gamma$-mediated stimulation of adenylyl cyclase (AC-II and AC-IV) requires the presence of active $G\alpha_s$, thus demonstrating a form of G protein crosstalk between $GABA_BRs$ and $G\alpha_s$-coupled GPCRs that augments cAMP production (Simonds 1999, Bowery et al. 2002, Calver et al. 2002). The $GABA_BR$-mediated bidirectional regulation of cAMP levels was confirmed in vivo using microdialysis in freely moving rats (Hashimoto and Kuriyama 2002).

Typical $GABA_BR$-induced reduction in cAMP and subsequent protein kinase A (PKA) activity influence several downstream processes. Presynaptic reductions in cAMP levels inhibit vesicle fusion and spontaneous neurotransmitter release (Sakaba and Neher 2003, Rost et al. 2011). Postsynaptic reductions in PKA activity alleviate an A-kinase anchoring protein (AKAP)-dependent tonic inhibition of TREK2 channels (Deng et al. 2009), decrease the Ca^{2+} permeability of NMDARs (Chalifoux and Carter 2010), enhance the magnitude of tonic $GABA_AR$ currents (Connelly et al. 2013), and influence gene expression (Ghorbel et al. 2005, Fukui et al. 2008, Schwirtlich et al. 2010). Taken together, $GABA_BR$-dependent regulation of adenylyl cyclase is poised to influence diverse cellular processes across short and long timeframes – by modifying neuronal excitability and synaptic transmission, altering levels of intracellular secondary messengers (cAMP, Ca^{2+}), and regulating gene expression.

4 Regulation of $GABA_BR$-Dependent Signaling in Neurons

Tight control over the timing and strength of $GABA_BR$-dependent signaling is crucial for establishing a proper inhibitory tone that balances excitation. In this regard, signaling through $GABA_BRs$ is subject to regulation via a myriad of mechanisms.

4.1 Desensitization

Desensitization is a common regulatory mechanism of GPCRs to prevent overstimulation. For many GPCRs, desensitization involves direct phosphorylation of the receptor by GPCR kinase (GRK), followed by arrestin binding and dynamin-dependent and clathrin-mediated endocytosis (Gurevich and Gurevich 2019).

Internalized receptors accumulate in endosomal sorting compartments where they may either be dephosphorylated and recycled back to the cell surface or targeted to lysosomes for degradation (Benke et al. 2012, Iacovelli and De Blasi 2013, Lefkowitz 2013).

While prolonged activation of $GABA_B$Rs induces desensitization of the receptor response, $GABA_B$R desensitization does not involve receptor internalization via the classical GRK phosphorylation and arrestin recruitment pathway. Rather, surface stability of $GABA_B$Rs is regulated through a variety of phosphorylation-independent and phosphorylation-dependent mechanisms (Fairfax et al. 2004, Grampp et al. 2008, Benke et al. 2012, Raveh et al. 2015). Although GRKs do not phosphorylate $GABA_B$Rs, GRK4 and GRK5 still promote agonist-induced desensitization of the $GABA_B$R response (Perroy et al. 2003, Fairfax et al. 2004, Kanaide et al. 2007). Since GRK4 and GRK5 directly associated with GB2, where they competed with the G protein for binding, they were proposed to induce desensitization by uncoupling the G protein from the $GABA_B$R (Benke et al. 2012, Raveh et al. 2015). This interaction may be highly cell specific, as biochemical analyses reveal that GRK4 and GRK5 are minimally expressed in tissues that have high $GABA_B$R expression, including the cerebral cortex and hippocampus (Sallese et al. 2000, Sato et al. 2015). GRK2 has also been reported to induce desensitization of several inhibitory GPCRs through a phosphorylation-independent mechanism involving the sequestration of $\beta\gamma$ (Raveh et al. 2010, 2015). In a heterologous system, GRK2 increased desensitization of $GABA_B$R-GIRK currents by ~30% (Turecek et al. 2014).

4.2 *Phosphorylation*

Unlike many GPCRs, $GABA_B$R activity is not correlated with the overall phosphorylation state of the receptor, as phosphorylation of different residues influences $GABA_B$R activity in distinct ways (Perroy et al. 2003, Terunuma 2018). There are five known phosphorylation sites on $GABA_B$Rs that regulate endocytosis, surface stability, and desensitization. These include serine 867 (S867) and S917/923 on GB1 and S783 and S892 on GB2. Several kinases mediate phosphorylation at these sites.

4.2.1 CaMKII

Ca^{2+}/calmodulin-dependent protein kinase II (CaMKII) phosphorylates S867 on primarily GB1b, leading to the dynamin-dependent endocytosis of $GABA_B$Rs that couple to GIRK channels (Guetg et al. 2010). Glutamatergic signaling downregulates $GABA_B$Rs (Guetg et al. 2010, Maier et al. 2010), in part through the activation of NMDARs that enhance CaMKII-mediated phosphorylation of S867 to promote $GABA_B$R internalization (Guetg et al. 2010). Indeed, blockade of either

CaMKII activity or phosphorylation of S867 was sufficient to prevent $GABA_BR$ internalization in hippocampal neurons (Guetg et al. 2010, Maier et al. 2010).

4.2.2 AMPK

AMP-activated protein kinase (AMPK) is a serine-threonine kinase that functions as an energy sensor that is activated by increased cellular levels of AMP due to high metabolic activity, or during anoxia or ischemia (Carling 2005, Carling et al. 2011). AMPK binds to the C-terminus of GB1, where it can phosphorylate two sites on GB1 (S917/923) and one site on GB2 (S783) (Kuramoto et al. 2007). The physiological relevance of AMPK-induced phosphorylation at all three sites was examined by measuring AMPK-mediated $GABA_BR$-GIRK coupling in HEK-293 cells. Phosphorylation of S783 on GB2 reduced desensitization of $GABA_BRs$ and enhanced $GABA_BR$-GIRK coupling by stabilizing receptors at the plasma membrane. Ischemic brain injury enhanced S783 phosphorylation in the hippocampus of rats, and S783 phosphorylation promoted neuronal survival of cultured hippocampal neurons after chemical anoxia (Kuramoto et al. 2007). Therefore, AMPK-mediated phosphorylation of S783 may play a neuroprotective role in limiting excitotoxicity by maintaining $GABA_BR$ inhibitory tone during times of high metabolic stress or ischemic injury (Kuramoto et al. 2007).

AMPK-mediated phosphorylation of S783 is bidirectionally regulated by glutamatergic signaling through NMDARs (Terunuma et al. 2010, Terunuma 2018). Transient activation of NMDARs enhances AMPK activity and promotes S783 phosphorylation, while prolonged NMDAR activation promotes S783 dephosphorylation (Terunuma et al. 2010). Prolonged NMDAR activity activates protein phosphatase 2A (PP2A), which dephosphorylates S783 and targets $GABA_BRs$ for lysosomal degradation, thus reducing surface expression and $GABA_BR$ function. Concurrent activation with $GABA_BRs$ prevents the NMDAR/PP2A-mediated reduction in $GABA_BR$ surface expression, likely via membrane hyperpolarization or decreased Ca^{2+} permeability of NMDARs (Terunuma et al. 2010). Thus, glutamatergic and GABAergic signaling delicately control the phosphorylation state of $GABA_BRs$ to regulate intracellular trafficking and cell surface stability. Interestingly, studies using S783A knock-in mice, in which mutation of serine to alanine decreases $GABA_BR$ degradation, revealed that the S783A mutation selectively enhanced postsynaptic, but not presynaptic, $GABA_BR$ activity (Terunuma et al. 2014). This suggests that presynaptic receptors are highly stable and have lower rates of endocytosis than postsynaptic receptors that have higher rates of phosphorylation-dependent internalization.

4.2.3 PKA

PKA phosphorylates the cytoplasmic tail of GB2 at S892, leading to increased $GABA_BR$ surface stability and reduced slow desensitization in HEK-293 and hippocampal cells (Couve et al. 2002). Prolonged activation of $GABA_B$Rs inhibits adenylyl cyclase to reduce PKA activity and S892 phosphorylation, which coincides with increased endocytosis-independent $GABA_B$R degradation (Fairfax et al. 2004). $GABA_B$R degradation induced by chronic exposure to baclofen is attenuated by either PKA activation or co-stimulation of $G\alpha_s$-coupled β-adrenergic receptors (Couve et al. 2002, Fairfax et al. 2004, Benke 2010). Thus, PKA-induced phosphorylation of S892 and $GABA_B$R surface stability are bidirectionally regulated by G protein signaling cascades that modulate PKA activity.

4.2.4 PKC

PKC has been reported to phosphorylate GB1 at an unknown site in Chinese hamster ovary cells. Activation of $GABA_B$Rs enhances PKC recruitment to the plasma membrane, induces phosphorylation of GB1, and disrupts the direct interaction between NEM-sensitive fusion (NSF) proteins and $GABA_B$Rs to facilitate agonist-induced, internalization-independent desensitization (Pontier et al. 2006).

4.3 Ubiquitination

Ubiquitination is a posttranslational modification that involves covalent attachment of ubiquitin to a target protein, generally directing the protein to proteasomes or lysosomes for degradation (Sarker et al. 2011). Many GPCRs undergo reversible ubiquitin modifications that regulate receptor degradation, among other functions (Dores and Trejo 2012, Cottrell 2013, Kommaddi and Shenoy 2013, Kennedy and Marchese 2015). Ubiquitination is reported to regulate $GABA_B$R trafficking from the endoplasmic reticulum to the cell surface via increased proteasomal degradation. Lys(48)-linked polyubiquitination of lysines 767/711 in the GB2 C-terminus promotes constitutive proteasomal degradation of $GABA_B$Rs in cultured cortical neurons, and inactivation of these sites increases cell surface receptor levels and enhances $GABA_B$R signaling (Zemoura et al. 2013).

Ubiquitination of GB1 is also reported to control lysosomal degradation of $GABA_B$Rs. GB1 is ubiquitinated at multiple sites by Mind bomb-2 (MIB2), a ubiquitin ligase catalyzing Lys(63)-linked ubiquitination, which promotes lysosomal degradation of $GABA_B$Rs (Zemoura et al. 2016). Mutational inactivation of putative GB1 ubiquitination sites prevented lysosomal degradation. Interestingly, MIB2-induced ubiquitination is believed to contribute to the glutamate-induced downregulation of $GABA_B$Rs (Zemoura et al. 2016). A recent follow-up study supports this notion, revealing that MIB2-induced GB1 ubiquitination is largely dependent on the phosphorylation state of S867 on GB1. CaMKIIβ-induced S867 phosphorylation promotes, while S867 dephosphorylation inhibits, Lys(63)-linked ubiquitination of GB1 (Zemoura et al. 2019).

PKC has also been reported to promote ubiquitination, internalization, and degradation of $GABA_BRs$. Cell surface $GABA_BRs$ undergo PKC-mediated constitutive ubiquitination and subsequent internalization, and deubiquitination of the receptor is catalyzed post-endocytically by USP14, a deubiquitinase that directly interacts with GB1 to target $GABA_BRs$ for lysosomal degradation (Lahaie et al. 2016).

4.4 KCTD Proteins

The four KCTD proteins (KCTD8, 12, 12b, 16) assemble as homo- or heteromeric pentamers on the C-terminus of GB2 (Fritzius et al. 2017, Pinkas et al. 2017, Fritzius and Bettler 2019, Zuo et al. 2019), where they stabilize G proteins at the receptor and regulate the kinetics of G protein-dependent signaling (Turecek et al. 2014, Zheng et al. 2019). KCTD proteins accelerate the onset of $GABA_BR$-GIRK currents, and KCTD12 and KCTD16 additionally increase agonist potency, as seen by a reduced EC_{50} value of baclofen-evoked GIRK currents (Schwenk et al. 2010). KCTD12 and KCTD12b also induce fast desensitization of $GABA_BR$-GIRK currents by directly binding receptor-activated Gβγ dimers to uncouple Gβγ from GIRK channels (Schwenk et al. 2010, Turecek et al. 2014, Fritzius et al. 2017, Fritzius and Bettler 2019, Zheng et al. 2019). Interestingly, PKA-mediated phosphorylation of S892 on GB2 can regulate KCTD12-induced fast desensitization (Adelfinger et al. 2014). PKA activation in hippocampal neurons slows, while PKA inhibition accelerates, KCTD12-induced fast desensitization of $GABA_BR$-GIRK currents. PKA fails to regulate desensitization in knock-in mice with a serine 892 to alanine mutation (S892A), demonstrating that phosphorylation of S892 slows KCTD12-induced fast desensitization in vivo (Adelfinger et al. 2014). In addition to regulating G protein signaling kinetics, KCTD proteins also scaffold effector channels and other proteins at the $GABA_BR$ (Table 1). For example, N-type Ca^{2+} channels, hyperpolarization-activated cyclic nucleotide-gated 2 (HCN2) channels, and 14-3-3 proteins associate with $GABA_BRs$ through direct interactions with KCTD16 (Schwenk et al. 2016).

4.5 RGS Proteins

Regulator of G protein signaling (RGS) proteins are GTPase-accelerating proteins (GAPs) that facilitate termination of G protein signaling by promoting hydrolysis of GTP on active Gα to enable reassembly of the heterotrimeric G protein complex (Anderson et al. 2009, Gerber et al. 2016). The mammalian RGS protein superfamily is divided into eight subfamilies (RZ, R4, R7, R12, RA, GED, GRK, SNX) based on amino acid sequence or structural similarity. While structural diversity among RGS proteins does explain the existence of noncanonical cell signaling roles, the RGS homology domain that is critical for accelerating GTPase activity is highly conserved among many members (Siderovski and Willard 2005, Gerber et al. 2016, Squires et al. 2018). RGS proteins across several subfamilies have been shown to regulate the kinetics of G protein-dependent signaling through $GABA_B$Rs.

4.5.1 R7 RGS/Gβ5

The R7 RGS protein family is composed of four members (RGS6, RGS7, RGS9, RGS11) that play critical roles in fundamental neuronal processes, including vision, motor control, reward behavior, and nociception (Anderson et al. 2009). R7 RGS proteins form obligate heterodimers with G protein β5 (Gβ5) through interactions at their Gγ-like domains (Snow et al. 1999, Witherow et al. 2000, Hollinger and Hepler 2002). RGS/Gβ5 heterodimers can then form reversible complexes with adaptor proteins, including R7-binding protein (R7BP) (Drenan et al. 2005, Martemyanov et al. 2005, Grabowska et al. 2008, Patil et al. 2018). When palmitoylated, R7BP anchors the heterodimeric complex at the plasma membrane and prevents RGS protein degradation (Drenan et al. 2005, 2006, Jia et al. 2011). R7BP also facilitates the functional association of RGS/Gβ5 with GIRK channels to promote deactivation of G proteins (Xie et al. 2010, Jia et al. 2014, Ostrovskaya et al. 2014). Indeed, genetic ablation of either RGS6, RGS7, Gβ5, or R7BP prolongs deactivation kinetics of $GABA_B$R-GIRK currents (Xie et al. 2010, Maity et al. 2012, Ostrovskaya et al. 2014). Ablation of RGS7 or R7BP also enhanced the coupling efficiency of $GABA_B$R-GIRK signaling, increasing the potency of baclofen-induced GIRK currents (Ostrovskaya et al. 2014).

In line with their functional association, biochemical, electrophysiological, and ultrastructural evidence support the existence of macromolecular complexes formed of RGS7/Gβ5, $GABA_B$Rs, and GIRK channels on dendritic spines of hippocampal CA1 pyramidal neurons (Xie et al. 2010, Fajardo-Serrano et al. 2013). Insights from the RGS7-Gβ5-R7BP crystal structure reveal that the orientation of the complex is compatible with macromolecular assemblies involving $GABA_B$Rs and GIRK channels (Patil et al. 2018).

In addition to forming complexes with R7BP, RGS7/Gβ5 can also assemble with G protein-coupled receptor 158 (GPR158) (Orlandi et al. 2012, Ostrovskaya et al. 2018). Formation of either complex is mutually exclusive and facilitates trafficking of RGS7 to the plasma membrane (Orlandi et al. 2012). The ability of RGS7 to negatively regulate $GABA_B$R signaling through GIRK channels or P/Q/N-type channels is enhanced by R7BP but opposed by GPR158 (Ostrovskaya et al. 2018). Interestingly, the RGS7/Gβ5-GPR158 complex has been reported to suppress homeostatic regulation of cAMP by $GABA_B$Rs (Orlandi et al. 2018). Altogether, this suggests that RGS7/Gβ5 dimers exist in two separate complexes at the plasma membrane that may guide RGS7-meditated regulation toward particular effector systems.

4.5.2 R4 RGS Proteins

Two members of the R4 RGS subfamily (RGS2 and RGS4) have been implicated in negatively regulating $GABA_BR$-GIRK signaling in neurons. RGS2 reduces the coupling efficiency of $GABA_BRs$ with heteromeric GIRK2/3 channels in VTA dopamine neurons. Evidence from immunoelectron microscopy and slice electrophysiology in GIRK subunit-specific knockout mice suggests that the effect of RGS2 on $GABA_BR$-GIRK signaling uniquely requires the GIRK3 subunit, and FRET analysis revealed direct interactions between RGS2 and GIRK3 (Labouèbe et al. 2007).

RGS4 has been proposed to form a signaling complex with $GABA_BRs$ to terminate $GABA_BR$-GIRK signaling. Double immunohistochemistry and immunoprecipitation assays revealed that RGS4 and $GABA_BRs$ associate together in the prefrontal cortex and hypothalamus, and FRET analysis in transfected HEK-293 cells indicated direct interactions between RGS4 and either GB1 or GB2 (Fowler et al. 2007, Kim et al. 2014). RGS4 enhances GIRK channel deactivation rates within a second of agonist application in vitro, and RGS4 expression in GIRK-transfected CHO cells mimics the fast deactivation kinetics observed in hippocampal neurons and atrial myocytes (Doupnik et al. 1997). RGS4 has also been reported to limit crosstalk between two $G\alpha_{i/o}$-coupled receptors, $GABA_BRs$ and A_2 adenosine receptors (A_2Rs), in pyramidal neurons of the prefrontal cortex. Within single dendritic spines, and through inhibition of PKA, $GABA_BR$ activation inhibits NMDARs, while A_2R activation inhibits AMPARs. RGS4 appears capable of limiting interference between the two receptors' neuromodulatory functions, as blocking RGS4 activity with either a small molecule inhibitor or an intracellular anti-RGS4 antibody enables crosstalk between pathways. This raises the intriguing possibility that RGS4 dysfunction in schizophrenia could disrupt pathway segregation and promote crosstalk that drives aberrant function (Gyorgy Lur et al. 2015).

4.5.3 RGS12

RGS12 negatively regulates presynaptic $GABA_BR$ signaling through N-type channels in chick dorsal root ganglion neurons (Schiff et al. 2000). In addition to its canonical GAP activity, RGS12 binds to the synprint region of N-type channels to directly modulate Ca^{2+} signaling (Schiff et al. 2000, Richman et al. 2005).

5 Plasticity of $GABA_BR$-Dependent Signaling

Native $GABA_BRs$ are multi-protein complexes with a remarkable diversity in protein composition across space and time (Malitschek et al. 1998, Fritschy et al. 1999, Schwenk et al. 2016). This endows functional diversity to $GABA_BR$-mediated signaling across brain regions and cell types (Cruz et al. 2004, Gassmann and Bettler 2012, Terunuma 2018) and even between sexes (Al-Dahan et al. 1994, Kelly et al. 2003, Marron Fernandez de Velasco et al. 2015). $GABA_B$R-mediated signaling is also plastic, changing throughout development (Moran et al. 2001, Luo et al. 2011, Dlouhá et al. 2013, Karls and Mynlieff 2015, Cai et al. 2017), during pathophysiological states (Fritzius and Bettler 2019), or in response to particular experiences (Terunuma 2018). Potential mechanisms underlying differences or changes in $GABA_B$R function can be remarkably diverse, as alterations in the expression/structure/function of either core and peripheral proteins that make up $GABA_B$R complexes, effector proteins, or regulatory proteins may have a dramatic influence at the molecular, cellular, and organismal level. In addition to examples of $GABA_B$R plasticity described earlier, listed below are a few proposed plasticity mechanisms, among the many that have been reported.

5.1 Phosphorylation-Dependent Plasticity

Phosphorylation-dependent changes in $GABA_B$R and GIRK channel surface availability underlie several reports of $GABA_B$R plasticity. As discussed earlier, sustained glutamatergic or GABAergic signaling can downregulate $GABA_B$Rs through several mechanisms, including receptor phosphorylation. Exposure to drugs of abuse can also induce phosphorylation-dependent changes in $GABA_B$R activity in vivo. For example, both acute and repeated exposure to cocaine suppressed $GABA_B$R-GIRK signaling in VTA dopamine neurons and prelimbic cortex pyramidal neurons, respectively, via phosphorylation-dependent internalization of $GABA_B$Rs and/or GIRK channels (Arora et al. 2011, Hearing et al. 2013). Exposure to inescapable footshocks similarly suppressed $GABA_B$R-GIRK signaling in the lateral habenula through increased PP2A activity and internalization of $GABA_B$Rs and GIRK channels (Lecca et al. 2016).

5.2 Plasticity of 14-3-3 Proteins

Multiple members of the 14-3-3 family of scaffolding proteins directly or indirectly interact with $GABA_B$Rs (Couve et al. 2001, Schwenk et al. 2016) and have been proposed to regulate $GABA_B$R structure and function. For example, 14-3-3ζ is overexpressed in the dorsal horn of neuropathic rats, where it inhibits $GABA_B$R signaling through K^+ channels and contributes to pain sensitization. Biochemical, electrophysiological, and ultrastructural evidence suggest that 14-3-3ζ disrupts $GABA_B$R-dependent signaling by dissociating $GABA_B$R heterodimers at the plasma membrane (Laffray et al. 2012). Exposure to NMDAR antagonists, acting as rapid antidepressants, increases the surface stability of 14-3-3η, which decouples $GABA_B$R signaling from GIRK channels in the hippocampus of socially defeated (model of depression) rodents (Workman et al. 2015). The existence of several other $GABA_BR$-interacting 14-3-3 isoforms raises the intriguing possibility of similar interactions throughout the CNS (Schwenk et al. 2016).

5.3 *Plasticity of RGS Regulation*

As mentioned earlier, RGS2 negatively regulates the $GABA_BR$-GIRK coupling efficiency in VTA dopamine neurons. Chronic exposure to either gamma hydroxybutyrate (GHB) or morphine reduced RGS2 transcription and enhanced the coupling efficiency of $GABA_BR$-GIRK signaling in VTA dopamine neurons of mice (Labouèbe et al. 2007). Another interesting example involves plasticity of GPR158, an RGS7/Gβ5-binding protein that suppresses homeostatic regulation of cAMP by $GABA_BRs$ (Orlandi et al. 2018). GPR158 is upregulated in the prefrontal cortex of both humans with major depressive disorder and mice exposed to chronic stress (Sutton et al. 2018). Viral overexpression of GPR158 in the mouse prefrontal cortex induced depressive-like behaviors, while constitutive GPR158 ablation produced an antidepressant effect (Sutton et al. 2018). Thus, GPR158 plasticity and $GABA_BR$-mediated dysregulation of cAMP levels in the prefrontal cortex may play a prominent role in stress-induced depression.

6 Concluding Remarks

$GABA_BRs$ are obligate heterodimers that interact with G proteins, effectors, and a wide variety of proteins to form spatially and temporally modular complexes that impart functional diversity to $GABA_BR$-dependent signaling throughout the brain. Given the vital roles $GABA_BRs$ play in regulating synaptic transmission and behavior, as well as their links to disease, it is critical to understand the dynamic structure and function of the $GABA_BR$ signalosome. Further research aiming to identify the functional roles of components within $GABA_BR$ complexes, as well as mechanisms underlying regulation/modulation of $GABA_BR$-dependent signaling, will occupy the field for many years. A deeper understanding of the relationships between the $GABA_BR$ signalosome and pathophysiological states will yield insights that are essential for drug discovery and development efforts. While drugs targeting $GABA_BRs$ are currently used on- and off-label to treat several disorders, broadening drug design to more selectively target components within $GABA_BR$ complexes, or interactions between components, will likely increase the therapeutic potential of new medicines (Berry-Kravis et al. 2018, Fritzius and Bettler 2019).

References

Adelfinger L, Turecek R, Ivankova K, Jensen AA, Moss SJ, Gassmann M, Bettler B (2014) GABAB receptor phosphorylation regulates KCTD12-induced K+ current desensitization. Biochem Pharmacol 91:369–379

Al-Dahan MI, Jalilian Tehrani MH, Thalmann RH (1994) Regulation of γ-aminobutyric acidB (GABAB) receptors in cerebral cortex during the estrous cycle. Brain Res 640:33–39

Anderson GR, Posokhova E, Martemyanov KA (2009) The R7 RGS protein family: multi-subunit regulators of neuronal G protein signaling. Cell Biochem Biophys 54:33–46

Andrade R, Malenka RC, Nicoll RA (1986) A G protein couples serotonin and GABAB receptors to the same channels in hippocampus. Science 234:1261

Arora D, Hearing M, Haluk DM, Mirkovic K, Fajardo-Serrano A, Wessendorf MW, Watanabe M, Lujan R, Wickman K (2011) Acute cocaine exposure weakens GABAB receptor-dependent G-protein-gated inwardly rectifying K+ signaling in dopamine neurons of the ventral tegmental area. J Neurosci 31:12251–12257

Badheka D, Yudin Y, Borbiro I, Hartle CM, Yazici A, Mirshahi T, Rohacs T (2017) Inhibition of transient receptor potential melastatin 3 ion channels by G-protein βγ subunits. eLife 6:e26147

Balasubramanian S, Fam SR, Hall RA (2006) GABAB receptor association with the PDZ scaffold Mupp1 alters receptor stability and function. J Biol Chem 282:4162–4171

Balasubramanian S, Teissére JA, Raju DV, Hall RA (2004) Hetero-oligomerization between $GABA_A$ and $GABA_B$ receptors regulates $GABA_B$ receptor trafficking. J Biol Chem 279:18840–18850

Bean BP (1989) Neurotransmitter inhibition of neuronal calcium currents by changes in channel voltage dependence. Nature 340:153–156

Becher A, White JH, McIlhinney RAJ (2008) The γ-aminobutyric acid receptor B, but not the metabotropic glutamate receptor type-1, associates with lipid rafts in the rat cerebellum. J Neurochem 79:787–795

Benians A, Nobles M, Hosny S, Tinker A (2005) Regulators of G-protein signaling form a quaternary complex with the agonist, receptor, and G-protein. J Biol Chem 280:13383–13394

Benke D (2010) Mechanisms of GABAB receptor exocytosis, endocytosis, and degradation. Adv Pharmacol 58:93–111

Benke D, Honer M, Michel C, Bettler B, Mohler H (1999) γ-Aminobutyric acid type B receptor splice variant proteins GBR1a and GBR1b are both associated with GBR2 in situ and display differential regional and subcellular distribution. J Biol Chem 274:27323

Benke D, Zemoura K, Maier PJ (2012) Modulation of cell surface GABA $_B$ receptors by desensitization, trafficking and regulated degradation. World J Biol Chem 3:61

Berry-Kravis EM, Lindemann L, Jønch AE, Apostol G, Bear MF, Carpenter RL, Crawley JN, Curie A, Des Portes V, Hossain F, Gasparini F, Gomez-Mancilla B, Hessl D, Loth E, Scharf SH, Wang PP, Von Raison F, Hagerman R, Spooren W, Jacquemont S (2018) Drug development for neurodevelopmental disorders: lessons learned from fragile X syndrome. Nat Rev Drug Discov 17:280–299

Bettler B, Kaupmann K, Mosbacher J, Gassmann M (2004) Molecular structure and physiological functions of $GABA_B$ receptors. Physiol Rev 84:835–867

Bettler B, Tiao JY-H (2006) Molecular diversity, trafficking and subcellular localization of GABAB receptors. Pharmacol Ther 110:533–543

Biermann B, Ivankova-Susankova K, Bradaia A, Abdel Aziz S, Besseyrias V, Kapfhammer JP, Missler M, Gassmann M, Bettler B (2010) The sushi domains of GABAB receptors function as axonal targeting signals. J Neurosci 30:1385–1394

Binet V, Brajon C, Le Corre L, Acher F, Pin J-P, Prézeau L (2004) The heptahelical domain of $GABA_{B2}$ is activated directly by CGP7930, a positive allosteric modulator of the $GABA_B$ receptor. J Biol Chem 279:29085–29091

Blednov YA, Stoffel M, Chang S, Adron HR (2001) GIRK2 deficient mice: evidence for hyperactivity and reduced anxiety. Physiol Behav 74:109–117

Blein S, Ginham R, Uhrin D, Smith BO, Soares DC, Veltel S, McIlhinney RAJ, White JH, Barlow PN (2004) Structural analysis of the complement control protein (CCP) modules of $GABA_B$ receptor 1a. J Biol Chem 279:48292–48306

Booker SA, Gross A, Althof D, Shigemoto R, Bettler B, Frotscher M, Hearing M, Wickman K, Watanabe M, Kulik Á, Vida I (2013) Differential GABAB-receptor-mediated effects in perisomatic- and dendrite-targeting parvalbumin interneurons. J Neurosci 33:7961–7974

Booker SA, Loreth D, Gee AL, Wyllie DJA, Watanabe M, Kind PC, Kos Kulik A, Vida I (2018) Postsynaptic GABABRs inhibit L-type calcium channels and abolish long-term potentiation in hippocampal somatostatin interneurons. Cell Rep 22:36–43

Bowery NG, Bettler B, Froestl W, Gallagher JP, Marshall F, Raiteri M, Bonner TI, Enna SJ (2002) International Union of Pharmacology. XXXIII. Mammalian gamma-aminobutyric acid (B) receptors: structure and function. Pharmacol Rev 54:247–264

Bowery NG, Brown DA (1997) The cloning of GABAB receptors. Nature 386:223–224

Bowery NG, Hudson AL (1979) Gamma-Aminobutyric acid reduces the evoked release of [3H]-noradrenaline from sympathetic nerve terminals [proceedings]. Br J Pharmacol 66:108P

Boyer SB, Clancy SM, Terunuma M, Revilla-Sanchez R, Thomas SM, Moss SJ, Slesinger PA (2009) Direct interaction of GABAB receptors with M2 muscarinic receptors enhances muscarinic signaling. J Neurosci 29:15796–15809

Boyes J, Bolam JP (2003) The subcellular localization of GABA(B) receptor subunits in the rat substantia nigra. Eur J Neurosci 18:3279–3293

Bray JG, Mynlieff M (2009) Influx of calcium through L-type calcium channels in early postnatal regulation of chloride transporters in the rat hippocampus. Dev Neurobiol 69:885–896; erratum 897-912

Brini M, Calì T, Ottolini D, Carafoli E (2014) Neuronal calcium signaling: function and dysfunction. Cell Mol Life Sci 71:2787–2814

Brock C, Boudier L, Maurel D, Blahos J, Pin J-P (2005) Assembly-dependent surface targeting of the heterodimeric GABAB receptor is controlled by COPI but not 14-3-3. Mol Biol Cell 16:5572–5578

Burke KJ, Bender KJ (2019) Modulation of ion channels in the axon: mechanisms and function. Front Cell Neurosci 13:221

Burmakina S, Geng Y, Chen Y, Fan QR (2014) Heterodimeric coiled-coil interactions of human GABAB receptor. Proc Natl Acad Sci U S A 111:6958–6963

Cai S, Fischer QS, He Y, Zhang L, Liu H, Daw NW, Yang Y (2017) GABAB receptor-dependent bidirectional regulation of critical period ocular dominance plasticity in cats. PLoS One 12: e0180162

Calebiro D, Rieken F, Wagner J, Sungkaworn T, Zabel U, Borzi A, Cocucci E, Zurn A, Lohse MJ (2013) Single-molecule analysis of fluorescently labeled G-protein-coupled receptors reveals complexes with distinct dynamics and organization. Proc Natl Acad Sci 110:743–748

Calver AR, Davies CH, Pangalos MN (2002) $GABA_B$ receptors: from monogamy to promiscuity. Neurosignals 11:299–314

Calver AR, Robbins MJ, Cosio C, Rice SQ, Babbs AJ, Hirst WD, Boyfield I, Wood MD, Russell RB, Price GW, Couve A, Moss SJ, Pangalos MN (2001) The C-terminal domains of the GABA (b) receptor subunits mediate intracellular trafficking but are not required for receptor signaling. J Neurosci 21:1203–1210

Campbell V, Berrow N, Dolphin AC (1993) GABAB receptor modulation of Ca2+ currents in rat sensory neurones by the G protein G(0): antisense oligonucleotide studies. J Physiol 470:1–11

Carling D (2005) AMP-activated protein kinase: balancing the scales. Biochimie 87:87–91

Carling D, Mayer FV, Sanders MJ, Gamblin SJ (2011) AMP-activated protein kinase: nature's energy sensor. Nat Chem Biol 7:512–518

Carter TJ, Mynlieff M (2004) γ-Aminobutyric acid type B receptors facilitate L-type and attenuate N-type Ca2+ currents in isolated hippocampal neurons. J Neurosci Res 76:323–333

Castelli MP, Gessa GL (2016) Distribution and localization of the GABAB receptor. Springer, Cham, pp 75–92

Chalifoux JR, Carter AG (2010) GABAB receptors modulate NMDA receptor calcium signals in dendritic spines. Neuron 66:101–113
Chalifoux JR, Carter AG (2011) GABAB receptor modulation of voltage-sensitive calcium channels in spines and dendrites. J Neurosci 31:4221–4232
Chang W, Tu C, Cheng Z, Rodriguez L, Chen T-H, Gassmann M, Bettler B, Margeta M, Jan LY, Shoback D (2007) Complex formation with the type B γ-aminobutyric acid receptor affects the expression and signal transduction of the extracellular calcium-sensing receptor. J Biol Chem 282:25030–25040
Charles K, Evans M, Robbins M, Calver A, Leslie R, Pangalos M (2001) Comparative immunohistochemical localisation of GABAB1a, GABAB1b and GABAB2 subunits in rat brain, spinal cord and dorsal root ganglion. Neuroscience 106:447–467
Chu DC, Albin RL, Young AB, Penney JB (1990) Distribution and kinetics of GABAB binding sites in rat central nervous system: a quantitative autoradiographic study. Neuroscience 34:341–357
Ciruela F, Fernández-Dueñas V, Sahlholm K, Fernández-Alacid L, Nicolau JC, Watanabe M, Luján R (2010) Evidence for oligomerization between GABAB receptors and GIRK channels containing the GIRK1 and GIRK3 subunits. Eur J Neurosci 32:1265–1277
Clancy SM, Fowler CE, Finley M, Suen KF, Arrabit C, Berton F, Kosaza T, Casey PJ, Slesinger PA (2005) Pertussis-toxin-sensitive Gα subunits selectively bind to C-terminal domain of neuronal GIRK channels: evidence for a heterotrimeric G-protein-channel complex. Mol Cell Neurosci 28:375–389
Colecraft HM, Brody DL, Yue DT (2001) G-protein inhibition of N- and P/Q-type calcium channels: distinctive elementary mechanisms and their functional impact. J Neurosci 21:1137–1147
Colecraft HM, Patil PG, Yue DT (2000) Differential occurrence of reluctant openings in G-protein–inhibited N- and P/Q-type calcium channels. J Gen Physiol 115:175–192
Comps-Agrar L, Kniazeff J, Nørskov-Lauritsen L, Maurel D, Gassmann M, Gregor N, Prézeau L, Bettler B, Durroux T, Trinquet E, Pin J-P (2011) The oligomeric state sets $GABA_B$ receptor signalling efficacy. EMBO J 30:2336–2349
Connelly WM, Fyson SJ, Errington AC, McCafferty CP, Cope DW, Di Giovanni G, Crunelli V (2013) GABAB receptors regulate extrasynaptic GABAA receptors. J Neurosci 33:3780–3785
Cottrell G (2013) Roles of proteolysis in regulation of GPCR function. Br J Pharmacol 168:576–590
Couve A, Filippov AK, Connolly CN, Bettler B, Brown DA, Moss SJ (1998) Intracellular retention of recombinant GABAB receptors. J Biol Chem 273:26361–26367
Couve A, Kittler JT, Uren JM, Calver AR, Pangalos MN, Walsh FS, Moss SJ (2001) Association of GABAB receptors and members of the 14-3-3 family of signaling proteins. Mol Cell Neurosci 17:317–328
Couve A, Restituito S, Brandon JM, Charles KJ, Bawagan H, Freeman KB, Pangalos MN, Calver AR, Moss SJ (2004) Marlin-1, a novel RNA-binding protein associates with GABA receptors. J Biol Chem 279:13934–13943
Couve A, Thomas P, Calver AR, Hirst WD, Pangalos MN, Walsh FS, Smart TG, Moss SJ (2002) Cyclic AMP–dependent protein kinase phosphorylation facilitates GABAB receptor–effector coupling. Nat Neurosci 5:415–424
Cruz HG, Ivanova T, Lunn M-L, Stoffel M, Slesinger PA, Lüscher C (2004) Bi-directional effects of GABAB receptor agonists on the mesolimbic dopamine system. Nat Neurosci 7:153–159
David M, Richer M, Mamarbachi AM, Villeneuve LR, Dupré DJ, Hebert TE (2006) Interactions between GABA-B1 receptors and Kir 3 inwardly rectifying potassium channels. Cell Signal 18:2172–2181
Deng P-Y, Xiao Z, Yang C, Rojanathammanee L, Grisanti L, Watt J, Geiger JD, Liu R, Porter JE, Lei S (2009) GABAB receptor activation inhibits neuronal excitability and spatial learning in the entorhinal cortex by activating TREK-2 K+ channels. Neuron 63:230–243

Dinamarca MC, Raveh A, Schneider A, Fritzius T, Früh S, Rem PD, Stawarski M, Lalanne T, Turecek R, Choo M, Besseyrias V, Bildl W, Bentrop D, Staufenbiel M, Gassmann M, Fakler B, Schwenk J, Bettler B (2019) Complex formation of APP with GABAB receptors links axonal trafficking to amyloidogenic processing. Nat Commun 10:1331

Dlouhá K, Kagan D, Roubalová L, Ujčíková H, Svoboda P (2013) Plasma membrane density of GABA B-R1a, GABA B-R1b, GABA-R2 and trimeric G-proteins in the course of postnatal development of rat brain cortex. Physiol Res 62:547–559

Doly S, Shirvani H, Gäta G, Meye FJ, Emerit M-B, Enslen H, Achour L, Pardo-Lopez L, Yang S-K, Armand V, Gardette R, Giros B, Gassmann M, Bettler B, Mameli M, Darmon M, Marullo S (2016) GABAB receptor cell-surface export is controlled by an endoplasmic reticulum gatekeeper. Mol Psychiatry 21:480–490

Dores MR, Trejo J (2012) Ubiquitination of G protein-coupled receptors: functional implications and drug discovery. Mol Pharmacol 82:563–570

Doupnik CA, Davidson N, Lester HA, Kofuji P (1997) RGS proteins reconstitute the rapid gating kinetics of gbetagamma-activated inwardly rectifying K+ channels. Proc Natl Acad Sci U S A 94:10461–10466

Drenan RM, Doupnik CA, Boyle MP, Muglia LJ, Huettner JE, Linder ME, Blumer KJ (2005) Palmitoylation regulates plasma membrane–nuclear shuttling of R7BP, a novel membrane anchor for the RGS7 family. J Cell Biol 169:623–633

Drenan RM, Doupnik CA, Jayaraman M, Buchwalter AL, Kaltenbronn KM, Huettner JE, Linder ME, Blumer KJ (2006) R7BP augments the function of RGS7·Gβ5 complexes by a plasma membrane-targeting mechanism. J Biol Chem 281:28222–28231

Duthey B, Caudron S, Perroy J, Bettler B, Fagni L, Pin J-P, Prézeau L (2002) A single subunit (GB2) is required for G-protein activation by the heterodimeric GABA(B) receptor. J Biol Chem 277:3236–3241

Fairfax BP, Pitcher JA, Scott MGH, Calver AR, Pangalos MN, Moss SJ, Couve A (2004) Phosphorylation and chronic agonist treatment atypically modulate GABAB receptor cell surface stability. J Biol Chem 279:12565–12573

Fajardo-Serrano A, Wydeven N, Young D, Watanabe M, Shigemoto R, Martemyanov KA, Wickman K, Luján R (2013) Association of Rgs7/Gβ5 complexes with girk channels and $GABA_B$ receptors in hippocampal CA1 pyramidal neurons. Hippocampus 23:1231–1245

Fan QR, Guo WY, Geng Y, Evelyn MG (2017) Class C GPCR: obligatory heterodimerization of GABAB receptor. In: G-protein-coupled receptor dimers. Springer, Cham, pp 307–325

Fernández-Alacid L, Aguado C, Ciruela F, Martín R, Colón J, Cabañero MJ, Gassmann M, Watanabe M, Shigemoto R, Wickman K, Bettler B, Sánchez-Prieto J, Luján R (2009) Subcellular compartment-specific molecular diversity of pre- and post-synaptic GABA-activated GIRK channels in Purkinje cells. J Neurochem 110:1363–1376

Fowler CE, Aryal P, Suen KF, Slesinger PA (2007) Evidence for association of GABA B receptors with Kir3 channels and regulators of G protein signalling (RGS4) proteins. J Physiol 580:51–65

Franek M, Pagano A, Kaupmann K, Bettler B, Pin JP, Blahos J (1999) The heteromeric GABA-B receptor recognizes G-protein alpha subunit C-termini. Neuropharmacology 38:1657–1666

Frangaj A, Fan QR (2018) Structural biology of GABAB receptor. Neuropharmacology 136:68–79

Freyd T, Warszycki D, Mordalski S, Bojarski AJ, Sylte I, Gabrielsen M (2017) Ligand-guided homology modelling of the GABAB2 subunit of the GABAB receptor. PLoS One 12:e0173889

Fritschy JM, Meskenaite V, Weinmann O, Honer M, Benke D, Mohler H (1999) GABAB-receptor splice variants GB1a and GB1b in rat brain: developmental regulation, cellular distribution and extrasynaptic localization. Eur J Neurosci 11:761–768

Fritzius T, Bettler B (2019) The organizing principle of GABA $_B$ receptor complexes: physiological and pharmacological implications. Basic Clin Pharmacol Toxicol. https://doi.org/10.1111/bcpt.13241

Fritzius T, Turecek R, Seddik R, Kobayashi H, Tiao J, Rem PD, Metz M, Kralikova M, Bouvier M, Gassmann M, Bettler B (2017) KCTD hetero-oligomers confer unique kinetic properties on hippocampal $GABA_B$ receptor-induced K^+ currents. J Neurosci 37:1162–1175

Fukui M, Nakamichi N, Yoneyama M, Ozawa S, Fujimori S, Takahata Y, Nakamura N, Taniura H, Yoneda Y (2008) Modulation of cellular proliferation and differentiation through GABAB receptors expressed by undifferentiated neural progenitor cells isolated from fetal mouse brain. J Cell Physiol 216:507–519

Gaillard S, Lo Re L, Mantilleri A, Hepp R, Urien L, Malapert P, Alonso S, Deage M, Kambrun C, Landry M, Low SA, Alloui A, Lambolez B, Scherrer G, Le Feuvre Y, Bourinet E, Moqrich A (2014) GINIP, a Gαi-interacting protein, functions as a key modulator of peripheral GABAB receptor-mediated analgesia. Neuron 84:123–136

Galvez T, Duthey B, Kniazeff J, Blahos J, Rovelli G, Bettler B, Prézeau L, Pin JP (2001) Allosteric interactions between GB1 and GB2 subunits are required for optimal GABA(B) receptor function. EMBO J 20(9):2152–2159

Garaycochea J, Slaughter MM (2016) GABAB receptors enhance excitatory responses in isolated rat retinal ganglion cells. J Physiol 594:5543–5554

Gassmann M, Bettler B (2012) Regulation of neuronal GABAB receptor functions by subunit composition. Nat Rev Neurosci 13:380–394

Gassmann M, Shaban H, Vigot R, Sansig G, Haller C, Barbieri S, Humeau Y, Schuler V, Müller M, Kinzel B, Klebs K, Schmutz M, Froestl W, Heid J, Kelly PH, Gentry C, Jaton A-L, Van der Putten H, Mombereau C, Lecourtier L, Mosbacher J, Cryan JF, Fritschy J-M, Lüthi A, Kaupmann K, Bettler B (2004) Redistribution of GABAB(1) protein and atypical GABAB responses in GABAB(2)-deficient mice. J Neurosci 24:6086–6097

Gerber KJ, Squires KE, Hepler JR (2016) Roles for regulator of G protein signaling proteins in synaptic signaling and plasticity programs in molecular and systems pharmacology. Mol Pharmacol 89:273–286

Ghorbel MT, Becker KG, Henley JM (2005) Profile of changes in gene expression in cultured hippocampal neurones evoked by the GABAB receptor agonist baclofen. Physiol Genomics 22:93–98

Gonchar Y, Pang L, Malitschek B, Bettler B, Burkhalter A (2001) Subcellular localization of GABA(B) receptor subunits in rat visual cortex. J Comp Neurol 431:182–197

Grabowska D, Jayaraman M, Kaltenbronn KM, Sandiford SL, Wang Q, Jenkins S, Slepak VZ, Smith Y, Blumer KJ (2008) Postnatal induction and localization of R7BP, a membrane-anchoring protein for RGS7 family-Gβ5 complexes in brain. Neuroscience 151:969

Grace CRR, Perrin MH, DiGruccio MR, Miller CL, Rivier JE, Vale WW, Riek R (2004) NMR structure and peptide hormone binding site of the first extracellular domain of a type B1 G protein-coupled receptor. Proc Natl Acad Sci U S A 101:12836–12841

Grampp T, Notz V, Broll I, Fischer N, Benke D (2008) Constitutive, agonist-accelerated, recycling and lysosomal degradation of GABAB receptors in cortical neurons. Mol Cell Neurosci 39:628–637

Guetg N, Aziz SA, Holbro N, Turecek R, Rose T, Seddik R, Gassmann M, Moes S, Jenoe P, Oertner TG, Casanova E, Bettler B (2010) NMDA receptor-dependent GABAB receptor internalization via CaMKII phosphorylation of serine 867 in GABAB1. Proc Natl Acad Sci 107:13924–13929

Gurevich VV, Gurevich EV (2019) GPCR signaling regulation: the role of GRKs and arrestins. Front Pharmacol 10:125

Gyorgy Lur A, Higley MJ, Lur G (2015) Glutamate receptor modulation is restricted to synaptic microdomains article glutamate receptor modulation is restricted to synaptic microdomains. Cell Rep 12:326–334

Halls ML, Cooper DMF (2017) Adenylyl cyclase signalling complexes – pharmacological challenges and opportunities. Pharmacol Ther 172:171–180

Hammond C, Mott D (2015) The metabotropic GABAB receptors. In: Cellular and molecular neurophysiology. Academic Press, Boston, pp 245–267

Hanack C, Moroni M, Lima WC, Wende H, Kirchner M, Adelfinger L, Schrenk-Siemens K, Tappe-Theodor A, Wetzel C, Kuich PH, Gassmann M, Roggenkamp D, Bettler B, Lewin GR,

Selbach M, Siemens J (2015) GABA blocks pathological but not acute TRPV1 pain signals. Cell 160:759–770
Hannan S, Gerrow K, Triller A, Smart TGG (2016) Phospho-dependent accumulation of GABABRs at presynaptic terminals after NMDAR activation. Cell Rep 16:1962–1973
Hannan S, Wilkins ME, Smart TG (2012) Sushi domains confer distinct trafficking profiles on GABAB receptors. Proc Natl Acad Sci U S A 109:12171–12176
Harrison NL, Lange GD, Barker JL (1988) (−)-Baclofen activates presynaptic GABAB receptors on GABAergic inhibitory neurons from embryonic rat hippocampus. Neurosci Lett 85:105–109
Hashimoto T, Kuriyama K (2002) In vivo evidence that GABAB receptors are negatively coupled to adenylate cyclase in rat striatum. J Neurochem 69:365–370
Havlickova M, Prezeau L, Duthey B, Bettler B, Pin J-P, Blahos J (2002) The intracellular loops of the GB2 subunit are crucial for G-protein coupling of the heteromeric gamma-aminobutyrate B receptor. Mol Pharmacol 62:343–350
Hearing M, Kotecki L, Marron Fernandez de Velasco E, Fajardo-Serrano A, Chung HJ, Luján R, Wickman K (2013) Repeated cocaine weakens GABAB-girk signaling in layer 5/6 pyramidal neurons in the prelimbic cortex. Neuron 80:159–170
Herlitze S, Garcia DE, Mackie K, Hille B, Scheuer T, Catterall WA (1996) Modulation of Ca2+ channels βγ G-protein py subunits. Nature 380:258–262
Hillenbrand M, Schori C, Schöppe J, Plückthun A (2015) Comprehensive analysis of heterotrimeric G-protein complex diversity and their interactions with GPCRs in solution. Proc Natl Acad Sci U S A 112:E1181–E1190
Hirono M, Yoshioka T, Konishi S (2001) GABAB receptor activation enhances mGluR-mediated responses at cerebellar excitatory synapses. Nat Neurosci 4:1207–1216
Hollinger S, Hepler JR (2002) Cellular regulation of RGS proteins: modulators and integrators of G protein signaling. Pharmacol Rev 54:527–559
Huang C-L, Feng S, Hilgemann DW (1998) Direct activation of inward rectifier potassium channels by PIP2 and its stabilization by Gβγ. Nature 391:803–806
Iacovelli L, De Blasi A (2013) Molecular mechanisms that desensitize metabotropic glutamate receptor signaling: an overview. Neuropharmacology 66:24–30
Iwakiri M, Mizukami K, Ikonomovic MD, Ishikawa M, Hidaka S, Abrahamson EE, DeKosky ST, Asada T (2005) Changes in hippocampal GABABR1 subunit expression in Alzheimer's patients: association with Braak staging. Acta Neuropathol 109:467–474
Jarvis SE, Magga JM, Beedle AM, Braun JE, Zamponi GW (2000) G protein modulation of N-type calcium channels is facilitated by physical interactions between syntaxin 1A and Gbetagamma. J Biol Chem 275:6388–6394
Jeong SW, Ikeda SR (1998) G protein alpha subunit G alpha z couples neurotransmitter receptors to ion channels in sympathetic neurons. Neuron 21:1201–1212
Jia L, Chisari M, Maktabi MH, Sobieski C, Zhou H, Konopko AM, Martin BR, Mennerick SJ, Blumer KJ (2014) A mechanism regulating G protein-coupled receptor signaling that requires cycles of protein palmitoylation and depalmitoylation. J Biol Chem 289:6249–6257
Jia L, Linder ME, Blumer KJ (2011) Gi/o signaling and the palmitoyltransferase DHHC2 regulate palmitate cycling and shuttling of RGS7 family-binding protein. J Biol Chem 286:13695–13703
Jiang X, Su L, Zhang Q, He C, Zhang Z, Yi P, Liu J (2012) GABAB receptor complex as a potential target for tumor therapy. J Histochem Cytochem 60:269–279
Kahanovitch U, Berlin S, Dascal N (2017) Collision coupling in the $GABA_B$ receptor-G protein-GIRK signaling cascade. FEBS Lett 591:2816–2825
Kanaide M, Uezono Y, Matsumoto M, Hojo M, Ando Y, Sudo Y, Sumikawa K, Taniyama K (2007) Desensitization of GABAB receptor signaling by formation of protein complexes of GABAB2 subunit with GRK4 or GRK5. J Cell Physiol 210:237–245
Kantamneni S, Corrêa SAL, Hodgkinson GK, Meyer G, Vinh NN, Henley JM, Nishimune A (2007) GISP: a novel brain-specific protein that promotes surface expression and function of $GABA_B$ receptors. J Neurochem 100:1003–1017

Karls A, Mynlieff M (2015) GABAB receptors couple to Gα q to mediate increases in voltage-dependent calcium current during development. J Neurochem 135:88–100

Karschin C, Dissmann E, Stühmer W, Karschin A (1996) IRK(1-3) and GIRK(1-4) inwardly rectifying K+ channel mRNAs are differentially expressed in the adult rat brain. J Neurosci 16:3559–3570

Kasten CR, Boehm SL II (2015) Identifying the role of pre-and postsynaptic GABA(B) receptors in behavior. Neurosci Biobehav Rev 57:70–87

Kaupmann K, Huggel K, Heid J, Flor PJ, Bischoff S, Mickel SJ, McMaster G, Angst C, Bittiger H, Froestl W, Bettler B (1997) Expression cloning of GABAB receptors uncovers similarity to metabotropic glutamate receptors. Nature 386:239–246

Kaupmann K, Malitschek B, Schuler V, Heid J, Froestl W, Beck P, Mosbacher J, Bischoff S, Kulik A, Shigemoto R, Karschin A, Bettler B (1998a) GABAB-receptor subtypes assemble into functional heteromeric complexes. Nature 396:683–687

Kaupmann K, Schuler V, Mosbacher J, Bischoff S, Bittiger H, Heid J, Froestl W, Leonhard S, Pfaff T, Karschin A, Bettler B (1998b) Human-aminobutyric acid type B receptors are differentially expressed and regulate inwardly rectifying K+ channels. Proc Natl Acad Sci 95:14991–14996

Kelly MJ, Qiu J, Rønnekleiv OK (2003) Estrogen modulation of G-protein-coupled receptor activation of potassium channels in the central nervous system. Ann N Y Acad Sci 1007:6–9

Kennedy JE, Marchese A (2015) Regulation of GPCR trafficking by ubiquitin. Prog Mol Biol Transl Sci 132:15–38

Kim G, Jung S, Son H, Kim S, Choi J, Lee DH, Roh GS, Kang SS, Cho GJ, Choi WS, Kim HJ (2014) The GABAB receptor associates with regulators of G-protein signaling 4 protein in the mouse prefrontal cortex and hypothalamus. BMB Rep 47:324–329

Kins S, Lauther N, Szodorai A, Beyreuther K (2006) Subcellular trafficking of the amyloid precursor protein gene family and its pathogenic role in Alzheimer's disease. Neurodegener Dis 3:218–226

Kitano J, Kimura K, Yamazaki Y, Soda T, Shigemoto R, Nakajima Y, Nakanishi S (2002) Tamalin, a PDZ domain-containing protein, links a protein complex formation of group 1 metabotropic glutamate receptors and the guanine nucleotide exchange factor cytohesins. J Neurosci 22:1280–1289

Kittler JT, McAinsh K, Moss SJ (2002) Mechanisms of GABA$_A$ receptor assembly and trafficking: implications for the modulation of inhibitory neurotransmission. Mol Neurobiol 26:251–268

Knight AR, Bowery NG (1996) The pharmacology of adenylyl cyclase modulation by GABAB receptors in rat brain slices. Neuropharmacology 35:703–712

Knott C, Maguire JJJ, Moratalla R, Bowery NGG (1993) Regional effects of pertussis toxin in vivo and in vitro on GABAB receptor binding in rat brain. Neuroscience 52:73–81

Kommaddi RP, Shenoy SK (2013) Arrestins and protein ubiquitination. Prog Mol Biol Transl Sci 118:175–204

Koppensteiner P, Melani R, Ninan I (2017) A cooperative mechanism involving Ca2+-permeable AMPA receptors and retrograde activation of GABAB receptors in interpeduncular nucleus plasticity. Cell Rep 20:1111–1122

Koyrakh L, Luján R, Colón J, Karschin C, Kurachi Y, Karschin A, Wickman K (2005) Molecular and cellular diversity of neuronal G-protein-gated potassium channels. J Neurosci 25:11468–11478

Kulik A, Vida I, Fukazawa Y, Guetg N, Kasugai Y, Marker CL, Rigato F, Bettler B, Wickman K, Frotscher M, Shigemoto R, Kulik Á, Vida I, Fukazawa Y, Guetg N, Kasugai Y, Marker CL, Rigato F, Bettler B, Wickman K, Frotscher M, Shigemoto R, Kulik A (2006) Compartment-dependent colocalization of Kir3.2-containing K+ channels and GABAB receptors in hippocampal pyramidal cells. J Neurosci 26:4289–4297

Kulik Á, Vida I, Luján R, Haas CA, López-Bendito G, Shigemoto R, Frotscher M (2003) Subcellular localization of metabotropic GABAB receptor subunits GABAB1a/b and GABAB2 in the rat hippocampus. J Neurosci 23:11026–11035

Kuramoto N, Wilkins ME, Fairfax BP, Revilla-Sanchez R, Terunuma M, Tamaki K, Iemata M, Warren N, Couve A, Calver A, Horvath Z, Freeman K, Carling D, Huang L, Gonzales C, Cooper E, Smart TG, Pangalos MN, Moss SJ (2007) Phospho-dependent functional modulation of GABAB receptors by the metabolic sensor AMP-dependent protein kinase. Neuron 53:233–247

Labouèbe G, Lomazzi M, Cruz HG, Creton C, Luján R, Li M, Yanagawa Y, Obata K, Watanabe M, Wickman K, Boyer SB, Slesinger PA, Lüscher C (2007) RGS2 modulates coupling between GABAB receptors and GIRK channels in dopamine neurons of the ventral tegmental area. Nat Neurosci 10:1559–1568

Ladera C, del Carmen Godino M, Cabañero MJ, Torres M, Watanabe M, Luján R, Sánchez-Prieto J (2008) Pre-synaptic $GABA_B$ receptors inhibit glutamate release through GIRK channels in rat cerebral cortex. J Neurochem 107:1506–1517

Laffray S, Bouali-Benazzouz R, Papon M-A, Favereaux A, Jiang Y, Holm T, Spriet C, Desbarats P, Fossat P, Le Feuvre Y, Decossas M, Héliot L, Langel U, Nagy F, Landry M (2012) Impairment of GABAB receptor dimer by endogenous 14-3-3ζ in chronic pain conditions. EMBO J 31:3239–3251

Lahaie N, Kralikova M, Prézeau L, Blahos J, Bouvier M (2016) Post-endocytotic deubiquitination and degradation of the metabotropic γ-aminobutyric acid receptor by the ubiquitin-specific protease. J Biol Chem 14:291

Laviv T, Vertkin I, Berdichevsky Y, Fogel H, Riven I, Bettler B, Slesinger PA, Slutsky I (2011) Compartmentalization of the GABAB receptor signaling complex is required for presynaptic inhibition at hippocampal synapses. J Neurosci 31:12523–12532

Leaney JL, Tinker A (2000) The role of members of the pertussis toxin-sensitive family of G proteins in coupling receptors to the activation of the G protein-gated inwardly rectifying potassium channel. Proc Natl Acad Sci 97:5651–5656

Lecca S, Pelosi A, Tchenio A, Moutkine I, Lujan R, Hervé D, Mameli M (2016) Rescue of GABAB and GIRK function in the lateral habenula by protein phosphatase 2A inhibition ameliorates depression-like phenotypes in mice. Nat Med 22:254–261

Lee C, Mayfield RD, Harris RA (2010) Intron 4 containing novel GABAB1 isoforms impair GABAB receptor function. PLoS One 5:e14044

Lefkowitz RJ (2013) Arrestins come of age: a personal historical perspective. Prog Mol Biol Transl Sci 118:3–18

Lehtinen MJ, Meri S, Jokiranta TS (2004) Interdomain contact regions and angles between adjacent short consensus repeat domains. J Mol Biol 344:1385–1396

Leung LS, Peloquin P (2006) GABAB receptors inhibit backpropagating dendritic spikes in hippocampal CA1 pyramidal cells in vivo. Hippocampus 16:388–407

Liu J, Maurel D, Etzol S, Brabet I, Ansanay H, Pin J-P, Rondard P (2004) Molecular determinants involved in the allosteric control of agonist affinity in the GABAB receptor by the GABAB2 subunit. J Biol Chem 279:15824–15830

Logothetis DE, Kurachi Y, Galper J, Neer EJ, Clapham DE (1987) The βγ subunits of GTP-binding proteins activate the muscarinic K+ channel in heart. Nature 325:321–326

Luján R, Aguado C (2015) Localization and targeting of GIRK channels in mammalian central neurons. Int Rev Neurobiol 123:161–200

Luján R, Aguado C, Ciruela F, Cózar J, Kleindienst D, de la Ossa L, Bettler B, Wickman K, Watanabe M, Shigemoto R, Fukazawa Y (2017) Differential association of GABAB receptors with their effector ion channels in Purkinje cells. Brain Struct Funct 223:1565–1587

Lujan R, Ciruela F (2012) GABAB receptors-associated proteins: potential drug targets in neurological disorders? Curr Drug Targets 13:129–144

Luján R, Marron Fernandez de Velasco E, Aguado C, Wickman K (2014) New insights into the therapeutic potential of Girk channels. Trends Neurosci 37:20–29

Luo B, Wang H-T, Su Y-Y, Wu S-H, Chen L (2011) Activation of presynaptic GABAB receptors modulates GABAergic and glutamatergic inputs to the medial geniculate body. Hear Res 280:157–165

Lüscher C, Jan LY, Stoffel M, Malenka RC, Nicoll RA (1997) G protein-coupled inwardly rectifying K+ channels (GIRKs) mediate postsynaptic but not presynaptic transmitter actions in hippocampal neurons. Neuron 19:687–695

Lüscher C, Slesinger PA (2010) Emerging roles for G protein-gated inwardly rectifying potassium (GIRK) channels in health and disease. Nat Rev Neurosci 11:301–315

Maguire G, Maple B, Lukasiewicz P, Werblin F (1989) Gamma-aminobutyrate type B receptor modulation of L-type calcium channel current at bipolar cell terminals in the retina of the tiger salamander. Proc Natl Acad Sci U S A 86:10144

Maier PJ, Marin I, Grampp T, Sommer A, Benke D (2010) Sustained glutamate receptor activation down-regulates GABAB receptors by shifting the balance from recycling to lysosomal degradation. J Biol Chem 285:35606–35614

Maity B, Stewart A, Yang J, Loo L, Sheff D, Shepherd AJ, Mohapatra DP, Fisher RA (2012) Regulator of G protein signaling 6 (RGS6) protein ensures coordination of motor movement by modulating $GABA_B$ receptor signaling. J Biol Chem 287:4972–4981

Malenka RC (1991) The role of postsynaptic calcium in the induction of long-term potentiation. Mol Neurobiol 5:289–295

Malitschek B, Rüegg D, Heid J, Kaupmann K, Bittiger H, Fröstl W, Bettler B, Kuhn R (1998) Developmental changes of agonist affinity at GABABR1 receptor variants in rat brain. Mol Cell Neurosci 12:56–64

Margeta-Mitrovic M, Jan YN, Jan LY (2000) A trafficking checkpoint controls GABA(B) receptor heterodimerization. Neuron 27:97–106

Marron Fernandez de Velasco E, Hearing M, Xia Z, Victoria NC, Luján R, Wickman K (2015) Sex differences in GABABR-GIRK signaling in layer 5/6 pyramidal neurons of the mouse prelimbic cortex. Neuropharmacology 95:353–360

Martemyanov KA, Yoo PJ, Skiba NP, Arshavsky VY (2005) R7BP, a novel neuronal protein interacting with RGS proteins of the R7 family. J Biol Chem 280:5133–5136

Maurel D, Comps-Agrar L, Brock C, Rives M-L, Bourrier E, Ayoub MA, Bazin H, Tinel N, Durroux T, Prézeau L, Trinquet E, Pin J-P (2008) Cell-surface protein-protein interaction analysis with time-resolved FRET and snap-tag technologies: application to GPCR oligomerization. Nat Methods 5:561–567

McCormick DA (1989) GABA as an inhibitory neurotransmitter in human cerebral cortex. J Neurophysiol 62:1018–1027

McMahon HT, Mills IG (2004) COP and clathrin-coated vesicle budding: different pathways, common approaches. Curr Opin Cell Biol 16:379–391

Mcpherson KB, Leff ER, Li M-H, Meurice C, Tai S, Traynor JR, Ingram SL (2018) Regulators of G protein signaling (RGS) proteins promote receptor coupling to G protein-coupled inwardly-rectifying potassium (GIRK) channels. J Neurosci 38:8737–8744

Menon-Johansson AS, Berrow N, Dolphin AC (1993) G(o) transduces GABAB-receptor modulation of N-type calcium channels in cultured dorsal root ganglion neurons. Pflugers Arch 425:335–343

Michaeli A, Yaka R (2010) Dopamine inhibits GABAA currents in ventral tegmental area dopamine neurons via activation of presynaptic G-protein coupled inwardly-rectifying potassium channels. Neuroscience 165:1159–1169

Milligan G, Kostenis E (2006) Heterotrimeric G-proteins: a short history. Br J Pharmacol 147:S46–S55

Misgeld U, Bijak M, Jarolimek W (1995) A physiological role for GABAB receptors and the effects of baclofen in the mammalian central nervous system. Prog Neurobiol 46:423–462

Møller TC, Moreno-Delgado D, Pin J-P, Kniazeff J (2017) Class C G protein-coupled receptors: reviving old couples with new partners. Biophys Rep 3:57–63

Monnier C, Tu H, Bourrier E, Vol C, Lamarque L, Trinquet E, Pin J-P, Rondard P (2011) *Trans*-activation between 7TM domains: implication in heterodimeric $GABA_B$ receptor activation. EMBO J 30:32–42

Moran JM, Enna SJ, McCarson KE (2001) Developmental regulation of GABA(B) receptor function in rat spinal cord. Life Sci 68:2287–2295

Morishita R, Kato K, Asano T (1990) GABAB receptors couple to G proteins Go, G*o and Gi1 but not to Gi2. FEBS Lett 271:231–235

Nehring RB, Horikawa HPM, El Far O, Kneussel M, Brandstätter JH, Stamm S, Wischmeyer E, Betz H, Karschin A (2000) The metabotropic $GABA_B$ receptor directly interacts with the activating transcription factor 4. J Biol Chem 275:35185–35191

Nicoll RA (2004) My close encounter with GABAB receptors. Biochem Pharmacol 68:1667–1674

Nishikawa M, Hirouchi M, Kuriyama K (1997) Functional coupling of Gi subtype with GABAB receptor/adenylyl cyclase system: analysis using a reconstituted system with purified GTP-binding protein from bovine cerebral cortex. Neurochem Int 31:21–25

Odagaki Y, Koyama T (2001) Identification of galpha subtype(s) involved in gamma-aminobutyric acid(B) receptor-mediated high-affinity guanosine triphosphatase activity in rat cerebral cortical membranes. Neurosci Lett 297:137–141

Odagaki Y, Nishi N, Koyama T (2000) Functional coupling of $GABA_B$ receptors with G proteins that are sensitive to N-ethylmaleimide treatment, suramin, and benzalkonium chloride in rat cerebral cortical membranes. J Neural Transm 107:1101–1116

Oh P, Schnitzer JE (2001) Segregation of heterotrimeric G proteins in cell surface microdomains. Mol Biol Cell 12:685–698

Orlandi C, Posokhova E, Masuho I, Ray TA, Hasan N, Gregg RG, Martemyanov KA (2012) GPR158/179 regulate G protein signaling by controlling localization and activity of the RGS7 complexes. J Cell Biol 197:711–719

Orlandi C, Sutton LP, Muntean BS, Song C, Martemyanov KA (2018) Homeostatic cAMP regulation by the RGS7 complex controls depression-related behaviors. Neuropsychopharmacology 44:642–653

Ostrovskaya O, Xie K, Masuho I, Fajardo-Serrano A, Lujan R, Wickman K, Martemyanov KA (2014) RGS7/Gβ5/R7BP complex regulates synaptic plasticity and memory by modulating hippocampal GABABR-GIRK signaling. eLife 3:e02053

Ostrovskaya OI, Orlandi C, Fajardo-Serrano A, Young SM, Lujan R, Martemyanov KA (2018) Inhibitory signaling to ion channels in hippocampal neurons is differentially regulated by alternative macromolecular complexes of RGS7. J Neurosci 38:10002–10015

Padgett CL, Lalive AL, Tan KR, Terunuma M, Munoz MB, Pangalos MN, Martínez-Hernández J, Watanabe M, Moss SJ, Luján R, Lüscher C, Slesinger PA (2012) Methamphetamine-evoked depression of GABAB receptor signaling in GABA neurons of the VTA. Neuron 73:978–989

Pagano A, Rovelli G, Mosbacher J, Lohmann T, Duthey B, Stauffer D, Ristig D, Schuler V, Meigel I, Lampert C, Stein T, Prezeau L, Blahos J, Pin J, Froestl W, Kuhn R, Heid J, Kaupmann K, Bettler B (2001) C-terminal interaction is essential for surface trafficking but not for heteromeric assembly of GABA(b) receptors. J Neurosci 21:1189–1202

Patil DN, Rangarajan ES, Novick SJ, Pascal BD, Kojetin DJ, Griffin PR, Izard T, Martemyanov KA (2018) Structural organization of a major neuronal G protein regulator, the RGS7-Gβ5-R7BP complex. eLife 7:e42150

Pérez-Garci E, Gassmann M, Bettler B, Larkum ME (2006) The GABAB1b isoform mediates long-lasting inhibition of dendritic Ca2+ spikes in layer 5 somatosensory pyramidal neurons. Neuron 50:603–616

Perroy J, Adam L, Qanbar R, Chénier S, Bouvier M (2003) Phosphorylation-independent desensitization of GABAB receptor by GRK4. EMBO J 22:3816–3824

Pfaff T, Malitschek B, Kaupmann K, Prézeau L, Pin JP, Bettler B, Karschin A (1999) Alternative splicing generates a novel isoform of the rat metabotropic GABA(B)R1 receptor. Eur J Neurosci 11:2874–2882

Pinkas DM, Sanvitale CE, Bufton JC, Sorrell FJ, Solcan N, Chalk R, Doutch J, Bullock AN (2017) Structural complexity in the KCTD family of Cullin3-dependent E3 ubiquitin ligases. Biochem J 474:3747–3761

Pontier SM, Lahaie N, Ginham R, St-Gelais F, Bonin H, Bell DJ, Flynn H, Trudeau L-E, McIlhinney J, White JH, Bouvier M (2006) Coordinated action of NSF and PKC regulates GABAB receptor signaling efficacy. EMBO J 25:2698–2709
Potier B, Dutar P (1993) Presynaptic inhibitory effect of baclofen on hippocampal inhibitory synaptic transmission involves a pertussis toxin-sensitive G-protein. Eur J Pharmacol 231:427–433
Proft J, Weiss N (2015) G protein regulation of neuronal calcium channels: back to the future. Mol Pharmacol 87:890–906
Puthiyedth N, Riveros C, Berretta R, Moscato P (2016) Identification of differentially expressed genes through integrated study of Alzheimer's disease affected brain regions. PLoS One 11: e0152342
Quallo T, Alkhatib O, Gentry C, Andersson DA, Bevan S (2017) G protein βγ subunits inhibit TRPM3 ion channels in sensory neurons. eLife 6:e26138
Raveh A, Cooper A, Guy-David L, Reuveny E (2010) Nonenzymatic rapid control of GIRK channel function by a G protein-coupled receptor kinase. Cell 143:750–760
Raveh A, Turecek R, Bettler B (2015) Mechanisms of fast desensitization of GABAB receptor-gated currents. Adv Pharmacol 73:145–165
Restituito S, Couve A, Bawagan H, Jourdain S, Pangalos MN, Calver AR, Freeman KB, Moss SJ (2005) Multiple motifs regulate the trafficking of GABAB receptors at distinct checkpoints within the secretory pathway. Mol Cell Neurosci 28:747–756
Reuveny E, Slesinger PA, Inglese J, Morales JM, Iñiguez-Lluhi JA, Lefkowitz RJ, Bourne HR, Jan YN, Jan LY (1994) Activation of the cloned muscarinic potassium channel by G protein βγ subunits. Nature 370:143–146
Rice HC, de Malmazet D, Schreurs A, Frere S, Van Molle I, Volkov AN, Creemers E, Vertkin I, Nys J, Ranaivoson FM, Comoletti D, Savas JN, Remaut H, Balschun D, Wierda KD, Slutsky I, Farrow K, De Strooper B, de Wit J, de Malmazet D, Schreurs A, Frere S, Van Molle I, Volkov AN, Creemers E, Vertkin I, Nys J, Ranaivoson FM, Comoletti D, Savas JN, Remaut H, Balschun D, Wierda KD, Slutsky I, Farrow K, De Strooper B, de Wit J (2019) Secreted amyloid-β precursor protein functions as a GABABR1a ligand to modulate synaptic transmission. Science 363:eaao4827
Richman RW, Strock J, Hains MD, Cabanilla NJ, Lau K-K, Siderovski DP, Diversé-Pierluissi M (2005) RGS12 interacts with the SNARE-binding region of the Cav2.2 calcium channel. J Biol Chem 280:1521–1528
Robbins MJ, Calver AR, Filippov AK, Hirst WD, Russell RB, Wood MD, Nasir S, Couve A, Brown DA, Moss SJ, Pangalos MN (2001) GABA(B2) is essential for g-protein coupling of the GABA(B) receptor heterodimer. J Neurosci 21:8043–8052
Rost BR, Nicholson P, Ahnert-Hilger G, Rummel A, Rosenmund C, Breustedt J, Schmitz D (2011) Activation of metabotropic GABA receptors increases the energy barrier for vesicle fusion. J Cell Sci 124:3066–3073
Sadana R, Dessauer CW (2009) Physiological roles for G protein-regulated adenylyl cyclase isoforms: insights from knockout and overexpression studies. Neurosignals 17:5–22
Saghatelyan AK, Snapyan M, Gorissen S, Meigel I, Mosbacher J, Kaupmann K, Bettler B, Kornilov AV, Nifantiev NE, Sakanyan V, Schachner M, Dityatev A (2003) Recognition molecule associated carbohydrate inhibits postsynaptic GABAB receptors: a mechanism for homeostatic regulation of GABA release in perisomatic synapses. Mol Cell Neurosci 24:271–282
Sakaba T, Neher E (2003) Direct modulation of synaptic vesicle priming by GABAB receptor activation at a glutamatergic synapse. Nature 424:775–778
Sallese M, Salvatore L, D'urbano E, Sala G, Storto M, Launey T, Nicoletti F, Knöpfel T, De Blasi A (2000) The G-protein-coupled receptor kinase GRK4 mediates homologous desensitization of metabotropic glutamate receptor 1. FASEB J 14:2569–2580
Sarker S, Xiao K, Shenoy SK (2011) A tale of two sites – how ubiquitination of a G protein-coupled receptor is coupled to its lysosomal trafficking from distinct receptor domains. Commun Integr Biol 4:528–531

Sato PY, Chuprun JK, Schwartz M, Koch WJ (2015) The evolving impact of g protein-coupled receptor kinases in cardiac health and disease. Physiol Rev 95:377–404

Sauter K, Grampp T, Fritschy J-M, Kaupmann K, Bettler B, Mohler H, Benke D (2005) Subtype-selective interaction with the transcription factor CCAAT/enhancer-binding protein (C/EBP) homologous protein (CHOP) regulates cell surface expression of GABA(B) receptors. J Biol Chem 280:33566–33572

Schiff ML, Siderovski DP, Jordan JD, Brothers G, Snow B, De Vries L, Ortiz DF, Divers-é-Pierluissi M (2000) Tyrosine-kinase-dependent recruitment of RGS12 to the N-type calcium channel. Nature 408:723–727

Schuler V, Lüscher C, Blanchet C, Klix N, Sansig G, Klebs K, Schmutz M, Heid J, Gentry C, Urban L, Fox A, Spooren W, Jaton AL, Vigouret J, Pozza M, Kelly PH, Mosbacher J, Froestl W, Käslin E, Korn R, Bischoff S, Kaupmann K, van der Putten H, Bettler B (2001) Epilepsy, hyperalgesia, impaired memory, and loss of pre- and postsynaptic GABA (B) responses in mice lacking GABA(B(1)). Neuron 31:47–58

Schwarz DA, Barry G, Eliasof SD, Petroski RE, Conlon PJ, Maki RA (2000) Characterization of γ-aminobutyric acid receptor $GABA_{B(1e)}$, a $GABA_{B(1)}$ splice variant encoding a truncated receptor. J Biol Chem 275:32174–32181

Schwenk J, Metz M, Zolles G, Turecek R, Fritzius T, Bildl W, Tarusawa E, Kulik A, Unger A, Ivankova K, Seddik R, Tiao JY, Rajalu M, Trojanova J, Rohde V, Gassmann M, Schulte U, Fakler B, Bettler B (2010) Native GABAB receptors are heteromultimers with a family of auxiliary subunits. Nature 465:231–235

Schwenk J, Pérez-Garci E, Schneider A, Kollewe A, Gauthier-Kemper A, Fritzius T, Raveh A, Dinamarca MC, Hanuschkin A, Bildl W, Klingauf J, Gassmann M, Schulte U, Bettler B, Fakler B (2016) Modular composition and dynamics of native GABAB receptors identified by high-resolution proteomics. Nat Neurosci 19:233–242

Schwindinger WF, Mirshahi UL, Baylor KA, Sheridan KM, Stauffer AM, Usefof S, Stecker MM, Mirshahi T, Robishaw JD (2012) Synergistic roles for G-protein γ3 and γ7 subtypes in seizure susceptibility as revealed in double knock-out mice. J Biol Chem 287:7121–7133

Schwirtlich M, Emri Z, Antal K, Máté Z, Katarova Z, Szabó G (2010) $GABA_A$ and $GABA_B$ receptors of distinct properties affect oppositely the proliferation of mouse embryonic stem cells through synergistic elevation of intracellular Ca^{2+}. FASEB J 24:1218–1228

Siderovski DP, Willard FS (2005) The GAPs, GEFs, and GDIs of heterotrimeric G-protein alpha subunits. Int J Biol Sci 1:51–66

Signorini S, Liao YJ, Duncan SA, Jan LY, Stoffel M (1997) Normal cerebellar development but susceptibility to seizures in mice lacking G protein-coupled, inwardly rectifying K+ channel GIRK2. Proc Natl Acad Sci U S A 94:923–927

Simonds WF (1999) G protein regulation of adenylate cyclase. Trends Pharmacol Sci 20:66–73

Smit MJ, Iyengar R (1998) Mammalian adenylyl cyclases. Adv Second Messenger Phosphoprotein Res 32:1–21

Snow BE, Betts L, Mangion J, Sondek J, Siderovski DP (1999) Fidelity of G protein β-subunit association by the G protein γ-subunit-like domains of RGS6, RGS7, and RGS11. Proc Natl Acad Sci 96:6489–6494

Squires KE, Montañez-Miranda C, Pandya RR, Torres MP, Hepler JR (2018) Genetic analysis of rare human variants of regulators of G protein signaling proteins and their role in human physiology and disease. Pharmacol Rev 70:446–474

Steiger JL, Bandyopadhyay S, Farb DH, Russek SJ (2004) cAMP response element-binding protein, activating transcription factor-4, and upstream stimulatory factor differentially control hippocampal GABABR1a and GABABR1b subunit gene expression through alternative promoters. J Neurosci 24:6115–6126

Stewart GD, Comps-Agrar L, Nørskov-Lauritsen LB, Pin J-P, Kniazeff J (2018) Allosteric interactions between GABAB1 subunits control orthosteric binding sites occupancy within GABAB oligomers. Neuropharmacology 136:92–101

Sutton LP, Orlandi C, Song C, Oh WC, Muntean BS, Xie K, Filippini A, Xie X, Satterfield R, Yaeger JDW, Renner KJ, Young SM, Xu B, Kwon H, Martemyanov KA, Martemyanov KA (2018) Orphan receptor GPR158 controls stress-induced depression. eLife 7:e33273

Tabata T, Araishi K, Hashimoto K, Hashimotodani Y, van der Putten H, Bettler B, Kano M (2004) Ca2+ activity at GABAB receptors constitutively promotes metabotropic glutamate signaling in the absence of GABA. Proc Natl Acad Sci 101:16952–16957

Takahashi T, Kajikawa Y, Tsujimoto T (1998) G-protein-coupled modulation of presynaptic calcium currents and transmitter release by a GABAB receptor. J Neurosci 18:3138–3146

Terunuma M (2018) Diversity of structure and function of GABAB receptors: a complexity of GABAB-mediated signaling. Proc Jpn Acad Ser B Phys Biol Sci 94:390–411

Terunuma M, Revilla-Sanchez R, Quadros IM, Deng Q, Deeb TZ, Lumb M, Sicinski P, Haydon PG, Pangalos MN, Moss SJ (2014) Postsynaptic GABAB receptor activity regulates excitatory neuronal architecture and spatial memory. J Neurosci 34:804–816

Terunuma M, Vargas KJ, Wilkins ME, Ramirez OA, Jaureguiberry-Bravo M, Pangalos MN, Smart TG, Moss SJ, Couve A (2010) Prolonged activation of NMDA receptors promotes dephosphorylation and alters postendocytic sorting of GABAB receptors. Proc Natl Acad Sci 107:13918–13923

Thinakaran G, Koo EH (2008) Amyloid precursor protein trafficking, processing, and function. J Biol Chem 283:29615–29619

Tian J, Zhu M (2018) GABAB receptors augment TRPC3-mediated slow excitatory postsynaptic current to regulate cerebellar purkinje neuron response to type-1 metabotropic glutamate receptor activation. Cells 7:90

Tiao JY, Bradaia A, Biermann B, Kaupmann K, Metz M, Haller C, Rolink AG, Pless E, Barlow PN, Gassmann M, Bettler B (2008) The sushi domains of secreted $GABA_{B1}$ isoforms selectively impair $GABA_B$ heteroreceptor function. J Biol Chem 283:31005–31011

Turecek R, Schwenk J, Fritzius T, Ivankova K, Zolles G, Adelfinger L, Jacquier V, Besseyrias V, Gassmann M, Schulte U, Fakler B, Bettler B (2014) Auxiliary GABAB receptor subunits uncouple G protein βγ subunits from effector channels to induce desensitization. Neuron 82:1032–1044

Valdés V, Valenzuela JI, Salas DA, Jaureguiberry-Bravo M, Otero C, Thiede C, Schmidt CF, Couve A (2012) Endoplasmic reticulum sorting and kinesin-1 command the targeting of axonal GABAB receptors. PLoS One 7:e44168

Vernon E, Meyer G, Pickard L, Dev K, Molnar E, Collingridge GL, Henley JM (2001) GABAB receptors couple directly to the transcription factor ATF4. Mol Cell Neurosci 17:637–645

Vertkin I, Styr B, Slomowitz E, Ofir N, Shapira I, Berner D, Fedorova T, Laviv T, Barak-Broner N, Greitzer-Antes D, Gassmann M, Bettler B, Lotan I, Slutsky I (2015) GABAB receptor deficiency causes failure of neuronal homeostasis in hippocampal networks. Proc Natl Acad Sci U S A 112:E3291–E3299

Vigot R, Barbieri S, Bräuner-Osborne H, Turecek R, Shigemoto R, Zhang Y-P, Luján R, Jacobson LH, Biermann B, Fritschy J-M, Vacher C-M, Müller M, Sansig G, Guetg N, Cryan JF, Kaupmann K, Gassmann M, Oertner TG, Bettler B (2006) Differential compartmentalization and distinct functions of GABAB receptor variants. Neuron 50:589–601

De Waard M, Liu H, Walker D, Scott VES, Gurnett CA, Campbell KP (1997) Direct binding of G-protein βλ complex to voltage-dependent calcium channels. Nature 385:446–450

Wettschureck N, Offermanns S (2005) Mammalian G proteins and their cell type specific functions. Physiol Rev 85:1159–1204

White JH, McIlhinney RAJ, Wise A, Ciruela F, Chan W-YY, Emson PC, Billinton A, Marshall FH (2000) The GABAB receptor interacts directly with the related transcription factors CREB2 and ATFx. Proc Natl Acad Sci 97:13967–13972

Whorton MR, MacKinnon R (2013) X-ray structure of the mammalian GIRK2–βγ G-protein complex. Nature 498:190–197

Wickman KD, Iñiguez-Lluhi JA, Davenport PA, Taussig R, Krapivinsky GB, Linder ME, Gilman AG, Clapham DE (1994) Recombinant G-protein βγ-subunits activate the muscarinic-gated atrial potassium channel. Nature 368:255–257

Witherow DS, Wang Q, Levay K, Cabrera JL, Chen J, Willars GB, Slepak VZ (2000) Complexes of the G protein subunit gbeta 5 with the regulators of G protein signaling RGS7 and RGS9. Characterization in native tissues and in transfected cells. J Biol Chem 275:24872–24880

Wojcik WJ, Neff NH (1984) Gamma-Aminobutyric acid B receptors are negatively coupled to adenylate cyclase in brain, and in the cerebellum these receptors may be associated with granule cells. Mol Pharmacol 25:24–28

Wong Y, Conklin B, Bourne H (1992) Gz-mediated hormonal inhibition of cyclic AMP accumulation. Science 255:339–342

Workman ER, Haddick PCG, Bush K, Dilly GA, Niere F, Zemelman BV, Raab-Graham KF (2015) Rapid antidepressants stimulate the decoupling of GABAB receptors from GIRK/Kir3 channels through increased protein stability of 14-3-3η. Mol Psychiatry 20:298–310

Workman ER, Niere F, Raab-Graham KF (2013) mTORC1-dependent protein synthesis underlying rapid antidepressant effect requires GABABR signaling. Neuropharmacology 73:192–203

Wright R, Newey SE, Ilie A, Wefelmeyer W, Raimondo JV, Ginham R, Mcllhinney RAJ, Akerman CJ (2017) Neuronal chloride regulation via KCC2 is modulated through a $GABA_B$ receptor protein complex. J Neurosci 37:5447–5462

Wu C, Sun D (2015) GABA receptors in brain development, function, and injury. Metab Brain Dis 30:367–379

Wu LG, Saggau P (1995) GABAB receptor-mediated presynaptic inhibition in guinea-pig hippocampus is caused by reduction of presynaptic Ca2+ influx. J Physiol 485:649–657

Xie K, Allen KL, Kourrich S, Colón-Saez J, Thomas MJ, Wickman K, Martemyanov KA (2010) Gβ5 recruits R7 RGS proteins to GIRK channels to regulate the timing of neuronal inhibitory signaling. Nat Neurosci 13:661

Xu C, Zhang W, Rondard P, Pin J-P, Liu J (2014) Complex GABAB receptor complexes: how to generate multiple functionally distinct units from a single receptor. Front Pharmacol 5:12

Xue L, Sun Q, Zhao H, Rovira X, Gai S, He Q, Pin J-P, Liu J, Rondard P (2019) Rearrangement of the transmembrane domain interfaces associated with the activation of a GPCR hetero-oligomer. Nat Commun 10:2765

Yang J, Jan YN, Jan LY (1995) Determination of the subunit stoichiometry of an inwardly rectifying potassium channel. Neuron 15:1441–1447

Zapata J, Moretto E, Hannan S, Murru L, Longatti A, Mazza D, Benedetti L, Fossati M, Heise C, Ponzoni L, Valnegri P, Braida D, Sala M, Francolini M, Hildebrand J, Kalscheuer V, Fanelli F, Sala C, Bettler B, Bassani S, Smart TG, Passafaro M (2017) Epilepsy and intellectual disability linked protein Shrm4 interaction with GABABRs shapes inhibitory neurotransmission. Nat Commun 8:14536

Zemoura K, Balakrishnan K, Grampp T, Benke D (2019) Ca2+/calmodulin-dependent protein kinase II (CaMKII) β-dependent phosphorylation of GABAB1 triggers lysosomal degradation of GABAB receptors via mind bomb-2 (MIB2)-mediated Lys-63-linked ubiquitination. Mol Neurobiol 56:1293–1309

Zemoura K, Schenkel M, Acuña MA, Yévenes GE, Ulrich Zeilhofer H, Benke D (2013) Endoplasmic reticulum-associated degradation controls cell surface expression of γ-aminobutyric acid, type B receptors. J Biol Chem 288:34897–34905

Zemoura K, Trümpler C, Benke D (2016) Lys-63-linked ubiquitination of γ-aminobutyric acid (GABA), type B1, at multiple sites by the E3 ligase mind bomb-2 targets $GABA_B$ receptors to lysosomal degradation. J Biol Chem 291:21682–21693

Zhang J, Tan L, Ren Y, Liang J, Lin R, Feng Q, Zhou J, Hu F, Ren J, Wei C, Yu T, Zhuang Y, Bettler B, Wang F, Luo M (2016) Presynaptic excitation via GABAB receptors in habenula cholinergic neurons regulates fear memory expression. Cell 166:716–728

Zhang Z, Zhang W, Huang S, Sun Q, Wang Y, Hu Y, Sun N, Zhang Y, Jiang Z, Minato N, Pin J-P, Su L, Liu J (2015) GABAB receptor promotes its own surface expression by recruiting a Rap1-dependent signaling cascade. J Cell Sci 128:2302–2313
Zheng S, Abreu N, Levitz J, Kruse AC (2019) Structural basis for KCTD-mediated rapid desensitization of GABAB signalling. Nature 567:127–131
Zuo H, Glaaser I, Zhao Y, Kurinov I, Mosyak L, Wang H, Liu J, Park J, Frangaj A, Sturchler E, Zhou M, McDonald P, Geng Y, Slesinger PA, Fan QR (2019) Structural basis for auxiliary subunit KCTD16 regulation of the GABA $_B$ receptor. Proc Natl Acad Sci 116:8370–8379

GABA$_B$ Receptor Chemistry and Pharmacology: Agonists, Antagonists, and Allosteric Modulators

A. Nieto, T. Bailey, K. Kaczanowska, and P. McDonald

Contents

Abstract The GABA$_B$ receptors are metabotropic G protein-coupled receptors (GPCRs) that mediate the actions of the primary inhibitory neurotransmitter, γ-aminobutyric acid (GABA). In the CNS, GABA plays an important role in behavior, learning and memory, cognition, and stress. GABA is also located throughout the gastrointestinal (GI) tract and is involved in the autonomic control of the intestine and esophageal reflex. Consequently, dysregulated GABA$_B$ receptor signaling is associated with neurological, mental health, and gastrointestinal disorders; hence, these receptors have been identified as key therapeutic targets and are the focus of multiple drug discovery efforts for indications such as muscle spasticity

A. Nieto, T. Bailey, and K. Kaczanowska contributed equally to this work.

A. Nieto, T. Bailey, and P. McDonald (✉)
Department of Cancer Physiology, Moffitt Cancer Center, Tampa, FL, USA
e-mail: Patsy.McDonald@moffitt.org

K. Kaczanowska
Department of Chemistry, The Scripps Research Institute, La Jolla, CA, USA

Curr Topics Behav Neurosci (2022) 52: 81–118
https://doi.org/10.1007/7854_2021_232
Published Online: 26 May 2021

disorders, schizophrenia, pain, addiction, and gastroesophageal reflex disease (GERD). Numerous agonists, antagonists, and allosteric modulators of the $GABA_B$ receptor have been described; however, Lioresal® (Baclofen; β-(4-chlorophenyl)-γ-aminobutyric acid) is the only FDA-approved drug that selectively targets $GABA_B$ receptors in clinical use; undesirable side effects, such as sedation, muscle weakness, fatigue, cognitive deficits, seizures, tolerance and potential for abuse, limit their therapeutic use. Here, we review $GABA_B$ receptor chemistry and pharmacology, presenting orthosteric agonists, antagonists, and positive and negative allosteric modulators, and highlight the therapeutic potential of targeting $GABA_B$ receptor modulation for the treatment of various CNS and peripheral disorders.

Keywords $GABA_B$ receptor pharmacology · Orthosteric and allosteric modulators · Therapeutic target

1 Introduction

γ-aminobutyric acid (GABA) is one of the most widely distributed amino acid neurotransmitters in the central nervous system (CNS), acting as the primary neurotransmitter responsible for neuronal inhibition. GABA activities are mediated through two distinct classes of receptors; ionotropic $GABA_A$ and $GABA_A$-ρ (formerly known as $GABA_C$, and prominently expressed in the retina (Naffaa et al. 2017)) and metabotropic $GABA_B$ receptors (Bowery et al. 2004). $GABA_A$ and $GABA_A$-ρ ionotropic receptor subunits form ion channels that are selectively permeable to anions like chloride and are responsible for the transient and rapid component of inhibitory postsynaptic potentials (Sigel and Steinmann 2012). Whereas, the metabotropic $GABA_B$ receptors belong to the superfamily of G-protein-coupled receptors (GPCRs) and mediate the slow and prolonged component of synaptic inhibition via indirect gating of neuronal K^+ and Ca^{2+} channels and lowering levels of other second messenger targets like cAMP (Bowery et al. 2002).

$GABA_B$ receptors are broadly expressed and distributed in the CNS. They are located pre- and postsynaptically; activation of presynaptic $GABA_B$ receptors by GABA located on GABAergic terminals (autoreceptors) inhibits the release of GABA, while activation of presynaptic $GABA_B$ receptors located on other nerve terminals (heteroreceptors) inhibits the release of several other neurotransmitters such as glutamate and bioactive peptides. In contrast, activation of postsynaptic receptors activate K^+ channels and induce slow inhibitory postsynaptic potentials (Benarroch 2012). $GABA_B$ receptors are also located in the periphery along the gastrointestinal (GI) tract where they regulate intestinal motility, gastric emptying, gastric acid secretion, and esophageal sphincter relaxation (Clarke et al. 2018; Lehmann et al. 2010; Ong and Kerr 1984). Dysregulated $GABA_B$ receptor function has been implicated in a variety of neurodegenerative diseases and pathophysiological disorders, including Parkinson's disease (Nambu 2012; Tyagi et al. 2015), Alzheimer's disease (Rice et al. 2019; Sun et al. 2020), Huntington's disease (Kim and Seo 2014), epilepsy (Billinton et al. 2001a; Teichgräber et al. 2009), spasticity (Francisco et al. 2001; Basmajian 1975; Korsgaard 1976), pain (Neto et al. 2006; Enna and McCarson 2006; Murai et al. 2019), anxiety (Kalinichev et al. 2017; Li et al. 2015) and depression (Cryan and Kaupmann 2005; Felice et al. 2012; Jacobson et al. 2018), schizophrenia (Glausier and Lewis 2017; Nair et al. 2020), narcolepsy (Black et al. 2014; Szabadi 2015), and addiction (Agabio and Colombo 2014, 2015; Agabio et al. 2018; Maccioni and Colombo 2019; Ranson et al. 2020). Owing to their presence in the gastrointestinal tract these receptors are also implicated in a variety of GI disorders such as gastroesophageal reflux disease (GERD) (Clarke et al. 2018; Lehmann et al. 2010; Ong and Kerr 1984; Lehmann 2009; Symonds et al. 2003).

2 Brief History

While GABA was discovered in the mammalian brain in 1950 (Awapara 1950; Roberts and Frankel 1950), it was not recognized as an inhibitory neurotransmitter until 1967 (Krnjević and Schwartz 1966; Dreifuss et al. 1969). Early attempts to interrogate the GABA system led to the development of the synthetic agonist, β-(4-chlorophenyl)-γ-aminobutyric acid, a poorly brain penetrant derivative of GABA better known as baclofen (Keberle et al. 1969). In 1968 the identification of the first GABA receptor antagonist "bicuculline" was reported (Curtis et al. 1970), and in 1987, bicuculline and GABA receptor agonists such as isoguvacine facilitated the cloning of the ionotropic $GABA_A$ receptor, a pentameric ligand gated ion channel (Schofield et al. 1987).

The existence of the $GABA_B$ receptors (so named to distinguish it from the $GABA_A$ receptor) first emerged in 1979. Dr. Norman Bowery and colleagues discovered that GABA blocks the release of neurotransmitters such as norepinephrine from nerve terminals but this effect was not blocked by bicuculline, instead it mimicked the effects of baclofen. It was also discovered that baclofen does not interact with the $GABA_A$ site (Bowery et al. 1979, 1980, 1981). A third GABA receptor with pharmacology distinct from $GABA_A$ and $GABA_B$ was identified in 1986 by virtue of the GABA response, "Cl-current blocked by picrotoxin," being both bicuculline and baclofen insensitive (Johnston 1986). This receptor was named $GABA_C$ (now referred to as $GABA_A$-ρ) and was later cloned in 1991 (Polenzani et al. 1991). However, it was almost 20 years since being identified that the $GABA_B$ receptor was cloned using expression cloning and radioligand binding of a high affinity antagonist (1997) by the Bettler group (Kaupmann et al. 1997). Thus, reagents that modulate the GABA receptors facilitated the cloning of, and have since defined those receptors; the ionotropic receptors $GABA_A$ and $GABA_A$-ρ are defined as "bicuculline-sensitive, isoguvacine-sensitive" and "bicuculline-insensitive, baclofen-insensitive" respectively, and the metabotropic $GABA_B$ receptor is defined as "bicuculline-insensitive, baclofen-sensitive."

3 Structure and Signaling

In common with other GPCRs, the $GABA_B$ receptor is an integral membrane protein that spans the cellular membrane with seven helices that are linked by three extracellular loops and three intracellular loops and possesses an extracellular N-terminus and an intracellular C-terminus. $GABA_B$ receptors are structurally related to metabotropic glutamate receptors (mGluRs), and together with the calcium-sensing receptor (CaSR), some pheromone and taste receptors, and orphan GPCRs (receptors with no known ligands), belong to the family C (or family III) of GPCRs (Bowery et al. 2002). Common to the members of family C GPCRs is the large extracellular N-terminus that contains a domain homologous to the periplasmic amino acid binding proteins (PBPs) found in bacteria. The X-ray structure of $GABA_B$ receptor PBP-like domains revealed an orthosteric ligand binding pocket that is made up of two globular lobes separated by a hinge region. The two lobes (LB1 and LB2) close upon ligand binding, much like a Venus flytrap does when touched by an insect, hence the globular domains in family C GPCRs are also referred to as "Venus flytrap" (VFT) domains (Galvez et al. 1999); an agonist binds and stabilizes the closed (active) conformation of the VFT, whereas an antagonist stabilizes and retains the VFT subunit in the open (inactive) configuration.

To date molecular cloning has identified two main $GABA_B$ receptor subunits, namely $GABA_{B1}$ and $GABA_{B2}$ which arise from distinct genes (Kaupmann et al. 1997, 1998). At the protein level $GABA_{B1}$ and $GABA_{B2}$ receptors share 35% identity and 54% similarity over their approximate length of 950 amino acid residues and both subunits are highly conserved across mammalian species, sharing 90–95% sequence homology between human, pig, rat, and mouse (Kaupmann et al. 1997). An active functional $GABA_B$ receptor with high affinity for agonist ligands depends upon the formation of a heterodimer between $GABA_{B1}$ and $GABA_{B2}$ receptor subunits (Kaupmann et al. 1998; Marshall et al. 1999; Jones et al. 1998). The association of the receptor subunits occurs, at least in part, through a coiled-coil motif found in the respective carboxyl termini of $GABA_{B1}$ and $GABA_{B2}$ subunits. It has been demonstrated in recombinant systems that $GABA_{B1}$ is unable to reach the cell surface in the absence of the $GABA_{B2}$ subunit because $GABA_{B1}$ contains endoplasmic retention motifs in its carboxy tail that are masked only upon heterodimerization with $GABA_{B2}$ subunit (Couve et al. 1998; Pagano et al. 2001). Interestingly, all orthosteric agonists and antagonists bind to the $GABA_{B1}$VFT and not to the $GABA_{B2}$ subunit VFT. Upon binding, an agonist induces conformational changes in the $GABA_{B1}$ subunit which by virtue of its physical interaction with the $GABA_{B2}$ subunit promotes conformational changes in the latter subunit allowing it to couple to its cognate G-protein promoting functional responses within the cell

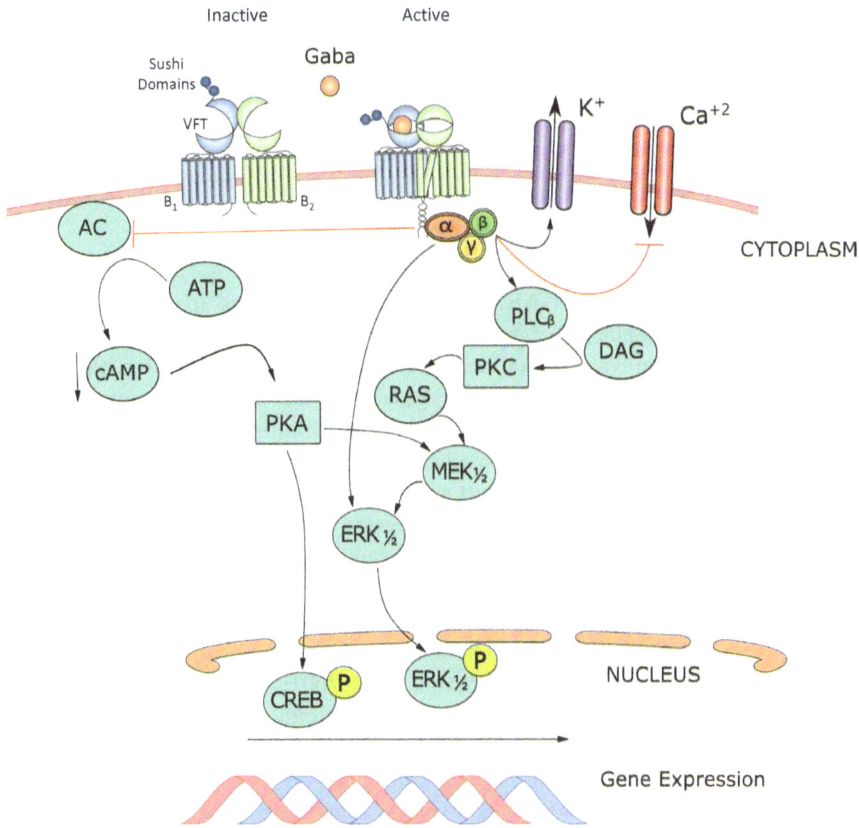

Fig. 1 Molecular diversity and signaling capacity of the GABA$_B$ receptor

(Galvez et al. 2001; Margeta-Mitrovic et al. 2001; Robbins et al. 2001; Duthey et al. 2002).

GABA$_B$ receptors provide a crucial component of inhibitory neurotransmission mainly via coupling to heterotrimeric $G_{i/o}$ type G-proteins, activation of which results in a Gα-mediated inhibition of cAMP production and a Gβγ-mediated modulation of the activity of ion channels such as high voltage-activated Ca^{2+} (Ca_v) channels and G protein-coupled inwardly rectifying Kir3-type potassium channels (GIRKs) (Morishita et al. 1990; Nishikawa et al. 1997). In rare cases and non-neuronal cells, GABA$_B$ receptor activation can promote increases in intracellular calcium either via activation of phospholipase C and store-operated channels or by inducing Ca^{2+} release from internal stores (Meier et al. 2008; New et al. 2006). Furthermore, GABA$_B$ receptor activation has been reported to induce phosphorylation of the Extracellular-signal Regulated protein Kinase 1/2 ($ERK_{1/2}$) in cerebellar neurons, as well as in the CA1 field of the mouse hippocampus (Tu et al. 2007; Vanhoose et al. 2002). Thus, GABA$_B$ receptor couples to multiple intracellular signal transduction pathways (Fig. 1) regulating ion homeostasis as well as MAPK signaling leading to downstream effects that include blocked neurotransmitter release and hyperpolarization of neurons (Bowery et al. 2002; Bettler et al. 2004), and the modulation of autonomic control of the intestine and esophageal reflex (Clarke et al. 2018; Ong and Kerr 1984; Lehmann 2009; Symonds et al. 2003).

4 Molecular Diversity and Complexity

Molecular diversity in the $GABA_B$ receptor system arises from expression of multiple $GABA_{B1}$ subunit isoforms of which 14 mammalian isoforms ($GABA_{B\ (1a\text{-}1n)}$) exist between various animal species and are generated by differential transcription or splicing (Bettler et al. 2004), whereas the $GABA_{B2}$ receptor encodes a singular form of the receptor (Bettler et al. 2004; Billinton et al. 2001b). The two predominant $GABA_{B1}$ isoforms, termed $GABA_{B(1a)}$ and $GABA_{B(1b)}$, are generated by use of alternative transcription start sites, whereas other less abundant isoforms such as $GABA_{B(1c)}$, and $GABA_{B(1e)}$ are generated by alternative splicing. Only the $GABA_{B(1a)}$ and $GABA_{B(1b)}$ variants have been identified as components of the native receptor $GABA_{B1}/GABA_{B2}$ complex. Although the identification of these variants is suggestive of pharmacologically distinct $GABA_B$ receptors, Ng and colleagues reported that the anticonvulsant gabapentin acts as an agonist at $GABA_{B(1a)}$ but not $GABA_{B(1b)}$ (Bertrand et al. 2001; Ng et al. 2001), this has been widely disputed as heterodimers comprised of either $GABA_{B(1a)}/GABA_{B2}$ or $GABA_{B(1b)}/GABA_{B2}$ are pharmacologically indistinguishable in heterologous systems (Jensen et al. 2002; Lanneau et al. 2001) and to date, no $GABA_B$ receptor ligand differentiates between these molecular variants. However, studies facilitated by the generation of $GABA_{B1}$ isoform-specific knockout mice (Vigot et al. 2006) demonstrated that $GABA_{B1a}$- and $GABA_{B1b}$-containing receptors have distinct functions owing to their different locations within neurons, where $GABA_{B1a}$ receptors are predominantly located presynaptically on axonal terminals and $GABA_{B1b}$ postsynaptically on dendritic spines. Consequently, global $GABA_{B1}$ receptor isoform knockout mice exhibit a wide spectrum of isoform-specific behaviors. For example, using the isoform-specific knockout mice, Vigot et al. showed that $GABA_{B1a}$ and not $GABA_{B1b}$ receptor was involved in impaired synaptic plasticity in hippocampus long-term potentiation (Vigot et al. 2006). It was also shown by Perez-Garci and colleagues that $GABA_{\ B1b}$ was responsible for mediating postsynaptic inhibition of Ca^{2+} spikes, whereas presynaptic inhibition of GABA release was mediated by $GABA_{B1a}$ (Pérez-Garci et al. 2006). Hence, based on numerous in vivo findings, the existence of pharmacologically distinct $GABA_B$ receptors has been proposed (Pinard et al. 2010).

$GABA_{B(1a)}$ and $GABA_{B(1b)}$ differ primarily in their extracellular amino-terminal domains by a pair of sushi domains only present in the $GABA_{B(1a)}$ subunit of the $GABA_{B1(a)}/GABA_{B2}$ heteromer (Bettler et al. 2004; Hawrot et al. 1998). Sushi domains, or short consensus repeats, are conserved protein domains commonly involved in protein–protein interactions mostly found in proteins involved in cell–

cell adhesion. In the context of the $GABA_B$ receptor, the sushi domains have been shown to play a role in targeting the $GABA_{B(1a)}$ receptor to specific subcellular regions by means of interaction of these motifs with proteins in the extracellular matrix or on the surface of neighboring cells (Hannan et al. 2012). The diversity in $GABA_{B1}$ isoforms may therefore provide a means for targeted subcellular localization and/or coupling to distinct intracellular signaling pathways while also providing, in part, an explanation for the complex and diverse physiology effects of the GABA/$GABA_B$ receptor axis observed in neuronal tissue and in vivo (Bettler and Tiao 2006).

The molecular complexity of the $GABA_B$ receptor is further enhanced through association of the receptor with numerous trafficking, effector, and regulatory proteins, as well as other membrane-bound receptors. For example, the extracellular matrix protein, fibulin-2, has been shown to bind to the first sushi domain of the $GABA_{B(1a)}$ and target this receptor to axon terminals of excitatory synapses (Blein et al. 2004). Likewise, amyloid precursor protein (APP), amyloid precursor protein-like 2 (APLP2), and adherence junction associated protein-1 (AJAP1) interact with the sushi domains and are also anticipated to direct axonal subcellular localization of the $GABA_{B(1a)}$/$GABA_{B2}$ receptor complex (Dinamarca et al. 2019). Whereas $GABA_{B(1b)}$-containing heteromers more frequently show dendritic localization (Vigot et al. 2006).

Furthermore, a subfamily of the potassium channel tetramerization domain (KCTD) proteins (KCTD 8, 12, 12b, and 16) has been shown to exclusively and constitutively interact with the $GABA_{B2}$ carboxy-terminus acting as auxiliary subunits of the receptor to regulate the kinetics and outcome of G-protein signaling (Bartoi et al. 2010; Schwenk et al. 2010). For example, the KCTD12 and 12b subunits mediate desensitization of the receptor, whereas KCTD8 and 16 regulate non-desensitizing activities. The receptor, KCTD subunits, and G-protein combined form the core receptor signaling complex required for normal function of inhibitory brain circuits. Recently, Zuo et al., reported a high-resolution crystal structure of the KCTD16 oligomerization domain in complex with a $GABA_{B2}$ C-terminal peptide and together with mutational analysis defined the interface between KCTD16 and $GABA_{B2}$ revealing a potential regulatory site that modulates $GABA_B$ receptor activity (Zuo et al. 2019).

Other proteins have been reported to transiently associate with the $GABA_B$ receptor either directly through $GABA_{B1}$ or $GABA_{B2}$ carboxy terminal domains, which include transcription factors (i.e., ATF-4 (CREB2) and CHOP (Gadd153) (Nehring et al. 2000; Ritter et al. 2004; Sauter et al. 2005)) and scaffolding and adaptor proteins (i.e., MUPP1, 14-3-3 protein, and NSF (Balasubramanian et al. 2007; Couve et al. 2001; Pontier et al. 2006)) or indirectly through multiprotein complexes, which include neuroligin-3, synaptotagmin-11, and calnexin (Schwenk et al. 2016). Novel functions of the $GABA_B$ receptor also arise through crosstalk with other membrane receptors such as $GABA_A$, mGluR1, NMDA, IGF-1, and TrkB receptors. For a more comprehensive description of the $GABA_B$ receptor interactome, see (Benke 2013; Fritzius and Bettler 2020).

Recent biophysical and structural studies have demonstrated that $GABA_B$ receptors can form higher-order multimeric receptor complexes and this has been shown to occur in both heterologous systems and in brain membranes. These multimers comprise oligomers of $GABA_{B1}$ and $GABA_{B2}$ heteromers that self-assemble through association of their $GABA_{B1}$ subunits into tetramers (dimers of dimers) and octamers (dimers of tetramers) (Comps-Agrar et al. 2011, 2012; Maurel et al. 2008). Tetramers were found to decrease $G\alpha_i$-protein coupling efficiency suggesting that the multimers exhibit negative cooperativity between heterodimers (Calebiro et al. 2013; Stewart et al. 2018). It has emerged that the core $GABA_{B1/B2}$ receptor not only assembles with itself (oligomerization) but can also form supercomplexes with other multiprotein complexes that are likely spatiotemporally regulated in response to neuronal and developmental cues (Fritzius and Bettler 2020). The role of higher-order receptor complexes in $GABA_B$ receptor function and physiology requires further investigation to determine the functional relevance of $GABA_B$ receptor oligomerization in native tissue.

5 Agonists

As mentioned previously, the synthesis of the GABA analogue baclofen (-β-(4-chlorophenyl)-GABA; Fig. 2) in 1962 as the prototypical $GABA_B$ receptor agonist (Keberle et al. 1964) has greatly facilitated the molecular and biochemical characterization of this receptor. Indeed, baclofen has served as an invaluable tool in elucidating the electrophysiological and behavioral responses linked to the $GABA_B$ receptor system revealing its versatility as a drug target to treat a wide variety of diseases (Bowery 1993; Froestl et al. 1995a, b). Owing to its extensive therapeutic potential, numerous attempts to improve baclofen's pharmacokinetic properties and

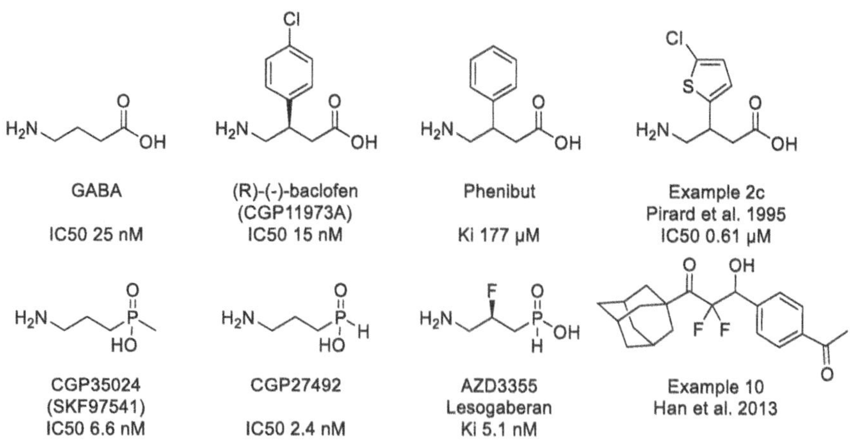

Fig. 2 Exemplar chemical structures of $GABA_B$ receptor full agonists

potency while maintaining selectivity have been pursued, but flat structure-activity relationships (SAR) around baclofen have resulted in very limited success. Of the 217 GABA$_B$ receptor-associated molecules reported in ChemBL Database (CHEMBL n.d.; Mendez et al. 2019), 55 compounds (42 agonists and 13 antagonists) are identified as being active at the GABA$_B$ receptor, most of which are chemically classified as analogues of either GABA or baclofen. However, the SAR investigations and the pharmacological properties of the resulting baclofen analogues have revealed important information regarding the chemical characteristics that endow baclofen with its activity at the GABA$_B$ receptor.

Following the resolution of baclofen in 1978 into the two enantiomers, (R)-(−)-baclofen and (S)-(+)-baclofen (Olpe et al. 1978; Weatherby et al. 1984) (CGP11973A and CGP11974A, respectively), in 1995, Froestl et al., demonstrated that the observed physiological effects of baclofen are stereoselective. They showed that the pharmacological action of baclofen is mediated by the R-(−)-enantiomer as R-(−)-baclofen (also known as Arbaclofen; Fig. 2) inhibits the binding of [^{3}H]-baclofen to GABA$_B$ receptors in cat cerebellum with an IC$_{50}$ of 15 nM, while the S-(+)-enantiomer and racemic mixture display >100-fold and 3-fold higher IC$_{50}$, respectively (Froestl et al. 1995a). Many analogues of (R)-(−)-baclofen have been generated to interrogate the role of the carboxylic acid, amine, and p-chlorophenyl groups in attempts to increase potency and improve pharmacokinetic properties; as a consequence, more agonists, partial agonists, and antagonists have been discovered (Froestl 2010).

The first analogues that proved to be more potent than baclofen were generated by replacing the carboxylic acid portion of GABA with phosphinic acid residues to generate full agonists, CGP35024 (SKF97541) (Froestl et al. 1995a) and CGP27492 (Chapman et al. 1993) (Fig. 2), which have greater or equal affinity than baclofen for the GABA$_B$ receptor and IC$_{50}$s of 2 nM and 5 nM (Froestl et al. 1995a; Patel et al. 2001; Bon and Galvan 1996; Seabrook et al. 1990), respectively. Later SAR efforts investigated the replacement of the p-chlorophenyl group of baclofen with heterocycles. The absence of the chlorine atom from baclofen produces another potent GABA$_B$ receptor agonist, phenibut, and like baclofen, the majority of the agonist activity at the GABA$_B$ receptor is attributed to (R)-phenibut. Substitution with a 2-chlorothienyl group also provides an active albeit weaker agonist (IC$_{50}$ ~ 0.6 μM) (Example 2c; Fig. 2) as determined in the [3H] baclofen displacement assay (Bolser et al. 1995). Further SAR and molecular modeling studies strongly implicated the p-chlorophenyl group (and its heteroaromatic substituents) as critical in the binding of baclofen and its analogues to the GABA$_B$ receptor (Costantino et al. 2001).

The phosphonous acid derivative, [(2R)-3-amino-2-fluoropropyl]phosphinic acid (AZD3355; Fig. 2) is a high affinity, non-brain penetrant analogue of baclofen that was developed by AstraZeneca and recently evaluated in clinical trials for the treatment of gastroesophageal reflux disease (GERD) under the generic name Lesogaberan® (Bredenoord 2009). AZD3355 has an EC$_{50}$ of 9 nM compared to GABA's EC$_{50}$ of 160 nM, and an increased binding affinity with a K$_i$ of 5 nM versus GABA's 110 nM for inhibition of [^{3}H]-GABA binding in rat brain (Niazi et al. 2011).

It has been suggested that the low structural diversity of the existing orthosteric $GABA_B$ receptor ligands may be due to the conformational space having not been fully explored (Evenseth et al. 2019). In the early 2000s extensive mutagenesis studies on the extracellular domain of the $GABA_{B1}$ receptor subunit identified critical residues in LB1 and LB2 that are key for both agonist and antagonist binding (Galvez et al. 1999, 2000; Kniazeff et al. 2002). More recently, the first X-ray crystal structures of the $GABA_B$ receptor heterodimeric complex between the extracellular VFT domains of $GABA_{B1}$ and $GABA_{B2}$, alone and in complex with bound agonists and antagonists were reported by Zuo et al., providing a more detailed understanding of how ligands act on the receptor (Geng et al. 2013). They demonstrated that in the inactive "apo" state and antagonist-bound state the VFT domains of both subunits adopt an open conformation whereas in the active "agonist-bound" state only the $GABA_{B1}$ subunit binds agonist and on doing so adopts a closed conformation.

Knowledge gained from these studies has since facilitated the development of a novel class of compounds that bind the orthosteric site of this receptor. In 2013, Colby et al. reported the discovery of $GABA_B$ receptor agonists comprised of β-hydroxy difluoromethyl ketones that represent the only structurally distinct $GABA_B$ receptor agonists as they lack the carboxylic acid or amino group of GABA (Example 10; Fig. 2) (Han et al. 2013). Additional analogues of the β-hydroxy difluoromethyl ketones have since been analyzed by the Colby laboratory, and docking models using the X-ray structures solved by Zuo et al. strongly suggest that these difluoromethyl ketones have similar binding modes to the orthosteric agonists (Sowaileh et al. 2018). Although some preliminary in vivo data suggest these compounds warrant further investigation as potential anxiolytic drugs (Han et al. 2013), their clinical utility has yet to be explored.

More recently, Mao and colleagues reported on Cryo-EM structures of the full-length inactive antagonist-bound and active agonist-bound in complex with $G\alpha_i$ protein of the $GABA_B$ receptor. This work further supports the findings that agonist binding stabilizes the closure of the $GABA_{B1}$ VFT domain (Geng et al. 2013). The Cryo-EM studies further revealed that agonist binding to $GABA_{B1}$ VFT domain induces rearrangement of the transmembrane (TM) interface between the $GABA_B$ subunits and this in turn promotes opening of the third intracellular loop in the $GABA_{B2}$ subunit allowing it to bind $G\alpha_i$ (Mao et al. 2020). Collectively, the structural studies of Zuo et al. and Mao et al. provide a deeper insight into $GABA_B$ receptor activation that will greatly assist in the design of novel modulators of the receptor.

6 Partial Agonists

Partial agonists are ligands that have varying degrees of intrinsic activities and affinity at their cognate receptors. They bind to and activate the receptor but elicit submaximal cell/tissue responses of the system relative to that produced by a full agonist. The naturally occurring GABA metabolite, γ-hydroxybutyric acid (GHB)

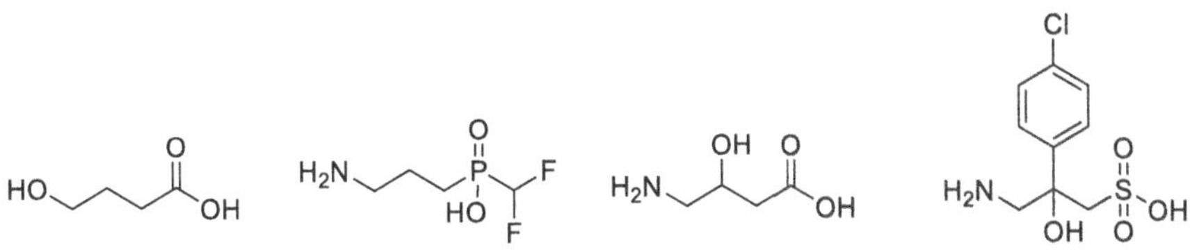

GHB
Ki 51 μM

CGP47656
IC50 89 nM

GABOB
1.6 μM

2-hydroxysaclofen*
IC50 5.1 μM

*Originally described as an antagonist.

Fig. 3 Exemplar structures of $GABA_B$ receptor partial agonists

(Fig. 3), exhibits partial agonism at the $GABA_B$ receptor and is used clinically to treat symptoms of narcolepsy, alcohol dependence and withdrawal, and also used illicitly as a drug of abuse. However, experiments performed in $GABA_{B1}$ receptor null mice clearly show that not all the in vivo effects of GHB are $GABA_B$ receptor-mediated (Wellendorph et al. 2005). GHB has both low and high affinity receptor targets in the brain. The high affinity binding site is well characterized but has yet to be incontrovertibly identified. Whereas, it is well established that the $GABA_B$ receptor is the low affinity binding site where GHB acts as a partial agonist (Wong et al. 2004). Several studies demonstrated this finding including those of Mathivet et al. (1997); using binding experiments GHB was shown to have K_i ~ 100 μM compared to baclofen K_i ~ 5 μM (Mathivet et al. 1997); and Lingenhoehl et al. (1999); using recombinant systems expressing $GABA_{B1}/GABA_{B2}$ heteromer together with Kir3 channels in xenopus oocytes showed that GHB activated these receptors with an EC_{50} ~ 5 mM and a maximal stimulation of 69% relative to baclofen. Furthermore, three $GABA_B$ receptor competitive antagonists, CGP5426A, 2-hydroxysaclofen, and CGP35348 each completely blocked the GHB-evoked response further supporting GHB is a weak, partial agonist (Lingenhoehl et al. 1999).

Returning to the baclofen analogues, as mentioned CGP35024/SKF97541 (-γ-aminopropyl(methyl)phosphinic acid) is a potent agonist harboring a methyl substituent on the phosphinic acid moiety. Exchanging the methyl group for a difluoromethyl group produces CGP47656 (γ-aminopropyl(difluoromethyl)-phosphinic acid) (Fig. 3), rendering the molecule a partial agonist at the $GABA_B$ receptor as demonstrated by measuring binding affinities (Urwyler et al. 2005), the release of GABA from rat cortex (Froestl et al. 1995a; Gemignani et al. 1994), or the cholinergic twitch contraction in guinea pig ileum (Marcoli et al. 2000). Replacing the aromatic substituent at the 3-position of baclofen with a hydroxyl group also produces partial agonistic activity as seen in 4-amino-3-hydroxybutanoic acid (GABOB) (Fig. 3), with (R)-(−)-GABOB being tenfold less potent than racemic baclofen in binding experiments from rat brain isolates (Hinton et al. 2008).

As noted above, CGP35348 (Fig. 4) and 2-hydroxysaclofen (Fig. 3) (Kerr et al. 1988) have previously been described as $GABA_B$ receptor neutral competitive

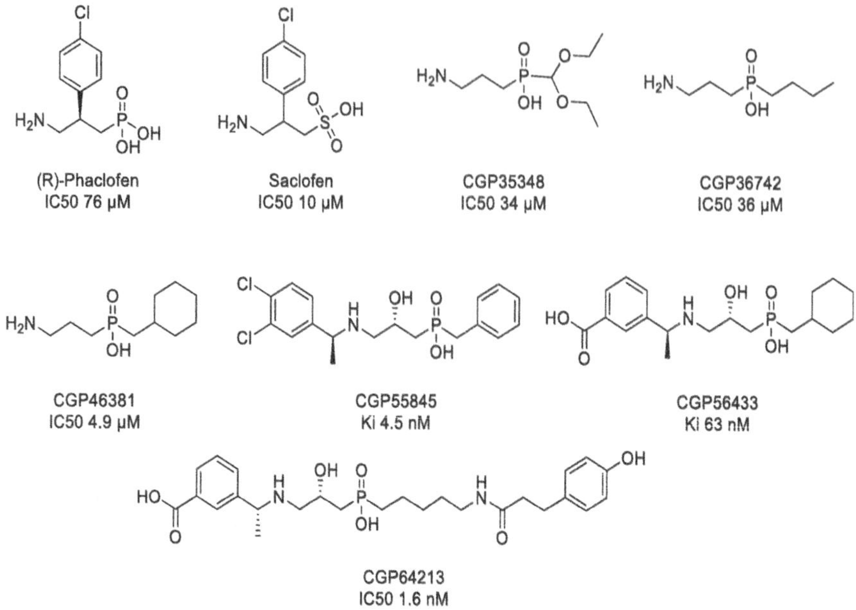

Fig. 4 Exemplar chemical structures of $GABA_B$ receptor antagonists

antagonists, having no intrinsic activity of their own and accordingly do not stimulate [^{35}S]-GTPγS-binding to membranes derived from CHO cells stably expressing the $GABA_B$ receptor. However, in the presence of CGP7930 or GS39783 (positive allosteric modulators, PAMs, of $GABA_B$ receptor) each "antagonist" stimulated [^{35}S]-GTPγS-binding to $GABA_B$ receptors with maximum efficiency of 31% and 35% of maximum GABA effect, respectively (Urwyler et al. 2005). A more sensitive assay measuring GABA/$GABA_B$ receptor-mediated inhibition of forskolin-stimulated cAMP accumulation revealed that CGP35348 and 2-hydroxysaclofen can have intrinsic partial agonistic activity in certain assay conditions that is enhanced by the PAMs. Thus, the PAMs revealed partial agonistic activity of compounds that otherwise appear to be devoid of intrinsic activity. Furthermore, the same experiments revealed that CGP7930 and GS39783 also possess intrinsic, low partial agonistic activity (Urwyler et al. 2005), an observation also reported by Binet et al. (Binet et al. 2004).

7 Antagonists

Following the 1979 discovery of a "bicuculline-insensitive, baclofen-sensitive" GABA receptor, efforts were immediately undertaken to design antagonists for this receptor. It was in the late 1980s that the first $GABA_B$ receptor antagonists were described. (R)-Phaclofen (Fig. 4), the phosphonic acid analogue of baclofen, was one of the first discovered antagonists and was shown to block the slow inhibitory postsynaptic potential in the rat hippocampus establishing the physiological importance of this receptor (Dutar and Nicoll 1988). This discovery was closely followed by the discovery of saclofen and (S)-2-hydroxysaclofen (Fig. 3), sulphonic analogues of baclofen (Kerr et al. 1987). (R)-Phaclofen has a low affinity (~130 μM) for the receptor in radioligand binding experiments using rat brain membranes (Kerr et al. 1987), whereas (S)-2-hydroxysaclofen is tenfold more potent an antagonist at the GABA$_B$ receptor than (R)-phaclofen in this assay.

In addition to their significant contributions in interrogating GABA$_B$ receptor function and pharmacological activity, as with agonists, preclinical studies strongly support GABA$_B$ receptor antagonists having clinical importance in the treatment of various CNS disorders. GABA$_B$ receptor antagonists have been shown to suppress absence seizures in preclinical animal models of epilepsy (Bernasconi et al. 1992; Ostojić et al. 2013; Marescaux et al. 1992; Snead 3rd 1992), improve learning and memory (Bianchi and Panerai 1993; Lasarge et al. 2009; Mondadori et al. 1993) and have also been widely shown to have antidepressant-like activity in animal models (Cryan and Kaupmann 2005; Felice et al. 2012; Jacobson et al. 2018; Cryan and Slattery 2010; Frankowska et al. 2007; Mombereau et al. 2004; Nowak et al. 2006) along with a rescue of withdrawal from drugs of abuse-induced stress (Vlachou et al. 2011). Anhedonia, a common symptom of both psychostimulant withdrawal and depression, appears to be the key to the role of GABA$_B$ receptor in these disorders, as previously described by Markou and colleagues (Markou et al. 1992, 1998). Furthermore, the GABA$_B$ receptor has been shown to play a role in the regulation of glucose homeostasis in vivo (Bonaventura et al. 2012), GABA$_B$ receptor antagonism as well as receptor knockout mice shows improved glucose-stimulated insulin secretion (Bonaventura et al. 2008; Braun et al. 2004).

Following the discovery of phaclofen and 2-hydroxysaclofen, additional antagonists were discovered leading to CGP35348 (3-aminopropyl(diethoxymethyl)-phosphinic acid), a potent GABA$_B$ receptor antagonist and the first shown to penetrate the blood-brain barrier; CGP36742 (3-aminopropyl(n-butyl)phosphinic acid), the first orally bioavailable antagonist; and CGP46381, the phosphinic acid bearing a methylcyclohexyl group. However, like their predecessors, these compounds have low affinity (high μM range) for the GABA$_B$ receptor as does the chemically distinct SCH50911 (Bolser et al. 1995) (Fig. 5). As a result of SAR studies during the generation of these compounds, it was discovered that the nature of the alkyl substituent on the phosphinic acid plays a critical role in ligand activity. For example, a methyl substituent is present on the potent agonist CGP35024 (Fig. 2); when this is replaced with the difluoromethyl group of CGP47656 (3-aminopropyl(difluoromethyl)phosphinic acid), a decrease in activity at the GABA$_B$ receptor is observed with CGP47656 (Fig. 3) acting as a partial agonist (Froestl et al. 1995a; Urwyler et al. 2005; Gemignani et al. 1994; Marcoli et al. 2000). Increases in size of the substituent as with the butyl group in CGP36742 (Fig. 4) result in a derivative that displays antagonist activity at the GABA$_B$ receptor. Hence, very modest structural modifications to the baclofen core can lead to

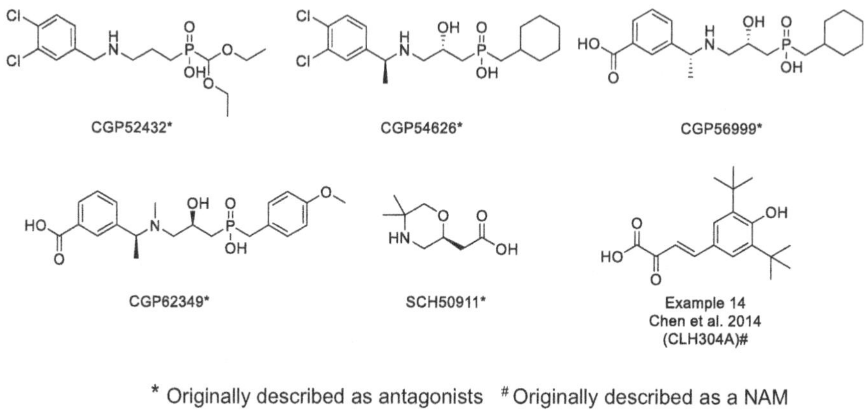

Fig. 5 Exemplar chemical structures of $GABA_B$ receptor inverse agonists

significant changes in ligand activity ranging from potent agonism to partial agonism to antagonism at the $GABA_B$ receptor (Pirard et al. 1995).

More potent antagonists have been developed since, displaying IC_{50} values in the nanomolar range. The radical shift in potency was achieved by substituting the amino group of existing $GABA_B$ modulators with benzyl substituents (3,4-dichlorobenzyl or 3-carboxybenzyl) as in CGP55845 and CGP56433, respectively. Other representatives of this generation of antagonists include CGP54626 (Fig. 4), and CGP62349, CGP52432, CGP56999, CGP54626, CGP64213 (Fig. 5); all highly potent antagonists and all demonstrating learning and memory-improving effects (Lasarge et al. 2009; Getova and Dimitrova 2007). These antagonists may also have significant clinical potential in absence epilepsy (Bernasconi et al. 1992; Marescaux et al. 1992; Snead 3rd 1992) as mice overexpressing the $GABA_{B1a}$ isoform exhibit characteristics associated with atypical absence epilepsy (Stewart et al. 2009).

8 Inverse Agonists

Given that GPCRs are believed to exist in equilibrium between inactive and active conformational states in which there is a continuum of structural conformations ranging from having no activity to being maximally active, these receptors have the potential to be active in the absence of an activating ligand, a phenomenon termed "constitutive activity." Ligands that stabilize the fully "inactive" conformation, thereby eliminating any intrinsic/constitutive activity the receptor may have, are referred to as "inverse agonists" (Berg and Clarke 2018; Kenakin 2004). Many GPCR-targeted drugs were initially characterized as "neutral" or "silent" antagonists as their discovery predated inverse agonism as a pharmacological concept. It is now estimated that at least 15% of compounds classified as antagonists have some intrinsic activity and that these drugs confer their therapeutic efficacy by reducing constitutive receptor activity (Urwyler et al. 2005; Grunewald et al. 2002; Hirst et al. 2003; Mukherjee et al. 2006).

In the context of the GABA$_B$ receptor, constitutive receptor activity has been demonstrated to modulate neurotransmitter release and neuronal excitability in the absence of GABA. For example, in cerebellar Purkinje cells, GABA$_B$ receptor has been shown to interact with extracellular calcium ions to increase the sensitivity of the glutamate receptor 1 (mGluR1) to its endogenous ligand, glutamate, by forming a complex with the mGlu1R (Tabata et al. 2004). The use of a selective GABA$_B$ receptor inverse agonist could serve to eliminate enhanced glutamate mediated mGluR1 activity which has been identified as an avenue with therapeutic potential for the treatment of fragile X syndrome (Niswender and Conn 2010).

As noted, compounds CGP52432, CGP54626, CGP56999, CGP62349 (Fig. 5) are closely related, sharing the same core structure, and were originally identified as competitive antagonists at the GABA$_B$ receptor. As antagonists, these compounds have the ability to block GABA/GABA$_B$ receptor-mediated inhibition of forskolin-stimulated cAMP in GABA$_B$ receptor expressing recombinant systems. However, following receptor desensitization resulting from sustained exposure to GABA, the activity of this family of compounds switches from antagonism to inverse agonism as demonstrated by the CGP54626-promoted increase in cAMP production. The atypical SCH50911 antagonist that lacks large hydrophobic substituents behaved in a similar manner (Gjoni and Urwyler 2009). Likewise, the structurally distinct CLH304a previously reported as a negative allosteric modulator (NAM; Fig. 5) (Chen et al. 2014) has also since been reported to exhibit inverse agonist properties in the absence of an agonist (Sun et al. 2016).

9 Allosteric Modulators

While endogenous neurotransmitter GABA agonists (i.e., baclofen) and antagonists (i.e., phaclofen) bind to the orthosteric site (VFT domain) in the GABA$_{B1}$ subunit, it is now widely accepted that the GABA$_B$ receptor modulators identified so far act at allosteric sites (binding sites topographically distinct from the orthosteric ligand binding site) and bind the transmembrane region of the GABA$_{B2}$ subunit. Allosteric modulators (AMs) are basically classified as either positive allosteric modulators (PAMs) or negative allosteric modulators (NAMs). PAMs that possess intrinsic agonist activity are referred to as "ago-PAMs." A third class of allosteric ligand has been described that binds to the receptor but has no intrinsic activity and no apparent effect on endogenous ligand activity, hence it is referred to as a "silent allosteric modulator" (SAM) (Burford et al. 2013). Importantly, since orthosteric and allosteric ligands bind to topographically distinct sites of the receptor, both ligands can interact with the receptor simultaneously and thus, each ligand can affect the binding (binding cooperativity) and the intrinsic activity (activation cooperativity) of the other. Theoretical models describing these interactions have been discussed extensively elsewhere (Keov et al. 2011; May et al. 2004).

Allosteric sites are attractive therapeutic targets because molecules that bind to these sites can act in concert with an orthosteric ligand, and in doing so are believed to offer several advantages over the use of orthosteric ligands alone (Kenakin and Miller 2010). As allosteric modulators typically rely on the presence of the endogenous ligand, they have the ability to modify receptor activity in a spatial and temporal manner by acting in concert with the endogenous receptor ligand (Kenakin and Miller 2010). Therefore, allosteric modulators are believed to have the potential to induce fewer side effects, as they simply modulate endogenous ligand-mediated receptor activation. In addition, upon prolonged exposure, allosteric modulators are less likely to induce GPCR desensitization compared to an orthosteric agonist, and as such, are less likely to induce drug tolerance.

Allosteric modulators of the $GABA_B$ receptor have generated significant attention for their therapeutic potential in the treatment of alcohol and drug addiction, anxiety, depression, muscle spasticity, epilepsy, pain, and gastrointestinal disorders (Urwyler 2011). It is postulated that the use of a PAM (or ago-PAM) will achieve a more desirable pharmacological signaling profiling and physiological responses by enhancing GABA-mediated receptor signaling rather than artificially stimulating the receptor with an exogenous agonist such as baclofen. Furthermore, $GABA_B$ receptor allosteric modulators hold the promise of more favorable pharmacokinetics compared to baclofen including improved bioavalability and brain exposure as well as cytotoxicity. Hence, the potential advantages of $GABA_B$ receptor allosteric modulation have led to the development of numerous small molecule allosteric modulators, the majority of which are PAMs.

While many of the described $GABA_B$ receptor PAMs are structurally distinct, based on the core structure they can be sorted into several groups (Fig. 6; each row representing a distinct structural class). The discovery of $GABA_B$ receptor PAMs was pioneered and first reported by Novartis scientists, Urwyler and colleagues, in 2001. These researchers demonstrated that small molecule CGP7930 (2,6-di-tert-butyl-4-(3-hydroxy-2,2-dimethyl-propyl)-phenol; discovered in a high throughput screening campaign) (Urwyler et al. 2001) potentiated GABA-stimulated [^{35}S]-GTPγS-accumulation in membrane preparations derived from CHO cells stably expressing the $GABA_B$ receptor. Using various combinations of wildtype and mutant $GABA_B$ subunits, Binet and colleagues investigated the mode of action of CGP7930, determining that the heptahelical domain (HD) of $GABA_{B2}$ was an absolute requirement for CGP7930 PAM action and that CGP7930 could also activate a truncated $GABA_{B2}$ subunit corresponding to the HD only (Binet et al. 2004).

In 2003, Novartis reported on another group of structurally distinct $GABA_B$ receptor PAMs, centered around GS39783 (*N,N′*-dicyclopentyl1–2-methylsulfanyl-5-nitro-pyrimidine-4,6-diamine). Like CGP7930, GS39783 was found to potentiate both affinity and maximal effects of GABA in biochemical and electrophysiological assay systems (Urwyler et al. 2003). Dupuis et al. studied point mutations in the TM region of $GABA_{B2}$ to identify the residues within the HD that

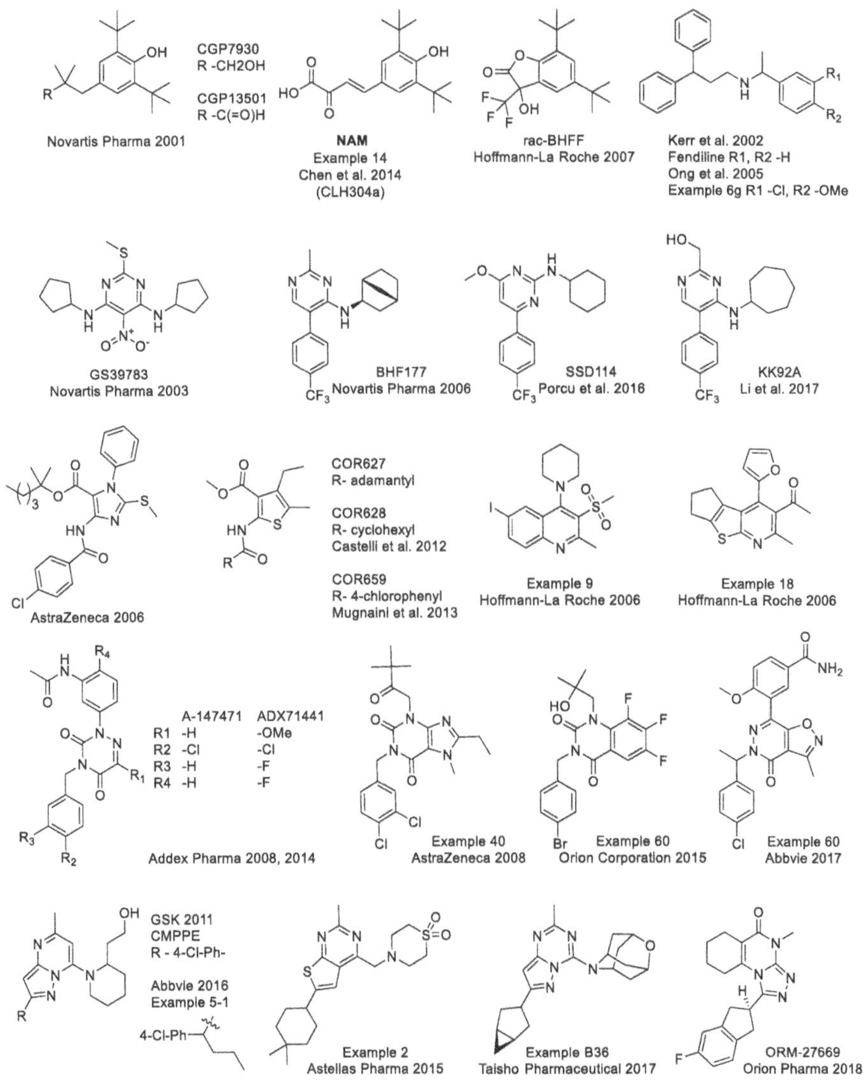

***NAM** CLH304a; an example of how subtle structural changes can change ligand activity

Fig. 6 Exemplar chemical structures of GABA$_B$ receptor allosteric modulators

interact with GS39783 and found that mutations G706T and A708P in TM6 were necessary and sufficient for GS39783 mediated agonist activation (Dupuis et al. 2006). Hence, both CGP7930 and GS39783 were found to bind to sites distinct from known agonist and antagonist receptor binding sites, and to require the presence of the GABA$_{B2}$ receptor subunit.

These findings prompted the pursuit of other molecules with similar PAM activities and consequently, numerous $GABA_B$ receptor allosteric modulators have been reported in the scientific and patent literature (accessible in SciFinder and Espacenet) over the past two decades. Roche scientists further developed the Novartis compounds by generating systematic modifications of CGP7930 structure and arrived at the bicyclic structure of rac-BHFF (Malherbe et al. 2008). Interestingly, it was found that both CGP7930 and rac-BHFF have intrinsic agonist activity, and distinct and differentiating ligand-induced signaling profiles compared to baclofen (Koek et al. 2013). Optimization of the genotoxic lead structure of the pyrimidine derivative of GS39783 led to the development of non-toxic $GABA_B$ receptor PAMs, such as BHF177 (N-[(1R,2R,4S)-bicyclo[2.2.1]heptan-2-yl]-2-methyl-5-[4-(trifluoromethyl)phenyl]-4-pyrimidinamine) reported by Novartis in 2006 (Floersheim et al. 2006). A decade later additional analogs from this series were reported by Porcu et al. (SSD114) (Porcu et al. 2016) and by our research group and collaborators Li et al. (KK-92A and approximately 100 additional analogs) (Li et al. 2017).

Substituted 5-membered heterocycles represent a substantial group of $GABA_B$ receptor PAMs, with the first examples of structurally novel modulators reported by AstraZeneca in patents aiming at the development of drugs for the treatment of gastrointestinal diseases (Bauer et al. 2005). Specific examples presented in the patents initially focused on imidazole derivatives that expanded in scope by scaffold hopping to cover other five-membered core heterocycles such as pyrazoles, oxazoles, and thiazoles. In 2011, a group led by Corelli identified COR627, COR628 (Castelli et al. 2012), and COR659 (Mugnaini et al. 2013) as $GABA_B$ receptor PAMs that displayed significant activity in vitro as $GABA_B$ receptor PAMs by potentiating $[^{35}S]$-GTPγS-binding induced by GABA while failing to exhibit intrinsic agonist activity. While the thiophene-based core of the active molecule differs from those reported by AstraZeneca, the substitution pattern resembles other representative molecules in this group.

Extensive work of Hoffman-La Roche resulted in the identification of additional classes of $GABA_B$ receptor PAMs disclosed in a series of patents published in 2006. The reported active molecules are based on a quinoline (Malherbe et al. 2006) or thieno[2,3-b]pyridine (Malherbe et al. 2007) as core heterocycles. A closely related set of $GABA_B$ receptor PAMs was reported in 2009 in an AstraZeneca patent (Cheng and Karle 2008). A separate group of $GABA_B$ receptor modulators represent a series of substituted triazinediones developed by Addex Pharma (Riguet et al. 2007). The Addex lead compound, ADX71441, is an orally available small molecule that demonstrated excellent preclinical efficacy and tolerability in several rodent models of pain, addiction, and overactive bladder (OAB) and has also proven efficacy in a genetic model of Charcot-Marie-Tooth Type 1A disease (CMT1A) (Cao and Zhang 2020).

In patents from 2008 and 2009 AstraZeneca scientists disclosed a new group of $GABA_B$ receptor PAMs based on bicyclic pyrimidinedione core, namely xanthines (Cheng et al. 2008a) and pteridine-2,4(1H,3H)-diones (Cheng et al. 2008b). Related structures were disclosed in 2015 by Orion Corporation (Prusis et al. 2015) and

Abbvie in 2017. In 2011, GlaxoSmithKline reported CMPPE, a novel moiety that positively modulated GABA-evoked in vitro [^{35}S]-GTPγS-binding signal with an EC_{50} value of 2.57 μM. The compound showed mild efficacy in a food consumption test in rats, modest in vivo potentiation of baclofen-induced muscle relaxation in mice, and poor metabolic stability in liver microsomal systems (Perdona et al. 2011). Other companies followed the CMPPE track with a range of modulators containing a modified core structure as the substitution pattern. Astellas Pharma reported a series of thieno[2,3-d]pyrimidines in 2015 (Shiraishi et al. 2014), whereas Abbvie (Faghih et al. 2016) and Richter Gedeon maintained pyrazolo[1,5-a]pyrimidinyl core in their series (WO 2018167630). In a single patent Taisho Pharmaceutical (Borza et al. 2018) covered analogs with the core heterocycle replaced by pyrazolo[1,5-a][1,3,5] triazine in addition to substituted pyrazolo[1,5-a]pyrimidines. ORM-27669, reported by Orion Pharma in 2017, with its tricyclic core structure containing [1,2,4]triazolo [4,3-a]pyrimidin-7(8H)-one represents a more original scaffold (de Miguel et al. 2019). Pretreatment with ORM-27669 reversed ethanol-induced neuroplasticity and attenuated ethanol drinking but had no effects on cocaine-induced neuroplasticity or self-administration.

Fendiline (Fig. 6) and its related arylalkylamines represent another unique structural class reported to be potential $GABA_B$ receptor PAMs. First reported as a non-selective calcium channel blocker and as a positive allosteric modulator of extracellular Ca^{2+} sensing receptors (CaSRs) (Nemeth et al. 1998), Fendiline is an FDA-approved (albeit obsolete) drug used in the treatment of coronary heart disease. Although not $GABA_B$ receptor specific, this compound is noteworthy as Ong and Kerr evaluated activity of Fendiline and its analogues as PAMs of $GABA_B$ receptors (Kerr et al. 2002, 2006) and demonstrated that the most potent analogue, (+)-N-1-(3-chloro-4-methoxyphenyl)ethyl-3,3-diphenylpropylamine) exhibited an EC_{50} of 30 nM in modulating baclofen-mediated function using grease gap recording in rat neocortical slices (Ong et al. 2005). However, direct action of Fendiline on $GABA_B$ receptor activity has been disputed (Urwyler et al. 2004) and further investigations are needed to determine the mechanism by which arylalkylamines enhance $GABA_B$ receptor-mediated responses.

$GABA_B$ receptor negative allosteric modulators (NAMs) have also been proposed as potential lead compounds for development into therapeutics for disorders such as CNS hyperexcitability-related disorders including epilepsy, anxiety, nerve damage, and low cognitive ability. Interestingly, modifications of $GABA_B$ receptor PAM CGP7930 (Fig. 6) led to discovery of the first $GABA_B$ receptor NAM, CLH304a, reported by Chen and colleagues in 2014 (Chen et al. 2014; Sun et al. 2016). CLH304a decreased agonist GABA-induced maximal effect of IP_3 production in HEK293 cells overexpressing $GABA_B$ receptor and Gαqi9 proteins without changing the EC_{50}. Moreover, it inhibited baclofen-induced $ERK_{1/2}$ phosphorylation and also blocked CGP7930-induced $ERK_{1/2}$ phosphorylation in HEK293 cells overexpressing $GABA_B$ receptor. This indicated that CLH304a (and some analogues) may be allosteric modulators, as orthosteric antagonists like CPG54626 are unable to attenuate PAM mediated signaling. Indeed, it was demonstrated that the compounds of interest bound to an allosteric site, negatively regulating orthosteric agonist mediated signaling (Chen et al. 2014).

10 Probe Dependency

An important aspect of allosteric modulation to be taken into consideration is that the extent and direction (positive or negative) of the interaction between the orthosteric and allosteric ligands depends on which orthosteric ligand is present; a phenomenon known as "probe-dependency," this is important as many individual GPCRs respond to multiple endogenous ligands (May et al. 2004; Kenakin 2005). For the $GABA_B$ receptor, only having one known endogenous ligand, probe-dependency might be considered irrelevant. However, the potential combination of an allosteric modulator with a synthetic therapeutic such as baclofen must also consider the possibility of probe-dependent effects on receptor signaling and function.

Indeed, it has been demonstrated that baclofen shows improved efficacy and an increased therapeutic window when administered in combination with $GABA_B$ receptor PAMs (Maccioni et al. 2012). In preclinical studies, treatment with $GABA_B$ receptor PAMs GS39783 and rac-BHFF potentiated the activity of low doses of baclofen in relation to alcohol seeking behaviors (Maccioni et al. 2015). Hence, the ability of PAMs to reduce the effective dose of baclofen not only has the potential to improve efficacy in disease relevant measures, but also to expand the therapeutic window of this drug by reducing the accompanying adverse side effects. Thus, leveraging the probe-dependent effects of treatment with multiple receptor ligands has the potential to "fine-tune" receptor signaling and facilitate the development of improved strategies to target the $GABA_B$ receptor.

11 Biased Agonism/Functional Selectivity

It is well established that any given ligand for a GPCR does not simply possess a single defined efficacy; rather, a ligand possesses multiple efficacies, depending on the specific downstream signal transduction pathway being investigated. This diversity is believed to be the result of conformational changes induced in the GPCR that are ligand-specific and hence receptors can adopt various conformations that preferentially activate/modulate one signaling pathway to the exclusion of others; a phenomenon referred to as "functional selectivity" or "ligand bias" (Kenakin 2017; Smith et al. 2018; Spangler and Bruchas 2017). Conceptually, as with allosteric modulation, functional selectivity is an appealing mechanism of therapeutic intervention, as modulating only a select subset receptor signaling pathway may allow for the development of drugs that demonstrate therapeutic efficacy without recruiting pathways that lead to downstream adverse side effects (Kenakin and Miller 2010).

Functional selectivity can be achieved by modulating the receptor with a single ligand or with multiple ligands. Our own studies identified a PAM, namely KK-92A (4-(cycloheptylamino)-5-(4-(trifluoromethyl) phenyl)pyrimidin-2-yl) methanol) that exhibits pathway-selective differential modulation of GABA$_B$ receptor signaling when compared to the structurally related allosteric modulator BHF177 (Sturchler et al. 2017). Using recombinant cell-based systems overexpressing the GABA$_B$ receptor, KK-92A exhibited similar activity to BHF177 in potentiating GABA-induced GABA$_B$ receptor-mediated inhibition of forskolin-stimulated cAMP production and GABA-induced increase in intracellular Ca^{2+} levels. However, in contrast to BHF177, in the absence of GABA, KK-92A exhibited intrinsic activity with regard to ERK$_{1/2}$ phosphorylation achieving ~70% maximum efficacy relative to GABA maximum efficacy (Li et al. 2017), demonstrating ago-PAM activity and pathway-selective effects.

12 GABA$_B$ Receptor-Targeted Pharmaceuticals

The presence of functional GABA$_B$ receptors in mammalian brain and the gastrointestinal tract has been known for more than 30 years. Given the widespread distribution of GABA and the GABA$_B$ receptor in the CNS and periphery, it is not surprising that activation of the GABA$_B$ receptor provokes a host of physiological responses, and as a consequence, dysregulation of GABA$_B$ receptor activity was proposed to be associated with various CNS diseases such as mood disorders (Kalinichev et al. 2017; Li et al. 2015; Felice et al. 2012; Jacobson et al. 2018; Cryan and Slattery 2010), epilepsy (Billinton et al. 2001a; Teichgräber et al. 2009), addiction (Agabio and Colombo 2014, 2015; Agabio et al. 2018; Ranson et al. 2020; Maccioni et al. 2015), Parkinson's disease (Nambu 2012; Tyagi et al. 2015), Alzheimer's disease (Rice et al. 2019; Sun et al. 2020), Huntington's disease (Kim and Seo 2014) as well as peripheral diseases such as gastroesophageal reflux disease (GERD) (Clarke et al. 2018; Lehmann et al. 2010). More recently, GABA has emerged as a tumor signaling molecule in the periphery that controls tumor cell proliferation (Young and Bordey 2009; Zhang et al. 2014; Jiang et al. 2012), and stimulation of GABA$_B$ receptor signaling has been proposed as a novel target for the treatment and prevention of pancreatic cancer (Schuller et al. 2008; Schuller 2018; Al-Wadei et al. 2012). Numerous studies have shown potential clinical benefit of targeting the GABA$_B$ receptor in the treatment of various CNS and peripheral disorders, yet there is still only one therapeutic agent used clinically that selectively activates the GABA$_B$ receptor, namely baclofen (Lioresal®).

As discussed, baclofen was originally synthesized in 1962 by chemists at Ciba, Switzerland in an attempt to generate a more lipophilic, brain penetrant GABA mimetic (Keberle et al. 1964). It was assessed in the treatment of epilepsy but failed to show sufficient efficacy in the clinic. However, as a consequence of an incidental finding in that it had positive effects on muscle spasticity (Hudgson and Weightman 1971), baclofen (Lioresal®) has been in clinical use since 1972, gaining FDA approval in 1977; long before its molecular target, the $GABA_B$ receptor was discovered and its mechanism of action identified. As previously mentioned, it has also shown therapeutic utility in a wide range of other off-label indications including addiction and was recently approved in Europe and Australia for the treatment of alcoholism (Agabio et al. 2018) but side effects such as sedation, nausea, muscle weakness, and rapid onset of tolerance limit its use (Kent et al. 2020).

With baclofen, an improvement over GABA regarding blood-brain barrier permeability was achieved, however, baclofen still has low brain penetration attributed to rapid efflux via the organic anion transporter (OAT3) (Ohtsuki et al. 2002). In parallel to Ciba's efforts in the 1960s to synthesize a GABA mimetic, a Russian team (Perekalin et al) synthesized a phenyl derivative of GABA, namely phenibut (-β-phenyl-γ-aminobutyric acid) that exhibits improved brain penetration over baclofen. Phenibut (Cirocard®) has been in clinical use in Russia and some Eastern European countries (not FDA-approved in USA) as a tranquilizer and cognition enhancer (nootropic) since the 1960s and is still used for these indications as well as for the treatment of mood and sleep disorders, PTSD, and a variety of neuropsychiatric diseases (Lapin 2001). However, phenibut suffers from many of the same liabilities as baclofen; sedation, muscle weakness, nausea, tolerance, and more recently has gained attention for its abuse potential (Jouney 2019).

Although baclofen has been in clinical use since 1972, it is far from an "ideal" drug; in addition to the unwanted side effects mentioned above, it also suffers from poor pharmacokinetic properties, including low brain penetration, limited absorption, short duration of action, rapid clearance from the blood, and narrow therapeutic window (Kent et al. 2020). Despite the lack of good "drug-like" qualities, the clinical success of baclofen has prompted numerous campaigns towards the identification and development of new and improved compounds that modulate the $GABA_B$ receptor and significant advances have been made. In 2009, XenoPort (now Arbor Pharmaceuticals) introduced Arbaclofen Placaril (XP19986), a transported prodrug of (R)-(−)-baclofen designed to possess a more favorable pharmacokinetic profile. Arbaclofen is absorbed throughout the intestinal tract and is rapidly converted to (R)-(−)-baclofen in tissues. It has been evaluated in Phase III clinical trials for GERD and multiple sclerosis, but these trials were discontinued in 2011 and 2013, respectively, due to lack of efficacy. It also reached Phase III trials in fragile X syndrome (FXS) but did not meet the primary outcome of improved social avoidance in FXS (Berry-Kravis et al. 2017). However, an extended release formula of Arbaclofen (Arbaclofen-ER; Ontinua®) developed by Osmotica is under FDA review as of July 2020 for the treatment of spasticity in multiple sclerosis. Also, two independent clinical trials evaluating benefit of Arbaclofen in children and adults with autism spectrum disorder (ASD) were initiated in 2019 (NCT03682978 and NCT03887676, respectively).

While XenoPort reported on Arbaclofen Placaril, AstraZeneca reported AZD3355 (Lesogaberan®; Fig. 2), a high affinity analogue of baclofen that was developed and evaluated in clinical trials for the treatment of GERD (Bredenoord 2009). AZD3355 is restricted peripherally and has a half-life of ~11 h in blood (Niazi et al. 2011). Unfortunately, Phase IIb clinical trials were terminated owing to lack of efficacy in GERD patients. As AZD3355 is not brain penetrant and devoid of unwanted CNS effects, with no other adverse effects reported, indicating that it is safe in humans, it has been proposed that Lesogaberan® could be repurposed for the treatment of type 1 diabetes; targeting the GABA$_B$ receptor in β cells to promote β-cell survival (Tian et al. 2017).

GHB (γ-hydroxybutyric acid) is approved in some countries and used clinically for the treatment of narcolepsy-related catalepsy (Xyrem®) (Szabadi 2015) and rarely alcoholism (Alcover®) (Keating 2014). GHB also has the potential for abuse and is used illicitly as a recreational drug and intoxicant (Busardò and Jones 2015). Although GHB itself is not FDA-approved for medical use, the first generic version of Xyrem®, sodium oxybate (the sodium salt of GHB), recently (2017) received FDA approval to treat symptoms of narcolepsy including excessive day-time sleepiness and narcolepsy with cataplexy.

The first (and only to the best of our knowledge) clinical investigation of GABA$_B$ receptor antagonists was an open trial with SGS742 (CGP36742; Fig. 4) (Bullock 2005) (Froestl et al. 2004). Even though its potency is low ($IC_{50} \approx 40$ μM (Froestl et al. 1995b)), many preclinical studies showed benefit with SGS742 for spatial memory improvement (Helm et al. 2005), the treatment of depression (Nowak et al. 2006), and arrest of cortical seizures (Mares and Kubova 2008). The initial Phase II clinical trial, conducted in mild cognitive impairment patients, showed that SGS742 significantly improved attention, in particular choice reaction time and visual information processing as well as working memory (Froestl et al. 2004). However, a second Phase II trial was undertaken in mild to moderate Alzheimer's disease patients and no statistically significant improvement was detected prompting the termination of the development program. The clinical implications of modulating the GABA$_B$ receptor are outlined in Table 1.

13 Concluding Remarks

The GABA$_B$ receptor and its physiological roles are extremely complex, consequently, dysregulation of this receptor is involved in a broad range of diseases, and as such the GABA$_B$ receptor is considered a highly attractive therapeutic target for the development of new anti-epileptic, antidepressant, analgesic, and anxiolytic drugs, as well as for the treatment of cognitive disorders, drug addiction, and depression. However, at present only one compound that targets the orthosteric site of GABA$_B$ receptor is in clinical use, namely baclofen (Lioresal™); used to treat muscle spasticity in multiple sclerosis, and more recently used off-label for alcohol addiction. Unfortunately, side effects such as sedation, muscle weakness, nausea, and the lack of efficacy observed in other indications, i.e., fragile X syndrome, limit its therapeutic use. In addition to unwanted side effects baclofen also suffers from low brain penetration, limited absorption, rapid tolerance, short duration of action, and narrow therapeutic window. As described earlier, numerous small molecule agonists, antagonists, and allosteric modulators of the GABA$_B$

Table 1 Current therapeutic use and potential clinical utility of $GABA_B$ receptor modulators

Pharmacology	[a]Therapeutic use/[b]clinical potential	Approved drug/[c]clinical trial	References
Agonists/positive allosteric modulators	[a]Muscle rigidity and spasticity	Baclofen (Lioresal®) Arbaclofen-ER (Ontinua®)	(Francisco et al. 2001; Basmajian 1975; Korsgaard 1976; Coffey et al. 1993)
	[a]GERD	Baclofen (Lioresal®)	(Clarke et al. 2018; Lehmann et al. 2010; Ong and Kerr 1984; Lehmann 2009; Symonds et al. 2003)
	[a]Charcot-Marie tooth type 1A		(Cao and Zhang 2020; Dyer 2013)
	[a]PTSD/tranquilizer/nootropic	Phenibut (Cirocard®) (Eastern Europe only)	(Lapin 2001; Drake et al. 2003)
	[a]Cough suppression		(Chung 2015; Martvon et al. 2020)
	[a]Alcoholism and addiction	Sodium oxybate/GHB/(Alcover®), Baclofen (Lioresal®)	(Agabio and Colombo 2014, 2015; Agabio et al. 2018; Maccioni and Colombo 2019; Ranson et al. 2020)
	[b]Anxiety		(Kalinichev et al. 2017; Li et al. 2015)
	[b]Epilepsy		(Billinton et al. 2001a; Teichgräber et al. 2009)
	[b]Cataplexy	Sodium oxybate/GHB (Xyrem®)	(Black et al. 2014; Szabadi 2015)
	[b]Binge eating disorder		(Broft et al. 2007; Tsunekawa et al. 2019)
	[b]Parkinson's disease		(Nambu 2012; Tyagi et al. 2015)
	[b]Schizophrenia		(Glausier and Lewis 2017; Nair et al. 2020)
	[b]Huntington's disease		(Kim and Seo 2014; Kleppner and Tobin 2001)
	[b]Spatial learning and memory		(Modaberi et al. 2019; Sahraei et al. 2019)
	[b]Autism spectrum disorder (ASD)	Arbaclofen [c]NCT03682978, [c]NCT03887676	(Veenstra-VanderWeele et al. 2017; Frye 2014)
	[b]Fragile X syndrome (FXS)		(Berry-Kravis et al. 2017; Zhang et al. 2015)
	[b]Alzheimer's disease		(Rice et al. 2019; Sun et al. 2020)
	[b]Analgesic (fibromyalgia)	[c]NCT03092726	(Neto et al. 2006; Enna and McCarson 2006; Murai et al. 2019)
	[b]Pancreatic cancer		(Young and Bordey 2009; Zhang et al. 2014; Jiang et al. 2012;

(continued)

Table 1 (continued)

Pharmacology	[a]Therapeutic use/[b]clinical potential	Approved drug/[c]clinical trial	References
			Schuller et al. 2008; Schuller 2018; Al-Wadei et al. 2012)
	[b]Type 1 diabetes		(Tian et al. 2017)
Antagonists/ negative allosteric modulators	[b]Depression/mood disorders		(Cryan and Kaupmann 2005; Felice et al. 2012; Jacobson et al. 2018)
	[b]Type 2 diabetes		(Bonaventura et al. 2008, 2012; Braun et al. 2004)
	[b]Absence epilepsy/ seizures		(Bernasconi et al. 1992; Ostojić et al. 2013)
	[b]Mild cognitive impairment and memory		(Lasarge et al. 2009; Mondadori et al. 1993)
	[b]Succinic semi-aldehyde dehydrogenase (SSADH) deficiency	[c]NCT02019667	(Cortez et al. 2004; Didiášová et al. 2020)

[a]Therapeutic use
[b]Clinical potential
[c]Clinical trail

receptor have been described in the scientific and patent literature that have been developed for their therapeutic potential; positive allosteric modulators, for example, have been proposed to mitigate the unwanted side effects and reduce tolerance but have yet to be approved for clinical use. Hence, identification of novel drugs targeting the GABA$_B$ receptor that display improved efficacy and pharmacokinetic properties and with a safer side effect profile is the subject of intense research and many industrial scale drug discovery efforts.

As mentioned, the multifaceted GABA$_B$ receptor is extremely complex. However, the same complexity that has historically hindered development of GABA$_B$ receptor-targeted therapeutics now provides the potential for discovery of GABA$_B$ receptor disease-specific therapeutics. For example, GABA$_B$ receptor subtype-selective ligands are highly desirable not only to dissect the physiological role of the predominant receptor subtypes, GABA$_{B1(a)/2}$ and GABA$_{B1(b)/2}$, but also to facilitate the development of more finely-tuned mode-of-action drugs to treat various diseases. From a drug discovery perspective, it may be possible to selectively modulate GABA$_{B(1a)}$ containing heteroreceptors by targeting their sushi domains, case in point; amyloid precursor protein (APP) binds to the N-terminal sushi domain of GABA$_{B(1a)}$ and acts as an axonal trafficking factor for GABA$_B$ receptors, it has been proposed that prevention of APP binding to this domain may interfere with GABA$_B$ receptor-mediated inhibition of glutamate release and thereby enhance cognitive function in patients with Alzheimer's disease and intellectual disabilities. Likewise, the discovery of functionally selective ligands for the different GABA$_B$ receptor effectors would provide powerful tools to identify a unique signaling profile that results in the desired in vivo effects without recruiting the adverse side effects.

Alternatively, recent biophysical and structural studies have greatly improved our understanding of the structural basis of $GABA_B$ receptor activation and modulation, and proteomic studies have identified receptor-associated proteins that work in concert with the receptor to orchestrate a variety of molecularly and functionally distinct multiprotein “signalosome” complexes, while providing spatiotemporal control of receptor activity. These findings also present new opportunities for drug discovery, modulating specific protein:protein interactions mediated through sushi domains of $GABA_{B1(a)}$ (as outlined above), C-terminal domain of $GABA_B1$ and/or $GABA_{B2}$; or KCTD subunits, all present potential target sites for designing drugs that selectively interfere with receptor function for disease-specific therapeutic intervention.

Thus, the successful collaboration between medicinal chemistry and pharmacology together with significant advances in our understanding of $GABA_B$ receptor structure and activation mechanisms has drug hunters well-poised for the discovery and development of chemically and mechanistically novel therapeutics targeting the multi-tasking $GABA_B$ receptor for the treatment of a wide variety of disease states.

Acknowledgements We were introduced to the GABA/$GABA_B$ receptor field by Dr. Athina Markou (UCSD). We are honored to have worked with Dr. Markou and forever grateful for the opportunities that she gave us to work with other experts in this field. It is on the shoulders of such giants that we write this chapter and hope that we have done their work justice. We apologize to those whose work we may have omitted.

References

Agabio R, Colombo G (2014) GABAB receptor ligands for the treatment of alcohol use disorder: preclinical and clinical evidence. Front Neurosci 8:140

Agabio R, Colombo G (2015) GABAB receptor as therapeutic target for drug addiction: from baclofen to positive allosteric modulators. Psychiatr Pol 49(2):215–223

Agabio R, Sinclair JM, Addolorato G, Aubin HJ, Beraha EM, Caputo F, Chick JD, de La Selle P, Franchitto N, Garbutt JC, Haber PS, Heydtmann M, Jaury P, Lingford-Hughes AR, Morley KC, Müller CA, Owens L, Pastor A, Paterson LM, Pélissier F, Rolland B, Stafford A, Thompson A, van den Brink W, de Beaurepaire R, Leggio L (2018) Baclofen for the treatment of alcohol use disorder: the Cagliari statement. Lancet Psychiatry 5(12):957–960

Al-Wadei HA, Al-Wadei MH, Ullah MF, Schuller HM (2012) Celecoxib and GABA cooperatively prevent the progression of pancreatic cancer in vitro and in xenograft models of stress-free and stress-exposed mice. PLoS One 7(8):e43376

Awapara J (1950) Occurrence of free gamma-aminobutyric acid in brain and its formation from L-glutamic acid. Tex Rep Biol Med 8(4):443–447

Balasubramanian S, Fam SR, Hall RA (2007) GABAB receptor association with the PDZ scaffold Mupp1 alters receptor stability and function. J Biol Chem 282(6):4162–4171

Bartoi T, Rigbolt KT, Du D, Köhr G, Blagoev B, Kornau HC (2010) GABAB receptor constituents revealed by tandem affinity purification from transgenic mice. J Biol Chem 285 (27):20625–20633

Basmajian JV (1975) Lioresal (baclofen) treatment of spasticity in multiple sclerosis. Am J Phys Med 54(4):175–177

Bauer U, Brailsford W, Chhajlani V, Egner B, Fjellström O, Gustafsson L, Mattsson J, Nilsson K, Olsson T (2005) Imidazole variants as modulators of gaba receptor for the treatment of GI disorders. U.S. Patent 20080269216

Benarroch EE (2012) GABAB receptors: structure, functions, and clinical implications. Neurology 78(8):578–584

Benke D (2013) GABAB receptor trafficking and interacting proteins: targets for the development of highly specific therapeutic strategies to treat neurological disorders? Biochem Pharmacol 86 (11):1525–1530

Berg KA, Clarke WP (2018) Making sense of pharmacology: inverse agonism and functional selectivity. Int J Neuropsychopharmacol 21(10):962–977

Bernasconi R, Lauber J, Marescaux C, Vergnes M, Martin P, Rubio V, Leonhardt T, Reymann N, Bittiger H (1992) Experimental absence seizures: potential role of gamma-hydroxybutyric acid and GABAB receptors. J Neural Transm Suppl 35:155–177

Berry-Kravis E, Hagerman R, Visootsak J, Budimirovic D, Kaufmann WE, Cherubini M, Zarevics P, Walton-Bowen K, Wang P, Bear MF, Carpenter RL (2017) Arbaclofen in fragile X syndrome: results of phase 3 trials. J Neurodev Disord 9:3

Bertrand S, Ng GY, Purisai MG, Wolfe SE, Severidt MW, Nouel D, Robitaille R, Low MJ, O'Neill GP, Metters K, Lacaille JC, Chronwall BM, Morris SJ (2001) The anticonvulsant, antihyperalgesic agent gabapentin is an agonist at brain gamma-aminobutyric acid type B receptors negatively coupled to voltage-dependent calcium channels. J Pharmacol Exp Ther 298(1):15–24

Bettler B, Tiao JY (2006) Molecular diversity, trafficking and subcellular localization of GABAB receptors. Pharmacol Ther 110(3):533–543

Bettler B, Kaupmann K, Mosbacher J, Gassmann M (2004) Molecular structure and physiological functions of GABA(B) receptors. Physiol Rev 84(3):835–867

Bianchi M, Panerai AE (1993) Reversal of scopolamine-induced amnesia by the GABAB receptor antagonist CGP 35348 in the mouse. Brain Res Cogn Brain Res 1(2):135–136

Billinton A, Baird VH, Thom M, Duncan JS, Upton N, Bowery NG (2001a) GABA(B) receptor autoradiography in hippocampal sclerosis associated with human temporal lobe epilepsy. Br J Pharmacol 132(2):475–480

Billinton A, Ige AO, Bolam JP, White JH, Marshall FH, Emson PC (2001b) Advances in the molecular understanding of GABA(B) receptors. Trends Neurosci 24(5):277–282

Binet V, Brajon C, Le Corre L, Acher F, Pin JP, Prezeau L (2004) The heptahelical domain of GABA(B2) is activated directly by CGP7930, a positive allosteric modulator of the GABA (B) receptor. J Biol Chem 279(28):29085–29091

Black SW, Morairty SR, Chen TM, Leung AK, Wisor JP, Yamanaka A, Kilduff TS (2014) GABAB agonism promotes sleep and reduces cataplexy in murine narcolepsy. J Neurosci 34 (19):6485–6494

Blein S, Ginham R, Uhrin D, Smith BO, Soares DC, Veltel S, McIlhinney RA, White JH, Barlow PN (2004) Structural analysis of the complement control protein (CCP) modules of GABA (B) receptor 1a: only one of the two CCP modules is compactly folded. J Biol Chem 279 (46):48292–48306

Bolser DC, Blythin DJ, Chapman RW, Egan RW, Hey JA, Rizzo C, Kuo SC, Kreutner W (1995) The pharmacology of SCH 50911: a novel, orally-active GABA-beta receptor antagonist. J Pharmacol Exp Ther 274(3):1393–1398

Bon C, Galvan M (1996) Electrophysiological actions of GABAB agonists and antagonists in rat dorso-lateral septal neurones in vitro. Br J Pharmacol 118(4):961–967

Bonaventura MM, Catalano PN, Chamson-Reig A, Arany E, Hill D, Bettler B, Saravia F, Libertun C, Lux-Lantos VA (2008) GABAB receptors and glucose homeostasis: evaluation in GABAB receptor knockout mice. Am J Physiol Endocrinol Metab 294(1):E157–E167

Bonaventura MM, Crivello M, Ferreira ML, Repetto M, Cymeryng C, Libertun C, Lux-Lantos VA (2012) Effects of GABAB receptor agonists and antagonists on glycemia regulation in mice. Eur J Pharmacol 677(1–3):188–196

Borza I, Román V, Éles J, Hadady Z, Huszár J (2018) Pharmacologically active aryl-substituted pyrazolo[1,5-a]pyrimidine derivatives. W.O. Patent 2018167629

Bowery NG (1993) GABAB receptor pharmacology. Annu Rev Pharmacol Toxicol 33:109–147

Bowery NG, Doble A, Hill DR, Hudson AL, Shaw JS, Turnbull MJ (1979) Baclofen: a selective agonist for a novel type of GABA receptor proceedings. Br J Pharmacol 67(3):444p–445p

Bowery NG, Hill DR, Hudson AL, Doble A, Middlemiss DN, Shaw J, Turnbull M (1980) (−) Baclofen decreases neurotransmitter release in the mammalian CNS by an action at a novel GABA receptor. Nature 283(5742):92–94

Bowery NG, Doble A, Hill DR, Hudson AL, Shaw JS, Turnbull MJ, Warrington R (1981) Bicuculline-insensitive GABA receptors on peripheral autonomic nerve terminals. Eur J Pharmacol 71(1):53–70

Bowery NG, Bettler B, Froestl W, Gallagher JP, Marshall F, Raiteri M, Bonner TI, Enna SJ (2002) International Union of Pharmacology. XXXIII. Mammalian gamma-aminobutyric acid (B) receptors: structure and function. Pharmacol Rev 54(2):247–264

Bowery N, Enna SJ, Olsen RW (2004) Six decades of GABA. Biochem Pharmacol 68 (8):1477–1478

Braun M, Wendt A, Buschard K, Salehi A, Sewing S, Gromada J, Rorsman P (2004) GABAB receptor activation inhibits exocytosis in rat pancreatic beta-cells by G-protein-dependent activation of calcineurin. J Physiol 559(Pt 2):397–409

Bredenoord AJ (2009) Lesogaberan, a GABA(B) agonist for the potential treatment of gastroesophageal reflux disease. IDrugs 12(9):576–584

Broft AI, Spanos A, Corwin RL, Mayer L, Steinglass J, Devlin MJ, Attia E, Walsh BT (2007) Baclofen for binge eating: an open-label trial. Int J Eat Disord 40(8):687–691

Bullock R (2005) SGS-742 Novartis. Curr Opin Investig Drugs 6(1):108–113

Burford NT, Clark MJ, Wehrman TS, Gerritz SW, Banks M, O'Connell J, Traynor JR, Alt A (2013) Discovery of positive allosteric modulators and silent allosteric modulators of the μ-opioid receptor. Proc Natl Acad Sci U S A 110(26):10830–10835

Busardò FP, Jones AW (2015) GHB pharmacology and toxicology: acute intoxication, concentrations in blood and urine in forensic cases and treatment of the withdrawal syndrome. Curr Neuropharmacol 13(1):47–70

Calebiro D, Rieken F, Wagner J, Sungkaworn T, Zabel U, Borzi A, Cocucci E, Zürn A, Lohse MJ (2013) Single-molecule analysis of fluorescently labeled G-protein-coupled receptors reveals complexes with distinct dynamics and organization. Proc Natl Acad Sci U S A 110(2):743–748

Cao W, Zhang R (2020) Research advance of underlying pathogenesis and target therapies in Charcot-Marie-tooth disease type 1A. Zhonghua Yi Xue Yi Chuan Xue Za Zhi 37(5):578–583

Castelli MP, Casu A, Casti P, Lobina C, Carai MA, Colombo G, Solinas M, Giunta D, Mugnaini C, Pasquini S, Tafi A, Brogi S, Gessa GL, Corelli F (2012) Characterization of COR627 and COR628, two novel positive allosteric modulators of the GABA(B) receptor. J Pharmacol Exp Ther 340(3):529–538

Chapman RW, Danko G, del Prado M, Egan RW, Kreutner W, Rizzo CA, Hey JA (1993) Further evidence for prejunctional GABA-B inhibition of cholinergic and peptidergic bronchoconstriction in guinea pigs: studies with new agonists and antagonists. Pharmacology 46(6):315–323

CHEMBL (n.d.) Database Release 27. https://www.ebi.ac.uk/chembl/

Chen LH, Sun B, Zhang Y, Xu TJ, Xia ZX, Liu JF, Nan FJ (2014) Discovery of a negative allosteric modulator of GABAB receptors. ACS Med Chem Lett 5(7):742–747

Cheng L, Karle M (2008) Quinoline compounds having an activity against the gabab receptor. W.O. Patent 2009041904A1

Cheng L, Holmqvist S, Raubacher F, Schell P (2008a) Xanthine compounds having a positive allosteric gabab receptor modulator effect. W.O. Patent 2008130314A1

Cheng L, Jonforsen M, Schell P (2008b) Pteridine compounds having activity on the gaba-receptors. W.O. Patent 2009041905

Chung KF (2015) NMDA and GABA receptors as potential targets in cough hypersensitivity syndrome. Curr Opin Pharmacol 22:29–36

Clarke JO, Fernandez-Becker NQ, Regalia KA, Triadafilopoulos G (2018) Baclofen and gastroesophageal reflux disease: seeing the forest through the trees. Clin Transl Gastroenterol 9 (3):137–137

Coffey JR, Cahill D, Steers W, Park TS, Ordia J, Meythaler J, Herman R, Shetter AG, Levy R, Gill B et al (1993) Intrathecal baclofen for intractable spasticity of spinal origin: results of a long-term multicenter study. J Neurosurg 78(2):226–232

Comps-Agrar L, Kniazeff J, Nørskov-Lauritsen L, Maurel D, Gassmann M, Gregor N, Prézeau L, Bettler B, Durroux T, Trinquet E, Pin JP (2011) The oligomeric state sets GABA(B) receptor signalling efficacy. EMBO J 30(12):2336–2349

Comps-Agrar L, Kniazeff J, Brock C, Trinquet E, Pin JP (2012) Stability of GABAB receptor oligomers revealed by dual TR-FRET and drug-induced cell surface targeting. FASEB J 26 (8):3430–3439

Cortez MA, Wu Y, Gibson KM, Snead OC 3rd (2004) Absence seizures in succinic semialdehyde dehydrogenase deficient mice: a model of juvenile absence epilepsy. Pharmacol Biochem Behav 79(3):547–553

Costantino G, Macchiarulo A, Entrena Guadix A, Pellicciari R (2001) QSAR and molecular modeling studies of baclofen analogues as GABA(B) agonists. Insights into the role of the aromatic moiety in GABA(B) binding and activation. J Med Chem 44(11):1827–1832

Couve A, Filippov AK, Connolly CN, Bettler B, Brown DA, Moss SJ (1998) Intracellular retention of recombinant GABAB receptors. J Biol Chem 273(41):26361–26367

Couve A, Kittler JT, Uren JM, Calver AR, Pangalos MN, Walsh FS, Moss SJ (2001) Association of GABA(B) receptors and members of the 14-3-3 family of signaling proteins. Mol Cell Neurosci 17(2):317–328

Cryan JF, Kaupmann K (2005) Don't worry 'B' happy!: a role for GABA(B) receptors in anxiety and depression. Trends Pharmacol Sci 26(1):36–43

Cryan JF, Slattery DA (2010) GABAB receptors and depression: current status. Adv Pharmacol 58:427–451

Curtis DR, Duggan AW, Felix D, Johnston GA (1970) GABA, bicuculline and central inhibition. Nature 226(5252):1222–1224

de Miguel E, Vekovischeva O, Kuokkanen K, Vesajoki M, Paasikoski N, Kaskinoro J, Myllymaki M, Lainiola M, Janhunen SK, Hyytia P, Linden AM, Korpi ER (2019) GABAB receptor positive allosteric modulators with different efficacies affect neuroadaptation to and self-administration of alcohol and cocaine. Addict Biol 24(6):1191–1203

Didiášová M, Banning A, Brennenstuhl H, Jung-Klawitter S, Cinquemani C, Opladen T, Tikkanen R (2020) Succinic semialdehyde dehydrogenase deficiency: an update. Cell 9(2):477

Dinamarca MC, Raveh A, Schneider A, Fritzius T, Früh S, Rem PD, Stawarski M, Lalanne T, Turecek R, Choo M, Besseyrias V, Bildl W, Bentrop D, Staufenbiel M, Gassmann M, Fakler B, Schwenk J, Bettler B (2019) Complex formation of APP with GABA(B) receptors links axonal trafficking to amyloidogenic processing. Nat Commun 10(1):1331

Drake RG, Davis LL, Cates ME, Jewell ME, Ambrose SM, Lowe JS (2003) Baclofen treatment for chronic posttraumatic stress disorder. Ann Pharmacother 37(9):1177–1181

Dreifuss JJ, Kelly JS, Krnjević K (1969) Cortical inhibition and gamma-aminobutyric acid. Exp Brain Res 9(2):137–154

Dupuis DS, Relkovic D, Lhuillier L, Mosbacher J, Kaupmann K (2006) Point mutations in the transmembrane region of GABAB2 facilitate activation by the positive modulator N,N′-dicyclopentyl-2-methylsulfanyl-5-nitro-pyrimidine-4,6-diamine (GS39783) in the absence of the GABAB1 subunit. Mol Pharmacol 70(6):2027–2036

Dutar P, Nicoll RA (1988) A physiological role for GABAB receptors in the central nervous system. Nature 332(6160):156–158

Duthey B, Caudron S, Perroy J, Bettler B, Fagni L, Pin JP, Prézeau L (2002) A single subunit (GB2) is required for G-protein activation by the heterodimeric GABA(B) receptor. J Biol Chem 277 (5):3236–3241

Dyer T (2013) Addex announces positive data with ADX71441 in a pre-clinical transgenic model of charcot-marie-tooth 1A disease. Addex Therapeutics, Geneva

Enna SJ, McCarson KE (2006) The role of GABA in the mediation and perception of pain. Adv Pharmacol 54:1–27

Evenseth LM, Warszycki D, Bojarski AJ, Gabrielsen M, Sylte I (2019) In silico methods for the discovery of orthosteric GABAB receptor compounds. Molecules 24(5):935

Faghih R, Moeller A, Ochse M, Pohlki F, Schmidt M, Sippy K, Turner S, van der Kam EL (2016) Substituted pyrazolopyrimidines and method of use. U.S. Patent 9828381B2

Felice D, O'Leary OF, Pizzo RC, Cryan JF (2012) Blockade of the GABA(B) receptor increases neurogenesis in the ventral but not dorsal adult hippocampus: relevance to antidepressant action. Neuropharmacology 63(8):1380–1388

Floersheim P, Froestl W, Guery S, Kaupmann K, Koller M (2006) Pyrimidine derivatives for the treatment of Gaba B mediated nervous system disorders. U.S Pantent 20100179127A1

Francisco GE, Kothari S, Huls C (2001) GABA agonists and gabapentin for spastic hypertonia. Phys Med Rehabil Clin N Am 12(4):875–888, viii

Frankowska M, Filip M, Przegalinski E (2007) Effects of GABAB receptor ligands in animal tests of depression and anxiety. Pharmacol Rep 59(6):645–655

Fritzius T, Bettler B (2020) The organizing principle of GABA(B) receptor complexes: physiological and pharmacological implications. Basic Clin Pharmacol Toxicol 126(Suppl 6):25–34

Froestl W (2010) Chemistry and pharmacology of GABAB receptor ligands. In: GABAB receptor pharmacology – a tribute to Norman Bowery, pp 19–62

Froestl W, Mickel SJ, Hall RG, von Sprecher G, Strub D, Baumann PA, Brugger F, Gentsch C, Jaekel J, Olpe HR et al (1995a) Phosphinic acid analogues of GABA. 1. New potent and selective GABAB agonists. J Med Chem 38(17):3297–3312

Froestl W, Mickel SJ, von Sprecher G, Diel PJ, Hall RG, Maier L, Strub D, Melillo V, Baumann PA, Bernasconi R et al (1995b) Phosphinic acid analogues of GABA. 2. Selective, orally active GABAB antagonists. J Med Chem 38(17):3313–3331

Froestl W, Gallagher M, Jenkins H, Madrid A, Melcher T, Teichman S, Mondadori CG, Pearlman R (2004) SGS742: the first GABA(B) receptor antagonist in clinical trials. Biochem Pharmacol 68 (8):1479–1487

Frye RE (2014) Clinical potential, safety, and tolerability of arbaclofen in the treatment of autism spectrum disorder. Drug Healthc Patient Saf 6:69–76

Galvez T, Parmentier ML, Joly C, Malitschek B, Kaupmann K, Kuhn R, Bittiger H, Froestl W, Bettler B, Pin JP (1999) Mutagenesis and modeling of the GABAB receptor extracellular domain support a venus flytrap mechanism for ligand binding. J Biol Chem 274 (19):13362–13369

Galvez T, Prezeau L, Milioti G, Franek M, Joly C, Froestl W, Bettler B, Bertrand HO, Blahos J, Pin JP (2000) Mapping the agonist-binding site of GABAB type 1 subunit sheds light on the activation process of GABAB receptors. J Biol Chem 275(52):41166–41174

Galvez T, Duthey B, Kniazeff J, Blahos J, Rovelli G, Bettler B, Prézeau L, Pin JP (2001) Allosteric interactions between GB1 and GB2 subunits are required for optimal GABA(B) receptor function. EMBO J 20(9):2152–2159

Gemignani A, Paudice P, Bonanno G, Raiteri M (1994) Pharmacological discrimination between gamma-aminobutyric acid type B receptors regulating cholecystokinin and somatostatin release from rat neocortex synaptosomes. Mol Pharmacol 46(3):558–562

Geng Y, Bush M, Mosyak L, Wang F, Fan QR (2013) Structural mechanism of ligand activation in human GABA(B) receptor. Nature 504(7479):254–259

Getova DP, Dimitrova DD (2007) Effects of GABAB receptor antagonists CGP63360, CGP76290A and CGP76291A on learning and memory processes in rodents. Cent Eur J Med 2:280–293

Gjoni T, Urwyler S (2009) Changes in the properties of allosteric and orthosteric GABAB receptor ligands after a continuous, desensitizing agonist pretreatment. Eur J Pharmacol 603(1–3):37–41

Glausier JR, Lewis DA (2017) GABA and schizophrenia: where we stand and where we need to go. Schizophr Res 181:2–3

Grunewald S, Schupp BJ, Ikeda SR, Kuner R, Steigerwald F, Kornau HC, Kohr G (2002) Importance of the gamma-aminobutyric acid(B) receptor C-termini for G-protein coupling. Mol Pharmacol 61(5):1070–1080

Han C, Salyer AE, Kim EH, Jiang X, Jarrard RE, Powers MS, Kirchhoff AM, Salvador TK, Chester JA, Hockerman GH, Colby DA (2013) Evaluation of difluoromethyl ketones as agonists of the γ-aminobutyric acid type B (GABAB) receptor. J Med Chem 56(6):2456–2465

Hannan S, Wilkins ME, Smart TG (2012) Sushi domains confer distinct trafficking profiles on GABAB receptors. Proc Natl Acad Sci U S A 109(30):12171–12176

Hawrot E, Xiao Y, Shi QL, Norman D, Kirkitadze M, Barlow PN (1998) Demonstration of a tandem pair of complement protein modules in GABA(B) receptor 1a. FEBS Lett 432 (3):103–108

Helm KA, Haberman RP, Dean SL, Hoyt EC, Melcher T, Lund PK, Gallagher M (2005) GABAB receptor antagonist SGS742 improves spatial memory and reduces protein binding to the cAMP response element (CRE) in the hippocampus. Neuropharmacology 48(7):956–964

Hinton T, Chebib M, Johnston GA (2008) Enantioselective actions of 4-amino-3-hydroxybutanoic acid and (3-amino-2-hydroxypropyl)methylphosphinic acid at recombinant GABA (C) receptors. Bioorg Med Chem Lett 18(1):402–404

Hirst WD, Babbs AJ, Green A, Minton JA, Shaw TE, Wise A, Rice SQ, Pangalos MN, Price GW (2003) Pharmacological characterisation of a cell line expressing GABA B1b and GABA B2 receptor subunits. Biochem Pharmacol 65(7):1103–1113

Hudgson P, Weightman D (1971) Baclofen in the treatment of spasticity. Br Med J 4(5778):15–17

Jacobson LH, Vlachou S, Slattery DA, Li X, Cryan JF (2018) The gamma-aminobutyric acid b receptor in depression and reward. Biol Psychiatry 83(11):963–976

Jensen AA, Mosbacher J, Elg S, Lingenhoehl K, Lohmann T, Johansen TN, Abrahamsen B, Mattsson JP, Lehmann A, Bettler B, Bräuner-Osborne H (2002) The anticonvulsant gabapentin (neurontin) does not act through gamma-aminobutyric acid-B receptors. Mol Pharmacol 61 (6):1377–1384

Jiang X, Su L, Zhang Q, He C, Zhang Z, Yi P, Liu J (2012) GABAB receptor complex as a potential target for tumor therapy. J Histochem Cytochem 60(4):269–279

Johnston GA (1986) Multiplicity of GABA receptors. In: Benzodiazepine/GABA receptors and chloride channels. Receptor biochemistry and methodology. A.R. Liss

Jones KA, Borowsky B, Tamm JA, Craig DA, Durkin MM, Dai M, Yao WJ, Johnson M, Gunwaldsen C, Huang LY, Tang C, Shen Q, Salon JA, Morse K, Laz T, Smith KE, Nagarathnam D, Noble SA, Branchek TA, Gerald C (1998) GABA(B) receptors function as a heteromeric assembly of the subunits GABA(B)R1 and GABA(B)R2. Nature 396 (6712):674–679

Jouney EA (2019) Phenibut (β-phenyl-γ-aminobutyric acid): an easily obtainable "dietary supplement" with propensities for physical dependence and addiction. Curr Psychiatry Rep 21(4):23

Kalinichev M, Girard F, Haddouk H, Rouillier M, Riguet E, Royer-Urios I, Mutel V, Lütjens R, Poli S (2017) The drug candidate, ADX71441, is a novel, potent and selective positive allosteric modulator of the GABA(B) receptor with a potential for treatment of anxiety, pain and spasticity. Neuropharmacology 114:34–47

Kaupmann K, Huggel K, Heid J, Flor PJ, Bischoff S, Mickel SJ, McMaster G, Angst C, Bittiger H, Froestl W, Bettler B (1997) Expression cloning of GABA(B) receptors uncovers similarity to metabotropic glutamate receptors. Nature 386(6622):239–246

Kaupmann K, Malitschek B, Schuler V, Heid J, Froestl W, Beck P, Mosbacher J, Bischoff S, Kulik A, Shigemoto R, Karschin A, Bettler B (1998) GABA(B)-receptor subtypes assemble into functional heteromeric complexes. Nature 396(6712):683–687

Keating GM (2014) Sodium oxybate: a review of its use in alcohol withdrawal syndrome and in the maintenance of abstinence in alcohol dependence. Clin Drug Investig 34(1):63–80
Keberle H, Faigle JW, Wilhelm M (1964) Gamma-amino-beta-(para-halophenyl)-butyric acids and their esters. U.S. Patent 3471548A
Keberle H, Faigle JW, Wilhelm M (1969) Gamma-amino-beta-(para-halophenyl)-butyric acids and their esters. Google Patents
Kenakin T (2004) Efficacy as a vector: the relative prevalence and paucity of inverse agonism. Mol Pharmacol 65(1):2–11
Kenakin T (2005) New concepts in drug discovery: collateral efficacy and permissive antagonism. Nat Rev Drug Discov 4(11):919–927
Kenakin T (2017) Signaling bias in drug discovery. Expert Opin Drug Discovery 12(4):321–333
Kenakin T, Miller LJ (2010) Seven transmembrane receptors as shapeshifting proteins: the impact of allosteric modulation and functional selectivity on new drug discovery. Pharmacol Rev 62 (2):265–304
Kent CN, Park C, Lindsley CW (2020) Classics in chemical neuroscience: baclofen. ACS Chem Neurosci 11(12):1740–1755
Keov P, Sexton PM, Christopoulos A (2011) Allosteric modulation of G protein-coupled receptors: a pharmacological perspective. Neuropharmacology 60(1):24–35
Kerr DI, Ong J, Prager RH, Gynther BD, Curtis DR (1987) Phaclofen: a peripheral and central baclofen antagonist. Brain Res 405(1):150–154
Kerr DI, Ong J, Johnston GA, Abbenante J, Prager RH (1988) 2-Hydroxy-saclofen: an improved antagonist at central and peripheral GABAB receptors. Neurosci Lett 92(1):92–96
Kerr DI, Ong J, Puspawati NM, Prager RH (2002) Arylalkylamines are a novel class of positive allosteric modulators at GABA(B) receptors in rat neocortex. Eur J Pharmacol 451(1):69–77
Kerr DIB, Khalafy J, Ong J, Perkins MV, Prager RH, Puspawati NM, Rimaz M (2006) Synthesis and biological activity of allosteric modulators of GABAB receptors, part 2. 3-(2,6-Bis-tert-butyl-4-hydroxyphenyl)propanols. Aust J Chem 59(7):457–462
Kim W, Seo H (2014) Baclofen, a GABAB receptor agonist, enhances ubiquitin-proteasome system functioning and neuronal survival in Huntington's disease model mice. Biochem Biophys Res Commun 443(2):706–711
Kleppner SR, Tobin AJ (2001) GABA signalling: therapeutic targets for epilepsy, Parkinson's disease and Huntington's disease. Emerg Thera Targets 5(2):219–239
Kniazeff J, Galvez T, Labesse G, Pin JP (2002) No ligand binding in the GB2 subunit of the GABA (B) receptor is required for activation and allosteric interaction between the subunits. J Neurosci 22(17):7352–7361
Koek W, Cheng K, Rice KC (2013) Discriminative stimulus effects of the GABAB receptor-positive modulator rac-BHFF: comparison with GABAB receptor agonists and drugs of abuse. J Pharmacol Exp Ther 344(3):553–560
Korsgaard S (1976) Baclofen (Lioresal) in the treatment ofneuroleptic-induced tardive dyskinesia. Acta Psychiatr Scand 54(1):17–24
Krnjević K, Schwartz S (1966) Is gamma-aminobutyric acid an inhibitory transmitter? Nature 211 (5056):1372–1374
Lanneau C, Green A, Hirst WD, Wise A, Brown JT, Donnier E, Charles KJ, Wood M, Davies CH, Pangalos MN (2001) Gabapentin is not a GABAB receptor agonist. Neuropharmacology 41 (8):965–975
Lapin I (2001) Phenibut (beta-phenyl-GABA): a tranquilizer and nootropic drug. CNS Drug Rev 7 (4):471–481
Lasarge CL, Banuelos C, Mayse JD, Bizon JL (2009) Blockade of GABA(B) receptors completely reverses age-related learning impairment. Neuroscience 164(3):941–947
Lehmann A (2009) GABAB receptors as drug targets to treat gastroesophageal reflux disease. Pharmacol Ther 122(3):239–245
Lehmann A, Jensen JM, Boeckxstaens GE (2010) GABAB receptor agonism as a novel therapeutic modality in the treatment of gastroesophageal reflux disease. Adv Pharmacol 58:287–313

Li X, Kaczanowska K, Finn MG, Markou A, Risbrough VB (2015) The GABA(B) receptor positive modulator BHF177 attenuated anxiety, but not conditioned fear, in rats. Neuropharmacology 97:357–364
Li X, Sturchler E, Kaczanowska K, Cameron M, Finn MG, Griffin P, McDonald P, Markou A (2017) KK-92A, a novel GABAB receptor positive allosteric modulator, attenuates nicotine self-administration and cue-induced nicotine seeking in rats. Psychopharmacology 234 (9–10):1633–1644
Lingenhoehl K, Brom R, Heid J, Beck P, Froestl W, Kaupmann K, Bettler B, Mosbacher J (1999) Gamma-hydroxybutyrate is a weak agonist at recombinant GABA(B) receptors. Neuropharmacology 38(11):1667–1673
Maccioni P, Colombo G (2019) Potential of GABA(B) receptor positive allosteric modulators in the treatment of alcohol use disorder. CNS Drugs 33(2):107–123
Maccioni P, Zaru A, Loi B, Lobina C, Carai MA, Gessa GL, Capra A, Mugnaini C, Pasquini S, Corelli F, Hyytia P, Lumeng L, Colombo G (2012) Comparison of the effect of the GABABeta receptor agonist, baclofen, and the positive allosteric modulator of the GABAB receptor, GS39783, on alcohol self-administration in 3 different lines of alcohol-preferring rats. Alcohol Clin Exp Res 36(10):1748–1766
Maccioni P, Vargiolu D, Thomas AW, Malherbe P, Mugnaini C, Corelli F, Leite-Morris KA, Gessa GL, Colombo G (2015) Inhibition of alcohol self-administration by positive allosteric modulators of the GABAB receptor in rats: lack of tolerance and potentiation of baclofen. Psychopharmacology 232(10):1831–1841
Malherbe P, Masciadri R, Norcross RD, Prinssen E (2006) 3-methanesulfonylquinolines as GABAB enhancers. U.S. Patent 7728142B2
Malherbe P, Masciadri R, Norcross RD, Ratni H, Thomas AW (2007) Thieno-pyridine derivatives as gaba-b allosteric enhancers. E.P.O. Patent 1828199
Malherbe P, Masciadri R, Norcross RD, Knoflach F, Kratzeisen C, Zenner MT, Kolb Y, Marcuz A, Huwyler J, Nakagawa T, Porter RH, Thomas AW, Wettstein JG, Sleight AJ, Spooren W, Prinssen EP (2008) Characterization of (R,S)-5,7-di-tert-butyl-3-hydroxy-3-trifluoromethyl-3H-benzofuran-2-one as a positive allosteric modulator of GABAB receptors. Br J Pharmacol 154(4):797–811
Mao C, Shen C, Li C, Shen DD, Xu C, Zhang S, Zhou R, Shen Q, Chen LN, Jiang Z, Liu J, Zhang Y (2020) Cryo-EM structures of inactive and active GABA(B) receptor. Cell Res 30(7):564–573
Marcoli M, Scarrone S, Maura G, Bonanno G, Raiteri M (2000) A subtype of the gamma-aminobutyric acid(B) receptor regulates cholinergic twitch response in the guinea pig ileum. J Pharmacol Exp Ther 293(1):42–47
Mares P, Kubova H (2008) What is the role of neurotransmitter systems in cortical seizures? Physiol Res 57(Suppl 3):S111–S120
Marescaux C, Vergnes M, Bernasconi R (1992) GABAB receptor antagonists: potential new anti-absence drugs. J Neural Transm Suppl 35:179–188
Margeta-Mitrovic M, Jan YN, Jan LY (2001) Ligand-induced signal transduction within heterodimeric GABA(B) receptor. Proc Natl Acad Sci U S A 98(25):14643–14648
Markou A, Hauger RL, Koob GF (1992) Desmethylimipramine attenuates cocaine withdrawal in rats. Psychopharmacology 109(3):305–314
Markou A, Kosten TR, Koob GF (1998) Neurobiological similarities in depression and drug dependence: a self-medication hypothesis. Neuropsychopharmacology 18(3):135–174
Marshall FH, White J, Main M, Green A, Wise A (1999) GABA(B) receptors function as heterodimers. Biochem Soc Trans 27(4):530–535
Martvon L, Kotmanova Z, Dobrolubov B, Babalova L, Simera M, Veternik M, Pitts T, Jakus J, Poliacek I (2020) Modulation of cough reflex by Gaba-ergic inhibition in medullary raphé of the cat. Physiol Res 69(Suppl 1):S151–s161
Mathivet P, Bernasconi R, De Barry J, Marescaux C, Bittiger H (1997) Binding characteristics of gamma-hydroxybutyric acid as a weak but selective GABAB receptor agonist. Eur J Pharmacol 321(1):67–75

Maurel D, Comps-Agrar L, Brock C, Rives ML, Bourrier E, Ayoub MA, Bazin H, Tinel N, Durroux T, Prézeau L, Trinquet E, Pin JP (2008) Cell-surface protein-protein interaction analysis with time-resolved FRET and snap-tag technologies: application to GPCR oligomerization. Nat Methods 5(6):561–567

May LT, Avlani VA, Sexton PM, Christopoulos A (2004) Allosteric modulation of G protein-coupled receptors. Curr Pharm Des 10(17):2003–2013

Meier SD, Kafitz KW, Rose CR (2008) Developmental profile and mechanisms of GABA-induced calcium signaling in hippocampal astrocytes. Glia 56(10):1127–1137

Mendez D, Gaulton A, Bento AP, Chambers J, De Veij M, Félix E, Magariños MP, Mosquera JF, Mutowo P, Nowotka M, Gordillo-Marañón M, Hunter F, Junco L, Mugumbate G, Rodriguez-Lopez M, Atkinson F, Bosc N, Radoux CJ, Segura-Cabrera A, Hersey A, Leach AR (2019) ChEMBL: towards direct deposition of bioassay data. Nucleic Acids Res 47(D1):D930–d940

Modaberi S, Heysieattalab S, Shahbazi M, Naghdi N (2019) Combination effects of forced mild exercise and GABA(B) receptor agonist on spatial learning, memory, and motor activity in striatum lesion rats. J Mot Behav 51(4):438–450

Mombereau C, Kaupmann K, Froestl W, Sansig G, van der Putten H, Cryan JF (2004) Genetic and pharmacological evidence of a role for GABA(B) receptors in the modulation of anxiety- and antidepressant-like behavior. Neuropsychopharmacology 29(6):1050–1062

Mondadori C, Jaekel J, Preiswerk G (1993) CGP 36742: the first orally active GABAB blocker improves the cognitive performance of mice, rats, and rhesus monkeys. Behav Neural Biol 60 (1):62–68

Morishita R, Kato K, Asano T (1990) GABAB receptors couple to G proteins Go, Go* and Gi1 but not to Gi2. FEBS Lett 271(1–2):231–235

Mugnaini C, Pedani V, Casu A, Lobina C, Casti A, Maccioni P, Porcu A, Giunta D, Lamponi S, Solinas M, Dragoni S, Valoti M, Colombo G, Castelli MP, Gessa GL, Corelli F (2013) Synthesis and pharmacological characterization of 2-(acylamino)thiophene derivatives as metabolically stable, orally effective, positive allosteric modulators of the GABAB receptor. J Med Chem 56(9):3620–3635

Mukherjee RS, McBride EW, Beinborn M, Dunlap K, Kopin AS (2006) Point mutations in either subunit of the GABAB receptor confer constitutive activity to the heterodimer. Mol Pharmacol 70(4):1406–1413

Murai N, Kondo Y, Akuzawa S, Mihara T, Shiraishi N, Kakimoto S, Matsumoto M (2019) A novel GABA(B) receptor positive allosteric modulator, ASP8062, exerts analgesic effects in a rat model of fibromyalgia. Eur J Pharmacol 865:172750

Naffaa MM, Hung S, Chebib M, Johnston GAR, Hanrahan JR (2017) GABA-ρ receptors: distinctive functions and molecular pharmacology. Br J Pharmacol 174(13):1881–1894

Nair PC, McKinnon RA, Miners JO, Bastiampillai T (2020) Binding of clozapine to the GABA (B) receptor: clinical and structural insights. Mol Psychiatry 25:1910–1919

Nambu A (2012) GABA-B receptor: possible target for Parkinson's disease therapy. Exp Neurol 233(1):121–122

Nehring RB, Horikawa HP, El Far O, Kneussel M, Brandstätter JH, Stamm S, Wischmeyer E, Betz H, Karschin A (2000) The metabotropic GABAB receptor directly interacts with the activating transcription factor 4. J Biol Chem 275(45):35185–35191

Nemeth EF, Steffey ME, Hammerland LG, Hung BCP, Van Wagenen BC, DelMar EG, Balandrin MF (1998) Calcimimetics with potent and selective activity on the parathyroid calcium receptor. Proc Natl Acad Sci 95(7):4040–4045

Neto FL, Ferreira-Gomes J, Castro-Lopes JM (2006) Distribution of GABA receptors in the thalamus and their involvement in nociception. Adv Pharmacol 54:29–51

New DC, An H, Ip NY, Wong YH (2006) GABAB heterodimeric receptors promote Ca2+ influx via store-operated channels in rat cortical neurons and transfected Chinese hamster ovary cells. Neuroscience 137(4):1347–1358

Ng GY, Bertrand S, Sullivan R, Ethier N, Wang J, Yergey J, Belley M, Trimble L, Bateman K, Alder L, Smith A, McKernan R, Metters K, O'Neill GP, Lacaille JC, Hébert TE (2001)

Gamma-aminobutyric acid type B receptors with specific heterodimer composition and post-synaptic actions in hippocampal neurons are targets of anticonvulsant gabapentin action. Mol Pharmacol 59(1):144–152
Niazi M, Skrtic S, Ruth M, Holmberg AA (2011) Pharmacokinetic profile of lesogaberan (AZD3355) in healthy subjects: a novel GABA(B)-receptor agonist reflux inhibitor. Drugs R&D 11(1):77–83
Nishikawa M, Hirouchi M, Kuriyama K (1997) Functional coupling of Gi subtype with GABAB receptor/adenylyl cyclase system: analysis using a reconstituted system with purified GTP-binding protein from bovine cerebral cortex. Neurochem Int 31(1):21–25
Niswender CM, Conn PJ (2010) Metabotropic glutamate receptors: physiology, pharmacology, and disease. Annu Rev Pharmacol Toxicol 50:295–322
Nowak G, Partyka A, Palucha A, Szewczyk B, Wieronska JM, Dybala M, Metz M, Librowski T, Froestl W, Papp M, Pilc A (2006) Antidepressant-like activity of CGP 36742 and CGP 51176, selective GABAB receptor antagonists, in rodents. Br J Pharmacol 149(5):581–590
Ohtsuki S, Asaba H, Takanaga H, Deguchi T, Hosoya K, Otagiri M, Terasaki T (2002) Role of blood-brain barrier organic anion transporter 3 (OAT3) in the efflux of indoxyl sulfate, a uremic toxin: its involvement in neurotransmitter metabolite clearance from the brain. J Neurochem 83 (1):57–66
Olpe HR, Demiéville H, Baltzer V, Bencze WL, Koella WP, Wolf P, Haas HL (1978) The biological activity of d- and l-baclofen (Lioresal). Eur J Pharmacol 52(1):133–136
Ong J, Kerr DI (1984) Evidence for a physiological role of GABA in the control of guinea-pig intestinal motility. Neurosci Lett 50(1–3):339–343
Ong J, Parker DA, Marino V, Kerr DI, Puspawati NM, Prager RH (2005) 3-Chloro,4-methoxyfendiline is a potent GABA(B) receptor potentiator in rat neocortical slices. Eur J Pharmacol 507(1–3):35–42
Ostojić ZS, Ilić TV, Vesković SM, Andjus PR (2013) GABAB receptors as a common target for hypothermia and spike and wave seizures: intersecting mechanisms of thermoregulation and absence epilepsy. Neuroscience 238:39–58
Pagano A, Rovelli G, Mosbacher J, Lohmann T, Duthey B, Stauffer D, Ristig D, Schuler V, Meigel I, Lampert C, Stein T, Prezeau L, Blahos J, Pin J, Froestl W, Kuhn R, Heid J, Kaupmann K, Bettler B (2001) C-terminal interaction is essential for surface trafficking but not for heteromeric assembly of GABA(b) receptors. J Neurosci 21(4):1189–1202
Patel S, Naeem S, Kesingland A, Froestl W, Capogna M, Urban L, Fox A (2001) The effects of GABA(B) agonists and gabapentin on mechanical hyperalgesia in models of neuropathic and inflammatory pain in the rat. Pain 90(3):217–226
Perdona E, Costantini VJ, Tessari M, Martinelli P, Carignani C, Valerio E, Mok MH, Zonzini L, Visentini F, Gianotti M, Gordon L, Rocheville M, Corsi M, Capelli AM (2011) In vitro and in vivo characterization of the novel GABAB receptor positive allosteric modulator, 2-{1-[2-(4-chlorophenyl)-5-methylpyrazolo[1,5-a]pyrimidin-7-yl]-2-piperidinyl}eth anol (CMPPE). Neuropharmacology 61(5–6):957–966
Pérez-Garci E, Gassmann M, Bettler B, Larkum ME (2006) The GABAB1b isoform mediates long-lasting inhibition of dendritic Ca2+ spikes in layer 5 somatosensory pyramidal neurons. Neuron 50(4):603–616
Pinard A, Seddik R, Bettler B (2010) GABAB receptors: physiological functions and mechanisms of diversity. Adv Pharmacol 58:231–255
Pirard B, Carrupt PA, Testa B, Tsai RS, Berthelot P, Vaccher C, Debaert M, Durant F (1995) Structure-affinity relationships of baclofen and 3-heteroaromatic analogues. Bioorg Med Chem 3(11):1537–1545
Polenzani L, Woodward RM, Miledi R (1991) Expression of mammalian gamma-aminobutyric acid receptors with distinct pharmacology in Xenopus oocytes. Proc Natl Acad Sci U S A 88 (10):4318–4322

Pontier SM, Lahaie N, Ginham R, St-Gelais F, Bonin H, Bell DJ, Flynn H, Trudeau LE, McIlhinney J, White JH, Bouvier M (2006) Coordinated action of NSF and PKC regulates GABAB receptor signaling efficacy. EMBO J 25(12):2698–2709

Porcu A, Lobina C, Giunta D, Solinas M, Mugnaini C, Castelli MP (2016) In vitro and in vivo pharmacological characterization of SSD114, a novel GABAB positive allosteric modulator. Eur J Pharmacol 791:115–123

Prusis P, Höglund L, Törmakangas O, Hietanen A, Arvela R, Vesalainen A, Heikkinen T (2015) Pharmacologically active quinazolinedione derivatives. W.O. Patent 2015169999-A1

Ranson DC, Ayoub SS, Corcoran O, Casalotti SO (2020) Pharmacological targeting of the GABA (B) receptor alters Drosophila's behavioural responses to alcohol. Addict Biol 25(2):e12725

Rice HC, de Malmazet D, Schreurs A, Frere S, Van Molle I, Volkov AN, Creemers E, Vertkin I, Nys J, Ranaivoson FM, Comoletti D, Savas JN, Remaut H, Balschun D, Wierda KD, Slutsky I, Farrow K, De Strooper B, de Wit J (2019) Secreted amyloid-β precursor protein functions as a GABA(B)R1a ligand to modulate synaptic transmission. Science 363(6423):eaao4827

Riguet E, Campo B, Gibelin A, Mhalla K (2007) Triazinedione derivatives as GABAB receptor modulators. U.S. Patent 8344138B2

Ritter B, Zschüntsch J, Kvachnina E, Zhang W, Ponimaskin EG (2004) The GABA(B) receptor subunits R1 and R2 interact differentially with the activation transcription factor ATF4 in mouse brain during the postnatal development. Brain Res Dev Brain Res 149(1):73–77

Robbins MJ, Calver AR, Filippov AK, Hirst WD, Russell RB, Wood MD, Nasir S, Couve A, Brown DA, Moss SJ, Pangalos MN (2001) GABA(B2) is essential for g-protein coupling of the GABA(B) receptor heterodimer. J Neurosci 21(20):8043–8052

Roberts E, Frankel S (1950) Gamma-aminobutyric acid in brain: its formation from glutamic acid. J Biol Chem 187(1):55–63

Sahraei H, Askaripour M, Esmaeilpour K, Shahsavari F, Rajabi S, Moradi-Kor N (2019) GABA (B) receptor activation ameliorates spatial memory impairments in stress-exposed rats. Neuropsychiatr Dis Treat 15:1497–1506

Sauter K, Grampp T, Fritschy JM, Kaupmann K, Bettler B, Mohler H, Benke D (2005) Subtype-selective interaction with the transcription factor CCAAT/enhancer-binding protein (C/EBP) homologous protein (CHOP) regulates cell surface expression of GABA(B) receptors. J Biol Chem 280(39):33566–33572

Schofield PR, Darlison MG, Fujita N, Burt DR, Stephenson FA, Rodriguez H, Rhee LM, Ramachandran J, Reale V, Glencorse TA et al (1987) Sequence and functional expression of the GABA A receptor shows a ligand-gated receptor super-family. Nature 328(6127):221–227

Schuller HM (2018) Regulatory role of G protein-coupled receptors in pancreatic cancer development and progression. Curr Med Chem 25(22):2566–2575

Schuller HM, Al-Wadei HA, Majidi M (2008) GABA B receptor is a novel drug target for pancreatic cancer. Cancer 112(4):767–778

Schwenk J, Metz M, Zolles G, Turecek R, Fritzius T, Bildl W, Tarusawa E, Kulik A, Unger A, Ivankova K, Seddik R, Tiao JY, Rajalu M, Trojanova J, Rohde V, Gassmann M, Schulte U, Fakler B, Bettler B (2010) Native GABA(B) receptors are heteromultimers with a family of auxiliary subunits. Nature 465(7295):231–235

Schwenk J, Pérez-Garci E, Schneider A, Kollewe A, Gauthier-Kemper A, Fritzius T, Raveh A, Dinamarca MC, Hanuschkin A, Bildl W, Klingauf J, Gassmann M, Schulte U, Bettler B, Fakler B (2016) Modular composition and dynamics of native GABAB receptors identified by high-resolution proteomics. Nat Neurosci 19(2):233–242

Seabrook GR, Howson W, Lacey MG (1990) Electrophysiological characterization of potent agonists and antagonists at pre- and postsynaptic GABAB receptors on neurones in rat brain slices. Br J Pharmacol 101(4):949–957

Shiraishi N, Hoshii H, Hamaguchi W, Honjo E, Takuwa T, Kondo Y, Goto T (2014) Sulfur-Containing Bicyclic Compound. W.O. Patent 2015056771

Sigel E, Steinmann ME (2012) Structure, function, and modulation of GABA(A) receptors. J Biol Chem 287(48):40224–40231

Smith JS, Lefkowitz RJ, Rajagopal S (2018) Biased signalling: from simple switches to allosteric microprocessors. Nat Rev Drug Discov 17(4):243–260

Snead OC 3rd (1992) Evidence for GABAB-mediated mechanisms in experimental generalized absence seizures. Eur J Pharmacol 213(3):343–349

Sowaileh MF, Salyer AE, Roy KK, John JP, Woods JR, Doerksen RJ, Hockerman GH, Colby DA (2018) Agonists of the γ-aminobutyric acid type B (GABA(B)) receptor derived from β-hydroxy and β-amino difluoromethyl ketones. Bioorg Med Chem Lett 28(16):2697–2700

Spangler S, Bruchas MR (2017) Tuning biased GPCR signaling for physiological gain. Cell 171 (5):989–991

Stewart LS, Wu Y, Eubanks JH, Han H, Leschenko Y, Perez Velazquez JL, Cortez MA, Snead OC 3rd (2009) Severity of atypical absence phenotype in GABAB transgenic mice is subunit specific. Epilepsy Behav 14(4):577–581

Stewart GD, Comps-Agrar L, Nørskov-Lauritsen LB, Pin JP, Kniazeff J (2018) Allosteric interactions between GABA(B1) subunits control orthosteric binding sites occupancy within GABA (B) oligomers. Neuropharmacology 136(Pt A):92–101

Sturchler E, Li X, de Lourdes Ladino M, Kaczanowska K, Cameron M, Griffin PR, Finn MG, Markou A, McDonald P (2017) GABA(B) receptor allosteric modulators exhibit pathway-dependent and species-selective activity. Pharmacol Res Perspect 5(2):e00288

Sun B, Chen L, Liu L, Xia Z, Pin JP, Nan F, Liu J (2016) A negative allosteric modulator modulates GABAB-receptor signalling through GB2 subunits. Biochem J 473(6):779–787

Sun Z, Sun L, Tu L (2020) GABAB receptor-mediated PI3K/Akt signaling pathway alleviates oxidative stress and neuronal cell injury in a rat model of Alzheimer's disease. J Alzheimers Dis 76(4):1513–1526

Symonds E, Butler R, Omari T (2003) The effect of the GABAB receptor agonist baclofen on liquid and solid gastric emptying in mice. Eur J Pharmacol 470(1–2):95–97

Szabadi E (2015) GHB for cataplexy: possible mode of action. J Psychopharmacol 29(6):744–749

Tabata T, Araishi K, Hashimoto K, Hashimotodani Y, van der Putten H, Bettler B, Kano M (2004) Ca2+ activity at GABAB receptors constitutively promotes metabotropic glutamate signaling in the absence of GABA. Proc Natl Acad Sci U S A 101(48):16952–16957

Teichgräber LA, Lehmann TN, Meencke HJ, Weiss T, Nitsch R, Deisz RA (2009) Impaired function of GABA(B) receptors in tissues from pharmacoresistant epilepsy patients. Epilepsia 50(7):1697–1716

Tian J, Dang H, Hu A, Xu W, Kaufman DL (2017) Repurposing lesogaberan to promote human islet cell survival and β-cell replication. J Diabetes Res 2017:6403539

Tsunekawa T, Banno R, Yaginuma H, Taki K, Mizoguchi A, Sugiyama M, Onoue T, Takagi H, Hagiwara D, Ito Y, Iwama S, Goto M, Suga H, Bettler B, Arima H (2019) GABA(B) receptor signaling in the mesolimbic system suppresses binge-like consumption of a high-fat diet. IScience 20:337–347

Tu H, Rondard P, Xu C, Bertaso F, Cao F, Zhang X, Pin JP, Liu J (2007) Dominant role of GABAB2 and Gbetagamma for GABAB receptor-mediated-ERK1/2/CREB pathway in cerebellar neurons. Cell Signal 19(9):1996–2002

Tyagi RK, Bisht R, Pant J, Kumar P, Majeed AB, Prakash A (2015) Possible role of GABA-B receptor modulation in MPTP induced Parkinson's disease in rats. Exp Toxicol Pathol 67 (2):211–217

Urwyler S (2011) Allosteric modulation of family C G-protein-coupled receptors: from molecular insights to therapeutic perspectives. Pharmacol Rev 63(1):59–126

Urwyler S, Mosbacher J, Lingenhoehl K, Heid J, Hofstetter K, Froestl W, Bettler B, Kaupmann K (2001) Positive allosteric modulation of native and recombinant gamma-aminobutyric acid (B) receptors by 2,6-Di-tert-butyl-4-(3-hydroxy-2,2-dimethyl-propyl)-phenol (CGP7930) and its aldehyde analog CGP13501. Mol Pharmacol 60(5):963–971

Urwyler S, Pozza MF, Lingenhoehl K, Mosbacher J, Lampert C, Froestl W, Koller M, Kaupmann K (2003) N,N′-Dicyclopentyl-2-methylsulfanyl-5-nitro-pyrimidine-4,6-diamine (GS39783) and

structurally related compounds: novel allosteric enhancers of gamma-aminobutyric acidB receptor function. J Pharmacol Exp Ther 307(1):322–330
Urwyler S, Gjoni T, Kaupmann K, Pozza MF, Mosbacher J (2004) Selected amino acids, dipeptides and arylalkylamine derivatives do not act as allosteric modulators at GABAB receptors. Eur J Pharmacol 483(2–3):147–153
Urwyler S, Gjoni T, Koljatic J, Dupuis DS (2005) Mechanisms of allosteric modulation at GABAB receptors by CGP7930 and GS39783: effects on affinities and efficacies of orthosteric ligands with distinct intrinsic properties. Neuropharmacology 48(3):343–353
Vanhoose AM, Emery M, Jimenez L, Winder DG (2002) ERK activation by G-protein-coupled receptors in mouse brain is receptor identity-specific. J Biol Chem 277(11):9049–9053
Veenstra-VanderWeele J, Cook EH, King BH, Zarevics P, Cherubini M, Walton-Bowen K, Bear MF, Wang PP, Carpenter RL (2017) Arbaclofen in children and adolescents with autism spectrum disorder: a randomized, controlled, phase 2 trial. Neuropsychopharmacology 42 (7):1390–1398
Vigot R, Barbieri S, Bräuner-Osborne H, Turecek R, Shigemoto R, Zhang YP, Luján R, Jacobson LH, Biermann B, Fritschy JM, Vacher CM, Müller M, Sansig G, Guetg N, Cryan JF, Kaupmann K, Gassmann M, Oertner TG, Bettler B (2006) Differential compartmentalization and distinct functions of GABAB receptor variants. Neuron 50(4):589–601
Vlachou S, Paterson NE, Guery S, Kaupmann K, Froestl W, Banerjee D, Finn MG, Markou A (2011) Both GABA(B) receptor activation and blockade exacerbated anhedonic aspects of nicotine withdrawal in rats. Eur J Pharmacol 655(1–3):52–58
Weatherby RP, Allan RD, Johnston GA (1984) Resolution of the stereoisomers of baclofen by high performance liquid chromatography. J Neurosci Methods 10(1):23–28
Wellendorph P, Høg S, Greenwood JR, de Lichtenberg A, Nielsen B, Frølund B, Brehm L, Clausen RP, Bräuner-Osborne H (2005) Novel cyclic gamma-hydroxybutyrate (GHB) analogs with high affinity and stereoselectivity of binding to GHB sites in rat brain. J Pharmacol Exp Ther 315 (1):346–351
Wong CG, Gibson KM, Snead OC 3rd (2004) From the street to the brain: neurobiology of the recreational drug gamma-hydroxybutyric acid. Trends Pharmacol Sci 25(1):29–34
Young SZ, Bordey A (2009) GABA's control of stem and cancer cell proliferation in adult neural and peripheral niches. Physiology 24:171–185
Zhang X, Du Z, Liu J, He J (2014) Gamma-aminobutyric acid receptors affect the progression and migration of tumor cells. J Recept Signal Transduct Res 34(6):431–439
Zhang W, Xu C, Tu H, Wang Y, Sun Q, Hu P, Hu Y, Rondard P, Liu J (2015) GABAB receptor upregulates fragile X mental retardation protein expression in neurons. Sci Rep 5(1):10468
Zuo H, Glaaser I, Zhao Y, Kurinov I, Mosyak L, Wang H, Liu J, Park J, Frangaj A, Sturchler E, Zhou M, McDonald P, Geng Y, Slesinger PA, Fan QR (2019) Structural basis for auxiliary subunit KCTD16 regulation of the GABA(B) receptor. Proc Natl Acad Sci U S A 116 (17):8370–8379

$GABA_B$ Receptors and Drug Addiction: Psychostimulants and Other Drugs of Abuse

Xiaofan Li and Paul A. Slesinger

Contents

Abstract Metabotropic $GABA_B$ receptors ($GABA_BRs$) mediate slow inhibition and modulate synaptic plasticity throughout the brain. Dysfunction of $GABA_BRs$ has been associated with psychiatric illnesses and addiction. Drugs of abuse alter $GABA_B$ receptor ($GABA_BR$) signaling in multiple brain regions, which partly contributes to the development of drug addiction. Recently, $GABA_BR$ ligands and positive allosteric modulators (PAMs) have been shown to attenuate the initial

X. Li (✉) and P. A. Slesinger (✉)
Nash Family Department of Neuroscience, Icahn School of Medicine at Mount Sinai, New York, NY, USA
e-mail: xiaofan.li@mssm.edu; paul.slesinger@mssm.edu

Curr Topics Behav Neurosci (2022) 52: 119–156
https://doi.org/10.1007/7854_2020_187
Published Online: 14 January 2021

rewarding effect of addictive substances, inhibit seeking and taking of these drugs, and in some cases, ameliorate drug withdrawal symptoms. The majority of the anti-addiction effects seen with $GABA_BR$ modulation can be localized to ventral tegmental area (VTA) dopamine neurons, which receive complex inhibitory and excitatory inputs that are modified by drugs of abuse. Preclinical research suggests that $GABA_BR$ PAMs are emerging as promising candidates for the treatment of drug addiction. Clinical studies on drug dependence have shown positive results with $GABA_BR$ ligands but more are needed, and compounds with better pharmacokinetics and fewer side effects are critically needed.

Keywords Dopamine · Positive allosteric modulators (PAMs) · Synaptic plasticity · Ventral tegmental area (VTA)

1 Introduction

γ-aminobutyric acid (GABA) is the primary inhibitory neurotransmitter in the brain that activates both ionotropic $GABA_A$ receptors ($GABA_ARs$) and metabotropic $GABA_B$ receptors ($GABA_BRs$) (Chebib and Johnston 1999). While $GABA_ARs$ are chloride channels that mediate fast inhibitory postsynaptic currents (IPSCs), $GABA_BRs$ mediate slower inhibition by activating the $G_{i/o}$ G proteins and downstream second messengers (Odagaki and Koyama 2001; Chebib and Johnston 1999). $GABA_BRs$ are dimeric proteins consisting of two subunits, $GABA_BR1$ which contains the ligand-binding site, and $GABA_BR2$ which contains a binding site for allosteric modulators and is responsible for G protein coupling (Liu et al. 2004; Galvez et al. 1999, 2000; Kniazeff et al. 2002; Binet et al. 2004; Dupuis et al. 2006). The $GABA_BR2$ C-terminal domain also associates with auxiliary subunits called K^+ channel tetramerization domain (KCTD) proteins, which modulate the activation and desensitization kinetics of $GABA_BRs$ (Schwenk et al. 2010; Turecek et al. 2014; Seddik et al. 2012). Activation of postsynaptic $GABA_BRs$ leads to opening of G protein-activated inwardly rectifying K^+ (GIRK) channels (Luscher et al. 1997; Gahwiler and Brown 1985), resulting in membrane hyperpolarization and shunting excitatory currents. Activation of presynaptic $GABA_BRs$ leads to inhibition of voltage-gated Ca^{2+} (Ca_V) channels, suppressing neurotransmitter release (Thompson and Gahwiler 1992; Takahashi et al. 1998). In addition, the $G\alpha_{i/o}$ subunits inhibit adenylyl cyclase and cAMP-dependent signaling (Enna 2001). $GABA_BR$ activation contributes to synaptic plasticity, either facilitating or inhibiting long-term potentiation (LTP) depending on whether the $GABA_BRs$ act pre- or postsynaptically (Ulrich and Bettler 2007; Heaney and Kinney 2016; Morrisett et al. 1991; Olpe et al. 1993; Davies and Collingridge 1996; Davies et al. 1991). Conversely, neuronal activity also dynamically regulates $GABA_BR$ surface expression and functional signaling (Bettler and Tiao 2006). A growing body of evidence implicates $GABA_BR$ dysfunction in various psychiatric illnesses, and supports the therapeutic potential of

$GABA_BR$ ligands in treating these conditions, as well as movement and neurodegenerative disorders (Kumar et al. 2013; Bowery 2006). While this is discussed in more detail in other chapters of this book, here we focus on the involvement of $GABA_BR$ in drug addiction. Drugs of abuse have been shown to alter $GABA_BR$ signaling in many brain regions, which can last for prolonged periods of time. Some of these changes contribute to the development of behavioral and psychological manifestations of addiction, while others may be compensatory mechanisms to restore normal circuit function. The significant role the $GABA_BR$ plays in the reward circuit offers the possibility of using pharmacological $GABA_BR$ modulation to rescue circuit malfunction in addiction.

2 $GABA_BR$ Signaling in the Reward Circuit

Dopamine (DA) neurons in the VTA are crucial players in reward and motivated behaviors (Ranaldi 2014; Baik 2013). VTA DA neurons mediate the initial reinforcing effects of drugs of abuse, and long-lasting adaptations in their synaptic inputs and intrinsic activity partially underlie various behavioral manifestations of drug addiction (Luscher and Malenka 2011; Self 2004; Wanat et al. 2009; Francis et al. 2019). Glutamatergic inputs onto VTA DA neurons are known to be potentiated by drugs of abuse (see Sect. 5.3) (Overton et al. 1999; Saal et al. 2003; Ungless et al. 2001; Kalivas 1995). VTA DA neurons also receive GABAergic inputs from both local VTA GABA interneurons, as well as afferents from many subcortical regions including the nucleus accumbens (NAc), ventral pallidum (VP), rostromedial tegmental nucleus (RMTg), and lateral hypothalamus (LH) (Fig. 1) (Blacktop et al. 2016; Soden et al. 2020). Recent studies combining optogenetics and pharmacology have attempted to resolve the specific receptors mediating these inhibitory inputs (Nieh et al. 2015, 2016; Polter et al. 2018; Matsui et al. 2014; Edwards et al. 2017), which will be discussed in more detail below. Some other brain regions involved in reward function such as the amygdala, prefrontal cortex (PFC), lateral habenula, dorsal and median raphe nuclei also express $GABA_BR$s (Margeta-Mitrovic et al. 1999), but the functional role of $GABA_BR$ signaling in these regions in reward processes is less well studied.

2.1 The VTA Microcircuit

Both $GABA_AR$s and $GABA_BR$s are expressed on VTA DA and GABA neurons (Ciccarelli et al. 2012). There are some indications that the two receptor types are activated by distinct inputs to the VTA, since the $GABA_A$ and $GABA_B$ IPSPs occur independently of each other and are differentially modulated by pharmacological agents (Sugita et al. 1992). VTA DA and GABA neurons have different sensitivities to $GABA_A$ and $GABA_B$ agonists as a result of differences in receptor subunit

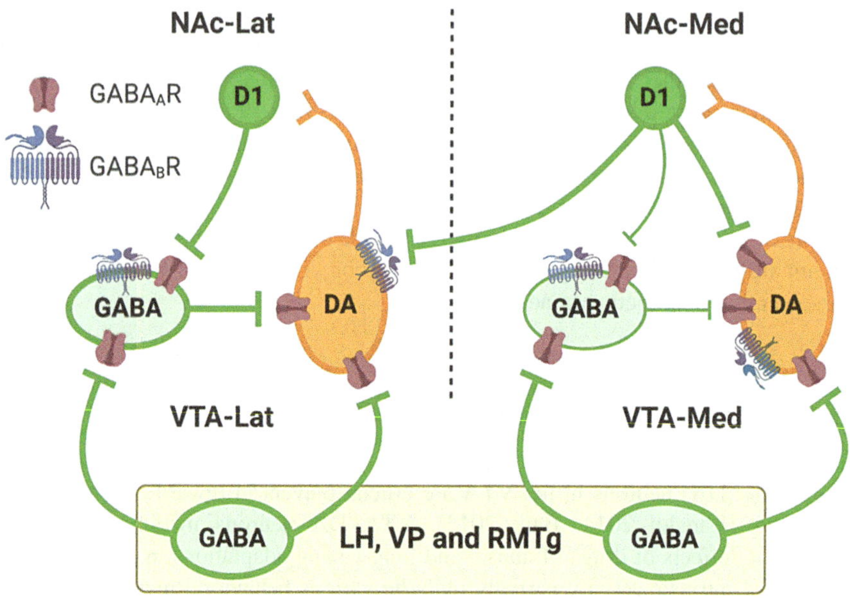

Fig. 1 Inhibitory inputs to VTA neurons. Main inhibitory inputs onto VTA dopamine neurons and the receptors that mediate inhibition are illustrated. D1: medium spiny neurons (MSN) expressing dopamine D1 receptor; GABA: VTA GABA interneurons; DA: VTA dopamine neurons; LH: lateral hypothalamus; VP: ventral pallidum; RMTg: rostromedial tegmental nucleus; Lat: lateral; Med: medial. Note that GABA$_A$R-mediated inhibitions from D1 MSNs to VTA GABA neurons and from VTA GABA neurons to VTA DA neurons are weaker in VTA-Med compared to VTA-Lat (Yang et al. 2018)

composition and coupling efficiency to GIRK channels, respectively (Tan et al. 2010; Cruz et al. 2004). Furthermore, VTA GABA interneurons inhibit DA neurons preferentially through GABA$_A$ receptors (Fig. 1) (Edwards et al. 2017; Polter et al. 2018). These facts should be taken into consideration when interpreting the effects of intra-VTA application of GABAergic drugs, as GABA$_A$R and GABA$_B$R agonists/antagonists can have bidirectional effects on DA neuron activity depending on whether the DA or GABA neurons are primarily affected (Laviolette and van der Kooy 2001; Laviolette et al. 2004; Cruz et al. 2004; Xi and Stein 1998). Nevertheless, in most in vivo studies GABA$_B$R ligands appear to bypass the VTA microcircuit and act directly on DA neurons. For example, infusion of baclofen (a GABA$_B$R agonist) locally into the VTA revealed decreases in DA release both within the VTA (Klitenick et al. 1992) and in the NAc (Westerink et al. 1996). DA release and locomotor activation in response to morphine, cocaine, and amphetamine were also blocked by intra-VTA baclofen (Kalivas et al. 1990; Klitenick et al. 1992; Leite-Morris et al. 2004). On the other hand, intra-VTA infusion of GABA$_B$R antagonist elicits an increase in DA level (Giorgetti et al. 2002). This GABA$_B$R-mediated inhibition of DA signaling serves as the basis for targeting GABA$_B$Rs in the treatment of drug addiction.

Recently, GABA$_B$R signaling in VTA DA neurons has been shown to be more nuanced than previously thought. Lateral VTA DA neurons that project to lateral NAc shell have significantly larger baclofen-induced currents than medial VTA DA neurons which project to medial NAc shell (Yang et al. 2018). Interestingly, GABA$_B$R1 expression detected by in situ hybridization is relatively uniform across different regions of the midbrain (Edwards et al. 2017), suggesting that other factors may determine the functional variability of GABA$_B$R signaling in subpopulations of VTA DA neurons.

2.2 GABAergic Inputs to the VTA

A wide range of subcortical regions send GABAergic projections to the VTA, and in most cases they preferentially synapse on local GABA neurons in the VTA (Soden et al. 2020). This pattern of innervation provides a potential mechanism of feedforward disinhibition of dopamine neurons which can switch them from tonic to burst firing to signal reward (Soden et al. 2020; Paladini and Tepper 1999). The major sources of GABAergic input to VTA with relevance to addiction are discussed below.

The RMTg is a GABAergic nucleus located just caudal to the VTA, and is therefore sometimes referred to as the tail of the VTA (tVTA) (Kaufling et al. 2009; Jhou et al. 2009b). GABAergic axons from RMTg neurons project to VTA/SNc DA neurons and provide tonic GABA$_A$R-mediated inhibition of DA neuron activity (Matsui and Williams 2011; Lecca et al. 2012). RMTg neurons are activated by noxious stimuli through innervation by glutamatergic lateral habenular neurons (Jhou et al. 2009a). Opioids and cannabinoids inhibit RMTg neurons and result in disinhibition of DA neurons (Lecca et al. 2012). The opioid sensitivity is specifically mediated by μ-opioid receptors, which are densely expressed on RMTg neurons (Jhou et al. 2009b; Matsui and Williams 2011). Interestingly, persistent induction of the transcription factor ΔFosB was observed in RMTg following chronic exposure to cocaine and amphetamine, but not morphine (Perrotti et al. 2005). Additional studies are needed to determine whether GABA release from RMTg afferents also activates GABA$_B$Rs.

The LH sends both glutamatergic and GABAergic axons to the VTA (Kallo et al. 2015; Nieh et al. 2015). While both DA and GABA neurons of the VTA receive excitatory input from LH, the GABAergic LH-VTA projection appears to preferentially target GABA neurons (Nieh et al. 2016). Activation of this GABAergic pathway results in disinhibition of VTA DA neurons and promotes approach behavior (Nieh et al. 2016). LH-VTA GABAergic projection also encodes cue-reward predictions to regulate learning (Sharpe et al. 2017). Picrotoxin-sensitive GABA$_A$R IPSCs have been observed in VTA neurons following LH activation, whereas the presence of GABA$_B$R IPSCs has not yet been tested (Nieh et al. 2016).

The VP has reciprocal connections with VTA (Faget et al. 2016; Taylor et al. 2014). GABAergic inputs from the VP provide a tonic inhibition on VTA DA neuron firing (Floresco et al. 2003), although this pathway can also be acutely activated by reward-associated cues (Smith et al. 2009). Silencing the GABAergic VP to VTA projection disrupts reinstatement of cocaine seeking, which is not recapitulated by $GABA_A$ antagonist (gabazine)-mediated direct disinhibition of VTA DA neurons, suggesting that the GABAergic VP afferents likely also inhibit non-DA neurons in the VTA (Mahler et al. 2014). Whether GABAergic VP afferents signal through $GABA_A$Rs or $GABA_B$Rs on VTA neurons has not been established.

In the NAc, dopamine D1 receptor-expressing medium spiny neurons (D1R-MSNs) send direct projections to the VTA, and comprise the "direct" pathway. Initial studies found that D1R-MSNs preferentially target non-DA neurons, in particular GABA neurons, even though a weak connection to DA neurons also exists (Bocklisch et al. 2013; Xia et al. 2011; Matsui et al. 2014). This pathway signals through $GABA_A$Rs and is inhibited by μ-OR activation (Xia et al. 2011; Matsui et al. 2014). Optogenetic stimulation of the D1R-MSN terminals failed to evoke $GABA_B$R IPSCs in VTA neurons, which can nevertheless be produced by electrical stimulation (Xia et al. 2011). A more recent study, however, identified a NAc to VTA DA connection that signals through $GABA_B$Rs (Edwards et al. 2017). The authors confirmed that NAc to VTA projection also inhibits GABA neurons via $GABA_A$Rs. Importantly, by combining immuno-electron microscopy and electrophysiological measurements on the spatio-temporal dynamics of $GABA_B$R activation, Edwards et al. (2017) showed that NAc axon terminals form symmetric synapses on DA cell bodies and dendrites, and that $GABA_B$Rs on DA neurons are activated by synaptic release of GABA. It will be important to determine whether the same population of NAc MSNs contact both DA and GABA neurons in the VTA. In contrast to the NAc input, VTA GABA neuron to DA neuron connection is mediated largely by $GABA_A$Rs, with only a small $GABA_B$R component (Edwards et al. 2017). Interestingly, VTA GABA neurons can inhibit presynaptic glutamate release onto DA neurons through $GABA_B$Rs, but not $GABA_A$Rs (Chen et al. 2015). $GABA_B$R-mediated inhibition of DA neuron terminal activity and DA efflux has also been reported in the NAc (Saigusa et al. 2012; Pitman et al. 2014; Xi et al. 2003).

A recent study by Yang et al. (2018) provided more precise anatomical resolution to the NAc-VTA connectivity on the medial-lateral axis (Fig. 1). Medial VTA (VTA-Med) DA neurons form reciprocal connections with NAc medial shell (NAc-Med), while lateral VTA (VTA-Lat) DA neurons are reciprocally connected with NAc lateral shell (NAc-Lat) (Yang et al. 2018). Interestingly, the two pathways show little overlap. Importantly, D1R-MSNs from NAc-Lat inhibit non-DA neurons (likely GABA neurons) more strongly than DA neurons in VTA-Lat, resulting in an overall excitation of VTA-Lat DA neurons (Yang et al. 2018). In contrast, NAc-Med D1R-MSNs equally target DA and non-DA neurons and when stimulated result in an overall inhibition of VTA-Med DA neurons. The inhibitory control by D1R-MSNs described above is mediated by $GABA_A$Rs, while $GABA_B$R-mediated inhibition is induced by NAc-Med, but not NAc-Lat stimulation in about half of NAc-Lat projecting DA neurons, but not in NAc-Med projecting DA neurons (Yang et al. 2018). The amplitudes of $GABA_B$R IPSCs (Yang et al. 2018) were smaller than in

Edwards et al. (2017), a difference that could partly be explained by a more widespread expression of ChR2 virus. What is the functional role of this NAc-Med to VTA-Lat $GABA_BR$ pathway? Intra-VTA antagonism of $GABA_BR$ with CGP 35348 removed the anxiety-like phenotype induced by NAc-Med D1R-MSN stimulation, and at the same time revealed a rewarding effect of the stimulation (Yang et al. 2018). This is consistent with $GABA_BRs$ having a reward-suppressing effect in the VTA (Willick and Kokkinidis 1995).

In summary, while a wide range of GABAergic inputs to the VTA has been identified, their full characterization remains incomplete. Currently, it is well established that these inputs can activate $GABA_ARs$ on VTA DA and GABA neurons to drive direct inhibition and disinhibition of DA neurons, respectively. However, with the exception of NAc, the ability of these inputs to activate $GABA_BRs$ in the VTA and contribute to drug addiction has not been explored, despite ample evidence that VTA neuron activity is modulated by $GABA_BR$ ligands. Future studies should address this gap in order to understand more completely GABAergic control of DA neuron activity.

3 Impact of Addictive Drugs on $GABA_BR$ Signaling in the Reward Circuit

$GABA_BR$ signaling undergoes activity-dependent plasticity and drug-evoked changes. A recent review by Lalive and Lüscher (2016) provides a good description of these processes and some underlying mechanisms. Here, we focus on the impact of these drug-induced alterations in $GABA_BR$ signaling on neurotransmission in the brain reward circuit.

3.1 Psychostimulants

Cocaine acutely modulates $GABA_BR$ signaling in the VTA in a bidirectional manner. In acutely prepared VTA slices, relatively low concentrations of cocaine (~0.1 μM) inhibit $GABA_BR$ inhibitory postsynaptic potentials (IPSPs) on VTA DA neurons via presynaptic 5-HT receptors, leading to disinhibition of DA neurons (Cameron and Williams 1994). On the other hand, cocaine at 1 μM or higher concentrations blocks dopamine reuptake and increases the level of extracellular dopamine, which activates D1 receptors on afferent GABA terminals to facilitate $GABA_BR$ IPSP (Cameron and Williams 1993, 1994; Lacey et al. 1990; Brodie and Dunwiddie 1990). Apart from immediate action, psychostimulants can also lead to long-lasting changes in $GABA_BR$ signaling, depending on the length of drug exposure (acute vs. chronic) and time of measurement (early vs. late withdrawal). A single injection of cocaine (15 or 30 mg/kg) decreases $GABA_BR$-GIRK signaling in VTA DA neurons for 3–4 days, possibly through a downregulation of surface GIRK channels (Arora et al. 2011). Furthermore, a single injection of cocaine (15 mg/kg) or amphetamine (10 mg/kg) increased AMPAR/NMDAR ratio in DA neurons, but not GABA neurons in the VTA (Saal et al. 2003; Ungless et al. 2001). This results in an altered excitation/inhibition balance on VTA DA neurons, enhancing DA signaling in the short term. On the other hand, a single cocaine or methamphetamine injection also diminished $GABA_BR$ signaling in VTA GABA neurons (Padgett et al. 2012), perhaps leading to more local GABA release as a compensatory mechanism to dampen VTA DA activity.

Chronic psychostimulant exposure results in altered $GABA_BR$ function within the VTA and other brain regions. $GABA_BR$-GIRK signaling in VTA DA neurons was diminished by either methamphetamine self-administration (Sharpe et al. 2014) or repeated non-contingent injections (Munoz et al. 2016). Suppression of $GABA_BR$-GIRK signaling was also observed in layer 5/6 pyramidal neurons of dorsal medial prefrontal cortex (mPFC) following repeated cocaine treatment (Hearing et al. 2013). This is consistent with the impaired ability of intra-mPFC baclofen to inhibit glutamate transmission (Jayaram and Steketee 2004) and cocaine-induced locomotor activity (Steketee and Beyer 2005). In the VTA, however, microdialysis analysis of somatodendritic DA and glutamate release in freely moving rats revealed a different story. Intra-VTA infusion of $GABA_BR$ antagonist CGP 55845A showed that at baseline, DA release is under tonic $GABA_BR$ inhibition while glutamate release is not (Giorgetti et al. 2002). After repeated amphetamine injections, both DA and glutamate release in the VTA was affected by increased $GABA_BR$ inhibitory tone (Giorgetti et al. 2002). Interestingly, this was seen during the early withdrawal period (3 days after last drug injection) but not at late withdrawal times (10–14 days after last drug injection), consistent with the finding that synaptic plasticity in the VTA that accompany psychostimulant sensitization is transient (Zhang et al. 1997). In the dorsolateral septal nucleus (DLSN), a brain region also implicated in reward function (Olds and Milner 1954), presynaptic $GABA_BR$ function was impaired after prolonged cocaine exposure, leading to enhancement of GABA and glutamate release in this area (Shoji et al. 1997). On the other hand, postsynaptic $GABA_BR$ function was not altered by chronic cocaine in DLSN (Shoji et al. 1997). In short, alterations in $GABA_BR$ function by psychostimulants can vary by brain region and timing.

Chronic cocaine exposure also decreased $G_{i\alpha}$ and $G_{o\alpha}$ G protein levels in VTA, NAc, and locus coeruleus (LC) (Nestler et al. 1990), and decreased functional coupling of $GABA_BRs$ to G proteins in VTA and NAc (Kushner and Unterwald 2001; Xi et al. 2003). Similarly, chronic amphetamine also attenuated $GABA_BR$ G protein coupling in NAc in late withdrawal (Zhang et al. 2000). Altogether, the diminished $GABA_BR$ signaling promotes behavioral sensitization to the drug, even though $GABA_BR$ expression may remain unaltered (Li et al. 2002a). Cocaine withdrawal can also indirectly reduce $GABA_BR$ transmission through impairing presynaptic GABA release, as seen at entopeduncular nucleus to lateral habenula (LHb) synapses (Tan et al. 2018). Reduced $GABA_BR$ transmission leads to higher excitability of LHb neurons, and in combination with enhanced excitatory input as a result of diminished $GABA_AR$ signaling, they contribute to the negative symptoms of cocaine withdrawal (Tan et al. 2018).

Animals that self-administer drugs of abuse can develop neuroadaptations associated with motivated responding and craving for the drug that are absent in animals passively receiving administered drug. Autoradiographic analysis using a $GABA_BR$ antagonist, [^{3}H]CGP 54626 revealed a large decrease in binding in a wide range of brain regions following a 10-day withdrawal in animals self-administering cocaine, but not in yoked animals that received non-contingent injections of cocaine (Frankowska et al. 2008). Decreases in GABABR binding were observed in the PFC, dorsal striatum, NAc, amygdala, hippocampus, VTA and substantia nigra, all regions important in the reward circuit. Whether the decreased $GABA_BR$ levels in these areas contribute to craving and relapse will require further study.

3.2 Nicotine

To date, there are few reports describing the impact of nicotine on $GABA_BR$ signaling. One group found that subcutaneous injection of 0.4 mg/kg nicotine for 14 days in rats abolished the baclofen-mediated inhibition of electrically evoked DA release from VTA slices (Amantea and Bowery 2004), and reduced $GABA_BR$ coupling to G proteins in the mPFC and NAc (Amantea et al. 2004). However, another study that chronically infused a much higher dose of nicotine (3.16 mg/kg/day for 7 days prior to and during testing) found that the ability of intra-VTA $GABA_BR$ agonist to elevate the threshold for intracranial self-stimulation (ICSS) is similar in vehicle and nicotine-treated rats (Paterson et al. 2005a, b). While the first two studies examined $GABA_BR$ function during nicotine withdrawal, the last study was done during nicotine administration, which may contribute at least partly to the discrepancies in these findings.

$GABA_BR1$ and $GABA_BR2$ RNA expression levels have been examined in rats chronically exposed to oral nicotine or cigarette smoke. A decrease in $GABA_BR1$ expression was observed in the hippocampus following both chronic nicotine and cigarette smoke (Li et al. 2002b). Interestingly the length of smoke exposure was negatively correlated with the degree of decrease in $GABA_BR1$ RNA. In the PFC, chronic nicotine led to a small decline in both $GABA_BR1$ and $GABA_BR2$ RNA expression, while cigarette smoke increased the level of both receptor subunits (Li et al. 2004). Whether the discrepancies are due to different route of exposure or ingredients other than nicotine in the cigarette smoke remains to be investigated.

3.3 Opioids

Acute morphine promotes glutamate release onto VTA DA neurons by removing $GABA_BR$ inhibition of glutamatergic inputs (Chen et al. 2015). Chronic morphine treatment, however, increases $GABA_BR$ signaling in the VTA during withdrawal, which in turn reduces glutamate release and leads to a decrease in DA neuron activity (Manzoni and Williams 1999). This may result from 1) enhanced GABA release during withdrawal (Bonci and Williams 1997) and 2) increased coupling efficiency of $GABA_BR$ to GIRK channels through downregulation of RGS2, a regulator of G protein signaling (Labouebe et al. 2007). Chronic infusion of morphine via osmotic pumps has also been reported to increase both $GABA_BR1$ and $GABA_BR2$ immunoreactivity in the globus pallidus (GP) and substantia nigra pars reticulata (SNr) although the functional implication of these changes is still unclear (Negrete-Diaz et al. 2019). On the other hand, chronic morphine treatment has also been shown to decrease the activity of inhibitory $G_{i/o}$ G proteins in the locus coeruleus (Selley et al. 1997). Since $GABA_BR$ and μ-opioid receptor both couple to $G_{i/o}$ G proteins, $GABA_BR$ signaling is potentially affected in this region and should be directly tested.

3.4 Cannabis

Tetrahydrocannabinol (THC) is the principal psychoactive cannabinoid in cannabis and is a partial agonist for the cannabinoid receptor CB1, which signals via $G_{i/o}$ as does $GABA_BR$ (Howlett et al. 1986; Bidaut-Russell et al. 1990). Chronic THC treatment resulted in CB1 downregulation and desensitization in most brain regions, whereas $GABA_BR$-stimulated G protein activation was not affected (Sim et al. 1996; Selley et al. 2004). However, a heterologous attenuation of adenylyl cyclase inhibition was observed in mouse cerebellum for $GABA_BR$ and adenosine A1 receptor following long-term THC treatment (Selley et al. 2004). This may contribute to the cross-tolerance to motor coordination deficits with cannabinoid, $GABA_BR$, and A1R agonists. Garcia-Gil et al. (1999) demonstrated that perinatal exposure to THC in male and female rats potentiates the motor inhibitory effect of baclofen in adult animals, despite the fact that baclofen-stimulated G protein activation was not changed, especially in the substantia nigra (Garcia-Gil et al. 1999). Whether $GABA_BR$ signaling in the reward circuit is altered by THC remains to be investigated.

3.5 Summary

In summary, drugs of abuse induce plasticity in $GABA_BR$ signaling in various parts of the brain, and these changes can take place at the level of RNA expression, receptor trafficking, G protein coupling, as well as the effector proteins. It appears that some of the changes contribute to the progression to addiction, while others may be compensatory responses to other changes in neurotransmission. Studies on psychostimulant-induced alterations in $GABA_BR$ signaling have been most abundant, whereas for other drugs we are still at the beginning of such investigations. It is important to figure out which of the plastic changes are universal mechanisms to all drugs of abuse, and which ones are drug-specific. Such knowledge will be beneficial for devising therapeutic interventions targeted at $GABA_BRs$.

4 Effects of Genetic Manipulations of $GABA_BR$ on Animal Models of Drug Addiction

Mice lacking either the $GABA_BR1$ or $GABA_BR2$ subunit have been generated and studied on the Balb/c genetic background (Gassmann et al. 2004; Schuler et al. 2001). Unfortunately, premature death in early adulthood was observed in knockout (KO) mice on other backgrounds such as C57BL6/J, preventing behavioral characterization (Prosser et al. 2001). Both $GABA_BR1$ and $GABA_BR2$ KO mice display hyperlocomotion, hyperalgesia, and spontaneous epileptic seizures. $GABA_BR1$ KO mice also have an anxious and antidepressant-like phenotype (Mombereau et al. 2004). Relatively few studies have examined the effects of $GABA_BR$ KO mice on drug addiction.

A series of studies conducted by Varani et al. on the $GABA_BR1$ KO in combination with pharmacological manipulations of $GABA_BRs$ have demonstrated a role for $GABA_BRs$ in mediating acute effects of nicotine as well as nicotine withdrawal symptoms with respect to reward and anxiety (Varani et al. 2012, 2014, 2015, 2018). $GABA_BRs$ are required for the acute locomotor, antinociceptive, and anxiolytic effects of nicotine, as well as the neurochemical changes that give rise to anxiety in nicotine withdrawal. On the other hand, $GABA_BR$ activity counteracts the rewarding effect of nicotine (Le Foll et al. 2008; Varani et al. 2018). While the specific brain regions where $GABA_BR$ activity is important for the nicotine effects are still elusive, nicotine-dependent changes in c-fos and BDNF expression in several brain areas involved in anxiety are absent in $GABA_BR1$ KO (Varani et al. 2012, 2014, 2015), and nicotine-induced increase in c-fos expression in NAc and VTA is potentiated in $GABA_BR1$ KO (Varani et al. 2018).

The two alternatively spliced isoforms of $GABA_BR1$ subunit, $GABA_BR1a$ and $GABA_BR1b$, localize to the presynaptic and postsynaptic membrane, respectively (Biermann et al. 2010; Vigot et al. 2006). $GABA_BR1b$ lacks the N-terminal sushi repeats that mediate axonal targeting in $GABA_BR1a$ (Kaupmann et al. 1997;

Biermann et al. 2010). In the hippocampus, $GABA_BR1a$-containing heteroreceptors mediate inhibition of glutamate release while $GABA_BR1b$-containing heteroreceptors mediate postsynaptic inhibition (Vigot et al. 2006). Isoform-specific knockouts showed differential effects on locomotor response to cocaine: While $GABA_BR1b$ KO mice had higher basal locomotion, they failed to develop sensitization; $GABA_BR1a$ KO had higher acute locomotor response as well as sensitization to cocaine (Jacobson et al. 2016).

Selective deletion of $GABA_BRs$ in VTA DA neurons has been carried out by injecting an adeno-associated virus expressing Cre recombinase under the tyrosine hydroxylase (TH) promoter (AAV-TH-iCre) into the VTA of mice with floxed $GABA_BR1$. This manipulation decreased baclofen currents in DA neurons and enhanced locomotor response to cocaine, while basal locomotion and morphine-induced locomotion were unaltered (Edwards et al. 2017). Rifkin et al. (2018) used a different approach to investigate the effect of reduced $GABA_BR$-activated GIRK currents on drug sensitivity. A conditional knockout of sorting nexin 27, which regulates surface expression of GIRK channels, led to reduced baclofen-induced currents in VTA DA neurons and increased sensitivity to cocaine-dependent locomotor sensitization (Rifkin et al. 2018).

5 Effects of Pharmacological Manipulations on Animal Models of Drug Addiction: Agonists, Antagonists, and Positive Allosteric Modulators (PAMs)

Research in the past few decades has yielded substantial evidence for the therapeutic potential of $GABA_BR$ ligands in treating a variety of neurological disorders (Bowery 2006; Heaney and Kinney 2016; Jacobson et al. 2018; Kumar et al. 2013). At the moment, baclofen is the only $GABA_BR$ agonist approved for human use as a muscle relaxant. In 2018, it was also approved in France for treating alcohol use disorder, despite controversy regarding its efficacy and safety (Braillon et al. 2020; de Beaurepaire et al. 2018). Recently several groups have attempted to validate its effect for treating abuse of addictive substances by using animal models. The common hypothesis is that activating $GABA_BRs$ on VTA DA neurons dampen their activity, thus reducing the rewarding effect of drugs of abuse and suppressing the motivation to work for them. Indeed, baclofen has been shown to dose-dependently reduce DA release in the NAc shell evoked by cocaine, nicotine, and morphine (Fadda et al. 2003). $GABA_BR$ agonists other than baclofen as well as antagonists have also been studied in the context of drug addiction. A thorough review of the chemistry and pharmacology of these ligands can be found in a recent book chapter (Froestl 2010). A potential issue with $GABA_BR$ orthosteric agonists (e.g., baclofen) is the induction of side effects, including sedation, weakness, vertigo, and headache (Agabio and Colombo 2015; Tyacke et al. 2010). Positive allosteric modulators (PAMs) may provide a remedy for this problem. In general,

Table 1 List of GABA$_B$R PAMs and chemical names

Name	Chemical name	Structure	Citation
BHF177	*N*-[(1*R*,2*R*,4*S*)-bicyclo[2.2.1]hept-2-yl]-2-methyl-5-[4-(trifluoromethyl)phenyl]-4-pyrimidinamine	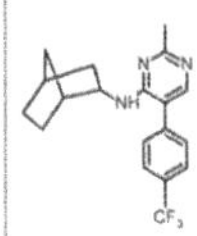	Guery et al. (2007)
KK-92A	(4-(cycloheptylamino)-5-(4-(trifluoromethyl)-phenyl)pyrimidin-2-yl) methanol (analog of BHF177)	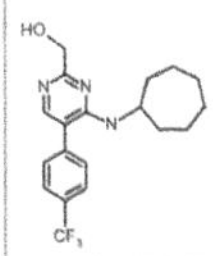	Li et al. (2017)
NVP998	(4-(bicyclo(2.2.1)heptylamino()-5-(4-(trifluoromethyl)phenyl)pyrimidin-2-yl) nitrile (analog of BHF177)	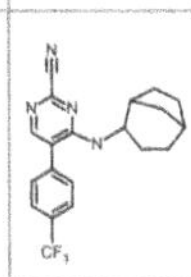	Sturchler et al. (2017)
CGP7930	3-(3′,5′-Di-*tert*-butyl-4′-hydroxy)phenyl-2,2-dimethylpropanol	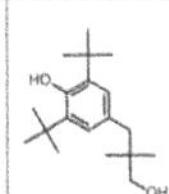	Urwyler et al. (2001)
CMPPE	2-{1-[2-(4-chlorophenyl)-5-methylpyrazolo[1,5-a] pyrimidin-7-yl]-2-piperidinyl}ethanol	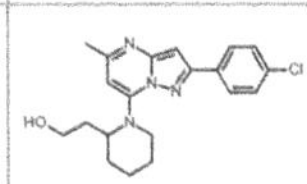	Perdona et al. (2011)
GS39783	*N*,*N*′-Dicyclopentyl-2-methylsulfanyl-5-nitro-pyrimidine-4,6-diamine	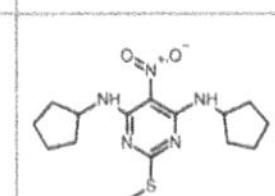	Urwyler et al. (2003)
rac-BHFF	(*R*,*S*)-5,7-di-*tert*-butyl-3-hydroxy-3-trifluoromethyl-3*H*-benzofuran-2-one	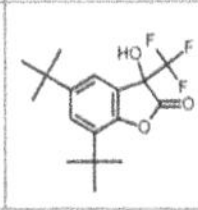	Malherbe et al. (2008)

allosteric modulators bind to a region of the GABA$_B$R outside of the ligand-binding site, where they induce conformational changes that either increase (i.e., positive, PAMs) or decrease (i.e., negative, NAMs) the effects of GABA. Therefore, they do not possess intrinsic agonistic activity, but can modulate the effect of endogenous GABA release on GABA$_B$Rs. As a result, they are expected to produce fewer side effects compared to orthosteric GABA$_B$R agonists (Filip et al. 2015). PAMs also appear to activate the receptor without inducing desensitization, which is partially responsible for the development of tolerance (Sturchler et al. 2017). CGP7930 (Urwyler et al. 2001) and GS39783 (Urwyler et al. 2003) were the first GABA$_B$R PAMs discovered through compound screening in a GTPγ^{35}S binding assay (Table 1). They increased both the potency and efficacy of GABA at GABA$_B$R, suggesting that they enhance not only agonist binding, but also receptor-G protein coupling (Urwyler 2011). The GABA$_B$R PAM binding site was mapped to the transmembrane domain of the GABA$_B$R2 subunit (Binet et al. 2004; Dupuis et al. 2006), Since then, new GABA$_B$R PAMs have been identified by screening compound libraries or structural analogs of existing PAMs in GTPγ^{35}S binding or functional $GABA_BR$ signaling assays (Table 1). In this section we will review studies that tested the effects of $GABA_BR$ ligands and PAMs in animal models of addiction for the common drugs of abuse. There are also a few other recent reviews on this topic (Filip et al. 2015; Filip and Frankowska 2008; Vlachou and Markou 2010; Phillips and Reed 2014).

5.1 Psychostimulants

5.1.1 Amphetamine and Methamphetamine

Baclofen given systemically (2–4 mg/kg) dose-dependently blocks the development and expression of locomotor sensitization to amphetamine (Bartoletti et al. 2004, 2005; Cedillo and Miranda 2013). A $GABA_BR$ PAM, CGP 7930, potentiated the effect of low dose (2 mg/kg) baclofen in blocking both the development and expression of amphetamine sensitization (Cedillo and Miranda 2013). GS39783, another $GABA_BR$ PAM, blocked expression of amphetamine conditioned place preference (CPP) without affecting locomotion (Halbout et al. 2011). In rats self-administering amphetamine, 1.8–5.6 mg/kg baclofen reduced responding under both fixed ratio (FR) and progressive ratio (PR) schedule of reinforcement, which was accompanied by an attenuation of amphetamine-induced DA level increase in the NAc (Brebner et al. 2005).

Baclofen (1.25–5 mg/kg) dose-dependently attenuated both the acquisition and expression of methamphetamine CPP (Li et al. 2001). Methamphetamine-induced CPP is resistant to extinction, but 2 mg/kg baclofen given after daily extinction sessions facilitated the extinction (Voigt et al. 2011a). Interestingly, two home cage injections of $GABA_BR$ PAMs (CGP 7930 or GS39783) in rats with established methamphetamine CPP abolished the expression of CPP in a subsequent test (Voigt et al. 2011a, b). This implicates $GABA_BRs$ in the short-term maintenance of memories associated with methamphetamine conditioning. Regarding methamphetamine self-administration, 2.5 and 5 mg/kg baclofen injection reduced break points for all doses of methamphetamine on a PR schedule, indicating diminished motivation to work for the drug without apparent motor impairment (Ranaldi and Poeggel 2002). There is also evidence that baclofen rescues cognitive deficits induced by methamphetamine. For example, methamphetamine-induced impairments of prepulse inhibition and novel object recognition were ameliorated by systemic baclofen at 2 mg/kg (Arai et al. 2009).

5.1.2 Cocaine

Baclofen (2 mg/kg) prevented cocaine-conditioned locomotion as well as stimulus-induced glutamate release in the NAc (Hotsenpiller and Wolf 2003). Both the development and expression of locomotor sensitization to cocaine were also reduced by baclofen as well as another $GABA_BR$ agonist, SKF 97541 (Frankowska et al. 2009). Intra-mPFC baclofen blocked acute locomotor activation by cocaine and the development of sensitization, but not the expression of sensitization (Steketee and Beyer 2005). The $GABA_BR$ PAM GS39783 modestly attenuated the development of cocaine sensitization, and also blocked cocaine-induced activation of DARPP-32 and CREB in the NAc (Lhuillier et al. 2007).

The impact of $GABA_BR$ ligands on cocaine self-administration has been extensively studied for more than two decades. Roberts et al. (1996) showed in rats trained to self-administer IV cocaine and tested on a progressive ratio task, that 2.5 mg/kg baclofen significantly reduced the break points for all cocaine doses (0.18 to 1.5 mg/kg/inj). In comparison, baclofen only slightly decreased responding for food reward. This suggests that at low dosages, baclofen can specifically dampen the reinforcing effect of cocaine without causing sedation or general disruption of performance, which typically emerges at 5 mg/kg (Roberts et al. 1996). In a subsequent study, the authors used a discrete trials procedure to probe baclofen's effect on the initiation of cocaine self-administration (Roberts and Andrews 1997). When each discrete trial was separated by 30 min, rats showed a clear diurnal pattern of cocaine taking where most of the infusions (1.5 mg/kg/inj) happen during the dark phase. The likelihood of cocaine taking is low but increasing at the beginning of the dark phase, and much higher in the middle of the dark phase. Baclofen (2.5 mg/kg) injection given at either time point strongly suppressed drug taking. A range of baclofen doses (1.25–5.0 mg/kg) were able to suppress cocaine taking while having no effect on food taking on a second lever.

A study from Campbell et al. (1999) showed that baclofen treatment prior to the 7-h cocaine self-administration session (FR1) suppressed cocaine intake in a dose-dependent manner. 2.5 and 5 mg/kg baclofen were effective in reducing cocaine infusions especially in the first 4 h of the session. The suppression effect was also stronger for the lower dose of cocaine (0.2 mg/kg/inj) compared to the higher dose (0.4 mg/kg/inj). In comparison, cocaine-primed reinstatement was more effectively blocked by baclofen at 1.25 & 2.5 mg/kg. Similarly, the Roberts group also found that baclofen had a greater effect in suppressing lower unit dose (0.75 mg/kg/inj) of cocaine intake on a FR1 schedule. Interestingly, the suppression can be attributed to a period of non-responding at the beginning of the session, instead of a reduced rate of responding (Brebner et al. 2000a). Regarding cue-induced reinstatement of cocaine seeking, 2.5 & 5 mg/kg baclofen was able to suppress responding without affecting food seeking. However, it is less effective on cocaine+cue primed reinstatement (Froger-Colleaux and Castagne 2016).

To gain more insights on the mechanism of baclofen-mediated reduction in drug intake/seeking, local infusions of baclofen into different brain regions have been carried out in several studies. One study found that intra-VTA baclofen injection was three times more potent than intra-NAc or intra-striatum baclofen in suppressing cocaine self-administration (1.5 mg/kg/inj) on a progressive ratio task (Brebner et al. 2000b). Another study showed that intra-NAc or intra-VTA injection of baclofen, but not intra-dorsal striatum injection, decreased responding for cocaine (0.66 mg/kg/inj) on an FR5 schedule (Shoaib et al. 1998). Interestingly, in a more recent report by Backes and Hemby (2008), intra-VTA injection of the $GABA_AR$ antagonist picrotoxin was found to inhibit cocaine self-administration to an extent comparable to that seen with baclofen, and that the effect of picrotoxin could be blocked by the $GABA_BR$ antagonist 2-hydroxysaclofen. This suggests that picrotoxin likely acts by disinhibiting VTA GABA neurons in the VTA, which in turn release GABA that act on $GABA_BRs$ on DA neurons.

CGP 44532, a highly selective and high-affinity $GABA_BR$ agonist, dose-dependently decreased cocaine self-administration on a progressive ratio schedule as well as on a discrete trials procedure with little effect on food self-administration (Brebner et al. 1999), similar to what has been reported for baclofen (Roberts and Andrews 1997; Brebner et al. 2000a). This is likely not due to an anhedonia state induced by CGP 44532, since it was found to be hedonically neutral in a brain stimulation reward (BSR) rate-frequency paradigm (Dobrovitsky et al. 2002). Cocaine-induced potentiation of BSR, however, was reduced by CGP 44532 in a dose-dependent manner (Dobrovitsky et al. 2002). When tested in baboons, CGP 44532 decreased cocaine-primed reinstatement of cocaine seeking to a degree comparable to that seen for baclofen (Weerts et al. 2007). However, both agonists reduced responding for food as much as for cocaine under an FR10 schedule (Weerts et al. 2005).

Several $GABA_BR$ PAMs have been tested on cocaine self-administration, usually in direct comparison with $GABA_BR$ agonists. $GABA_BR$ PAM CGP 7930 has been shown to reduce cocaine self-administration without affecting food responding across a range of doses (10–100 mg/kg), while high doses of the $GABA_BR$ agonists baclofen (5 mg/kg) and SKF 97541 (0.3 mg/kg) decreased food responding (Filip et al. 2007). A similar pattern was found for cocaine or cue-induced reinstatement, where CGP 7930 dose-dependently reduced cocaine seeking but not food seeking, while both agonists reduced cocaine but also food seeking at high doses (Filip and Frankowska 2007). Intriguingly, the same group found that while SCH 50911 ($GABA_BR$ antagonist) blocked the effects of $GABA_BR$ agonists and PAM on cocaine self-administration, it also decreased cue-induce reinstatement for cocaine but not for food (Filip and Frankowska 2007; Filip et al. 2007). Whether the latter phenomenon may be due to a possible rewarding property of SCH 50911 or its partial agonistic properties at the $GABA_BRs$ is yet uncertain. Another $GABA_BR$ PAM GS39783 reduced responding for cues previously paired with cocaine when administered i.p. at 30 and 100 mg/kg, and did not suppress locomotor activity at the lower dose (30 mg/kg) (Halbout et al. 2011). In contrast, baclofen significantly decreased locomotion at the minimal effective dose (2.5 mg/kg) for reducing cocaine seeking (Halbout et al. 2011). GS39783 also attenuated cocaine's rewarding effect in that it reduced the threshold lowering effect of cocaine on ICSS (Slattery et al. 2005). Unlike baclofen which elevated ICSS threshold, GS39783 has no effect on the threshold, suggesting that it is hedonically neutral. When compared side-by-side, CGP 7930 is somewhat more effective than GS39783 in reducing responding for cocaine across different self-administration schedules, including PR, FR1 as well as discrete trials (DT), a difference that may stem from the difference in drug bioavailability (Smith et al. 2004). CMPPE, a relatively new $GABA_BR$ PAM (Perdona et al.

2011) abolished cue-induced cocaine reinstatement with no sign of sedation or body weight loss, which would be observed with a high dose (3 mg/kg) of baclofen (Vengeliene et al. 2018). Non-sedative doses of the $GABA_BR$ PAM rac-BHFF (Malherbe et al. 2008) attenuated cocaine self-administration and prevented the cocaine-induced increase in AMPAR/NMDAR ratio in VTA DA neurons (de Miguel et al. 2018).

An important caveat of these self-administration studies is that they were performed on male animals only, even though in humans, women are more likely to progress from initial drug use to addiction (Van Etten and Anthony 1999). So far only a few papers have touched on the sex differences in the effect of $GABA_BR$ ligands on cocaine self-administration. It was noted that baclofen has a greater effect in decreasing the acquisition rate and percentage of cocaine self-administration in female rats than in males (Campbell et al. 2002). The relationship between cocaine use in adolescent and adulthood is also sexually dimorphic: males that display escalating self-administration in adolescence develop habit-based inflexible behaviors in adulthood, while in females it was not the escalating animals, but ones with low response rates that later develop behavioral inflexibility (DePoy et al. 2016). Baclofen similarly reduced drug seeking in these animals when they were re-exposed to cocaine-associated context in adulthood (DePoy et al. 2016). When female rats were selectively bred for high (HiS) versus low (LoS) saccharin intake, a pattern emerged in that HiS rats exhibited more cocaine self-administration and higher reinstatement than LoS rats. While baclofen was effective in reducing cocaine-primed reinstatement in both groups, it potentiated cocaine intake escalation in HiS rats but attenuated it in LoS rats during long-access sessions (Holtz and Carroll 2011). This highlights the potential complication of individual variability which is especially relevant in human clinical trials in interpreting the effects of $GABA_BR$ modulation on drug addiction. More studies are needed in the future on sex differences for drug addiction and treatment.

5.2 *Nicotine*

A recent review by Jacobson et al. (2018) provides an excellent summary of preclinical studies on the effects of $GABA_BR$ compounds on nicotine addiction. Nicotine can directly activate VTA DA neurons through nicotinic acetylcholine receptors, or enhance glutamate transmission onto DA neurons (Mansvelder et al. 2002; Zhao-Shea et al. 2011). Therefore, it is conceivable that $GABA_BR$ activation on VTA DA neurons should counteract the reinforcing effect of nicotine. Indeed, intra-VTA infusion of baclofen as well as the $GABA_AR$ agonist muscimol reduced nicotine self-administration in rats (Corrigall et al. 2000). Systemic baclofen and CGP 44532 have been reported to decrease nicotine self-administration on both FR and PR schedules, block nicotine CPP, and suppress cue-induced reinstatement although high doses of these drugs frequently affect responding for food as well (Paterson et al. 2004; Paterson et al. 2005b; Le Foll et al. 2008). Interestingly, the anxiolytic and anxiogenic effects produced by low and high doses of nicotine require $GABA_BR$ activity, since they are blocked by saclofen (Varani and Balerio 2012), consistent with the absence of nicotine's anxiolytic and anxiogenic responses in the $GABA_BR1$ KO (Varani et al. 2012, 2014).

$GABA_BR$ PAMs generally show more specificity than $GABA_BR$ agonists in attenuating nicotine's stimulant and rewarding effects or nicotine self-administration without nonspecific effects on locomotion or food seeking. GS39783 was able to block the acquisition of nicotine CPP and nicotine-induced ΔFosB in the NAc (Mombereau et al. 2007); it abolished nicotine-induced hyperlocomotion without affecting basal locomotion (Lobina et al. 2011), and it potentiated the effect of a sub-effective dose of the $GABA_BR$ agonist CGP 44532 on nicotine reinforcement without affecting food responding (Paterson et al. 2008). CGP 7930 also specifically blocked nicotine's locomotor and rewarding effects (Lobina et al. 2011; Paterson et al. 2008). BHF177, a $GABA_BR$ PAM characterized by Guery et al. (2007) (Table 1), decreased responding for nicotine under FR and PR schedules without affecting food responding, blocked cue-induced nicotine but not food seeking, and abolished the reward-enhancing effect of nicotine on ICSS (Paterson et al. 2008; Vlachou et al. 2011). A novel analog of BHF177 named KK-92A exhibited even better selectivity for inhibiting nicotine taking and seeking over food responding (Li et al. 2017). A recent study by Sturchler et al. compared the efficacies of GS39783, BHF177 and a structural analog NVP998 both in vitro and in vivo (Sturchler et al. 2017). Interestingly these PAMs exhibit different functional influence on intracellular signaling pathways and their effects show species selectivity. NVP998 displays highest PAM activity at the human receptor while it has no effect on nicotine self-administration in rats. This underscores the importance of testing potential drugs in nonhuman primate models before moving on to clinical research.

5.3 *Opioids*

The main mechanism of opioid reward is through inhibition of midbrain GABA neurons via μ-opioid receptors, and resulting disinhibition of VTA DA neuron activity (Johnson and North 1992; Corre et al. 2018; Margolis et al. 2014; Matthews and German 1984; Gysling and Wang 1983). Not surprisingly, $GABA_BR$ agonists have been reported to block the initial rewarding effect of opioids, and the effect can be localized to the VTA. For example, morphine-induced CPP was dose-dependently inhibited by intra-VTA infusion of baclofen prior to morphine injection during acquisition, an effect that was suppressed by the $GABA_BR$ antagonist saclofen (Tsuji et al. 1996). Morphine-induced locomotor activation and sensitization, as well as NAc c-fos activation were also dose-dependently blocked by intra-VTA baclofen (Leite-Morris et al. 2004). Baclofen given systemically (2.5 & 5 mg/kg) likewise inhibited the development of morphine sensitization and morphine-induced DA release in the NAc (Fu et al. 2012). In animals that have developed morphine CPP, systemic baclofen treatment after, but not before extinction sessions, dose-dependently facilitated the extinction process, presumably by affecting the reconsolidation of morphine-related memories (Heinrichs et al. 2010). Injection of baclofen into the cerebral ventricles decreased morphine self-administration while the $GABA_BR$ antagonist phaclofen increased self-infusions (Ramshini et al. 2013). Another group also showed that systemic baclofen (1.8 mg/kg) decreased morphine self-administration under FR1 schedule, and that $GABA_BR$ antagonist SCH 50911 blocked this effect (Yoon et al. 2007). Similarly for heroin, systemic baclofen dose-dependently decreased heroin self-administration, and the higher dose (1 mg/kg) of baclofen completely prevented acquisition of self-administration, even when baclofen treatment was discontinued after a week (Xi and Stein 1999). Baclofen also abolished heroin-induced DA release in the NAc, an effect partially blocked by intra-VTA saclofen. Intra-VTA infusion of baclofen (2 μg), however, increased heroin self-administration in well-trained animals, which likely indicates a compensatory response to reduced heroin reward at low baclofen dosage (Xi and Stein 1999). Intra-NAc baclofen on the other hand did not affect heroin self-administration. The effects of $GABA_BR$ PAMs on opioid addiction have so far not been reported.

Opioid withdrawal results in a wide range of physiological and psychological symptoms, which may precipitate relapse. Rodents also display withdrawal signs following naloxone-precipitated withdrawal from chronic morphine treatment. Systemic baclofen (2 or 20 mg/kg) treatment before naloxone injection attenuated behavioral signs of morphine withdrawal in multiple studies (Bexis et al. 2001; Pedron et al. 2016; Diaz et al. 2006). Infusion of baclofen into the locus coeruleus also dose-dependently decreased withdrawal signs (Riahi et al. 2009).

5.4 Other Drugs/Reinforcers

Propofol is a PAM of the $GABA_AR$ and is widely used as an intravenous anesthetic. It has abuse potential in humans and is self-administered in rats (LeSage et al. 2000). The rewarding effect of propofol is postulated to be the result of its activation of $GABA_ARs$ on VTA GABA neurons, leading to disinhibition of VTA DA neurons. Baclofen, either systemic (3 mg/kg) or intra-VTA (50 or 100 ng/side) suppressed propofol self-administration under FR1 schedule without affecting motor activities or food-maintained responses (Yang et al. 2011).

In rats trained to self-stimulate the medial forebrain bundle (MFB) upon discriminative cue presentation, $GABA_BR$ agonist baclofen, but not a $GABA_AR$ agonist or NMDAR agonist, infused into the VTA resulted in a rightward shift of the current intensity-operant response curve, indicating a dampening of the rewarding effect of MFB stimulation (Willick and Kokkinidis 1995). In another study, $GABA_BR$ agonist CGP 44532, $GABA_BR$ antagonists CGP 56433A and CGP 51176 all increased thresholds for brain stimulation reward (Macey et al. 2001). In addition, the antagonists had additive effects when given with the agonist, instead of blocking its effect. This paradoxical result could be attributed to $GABA_BR$ action at pre- versus post-synaptic sites (Macey et al. 2001).

5.5 Stress and Addiction

Stress is a common trigger of relapse in drug addiction (Sinha 2008). The role of $GABA_BRs$ in stress-induced reinstatement of drug seeking has been investigated in animal models. Intra-VTA infusion of the $GABA_BR$ antagonist 2-hydroxysaclofen prevented reinstatement of cocaine seeking induced either by stress or intra-VTA injection of corticotropin-releasing factor (CRF) (Blacktop et al. 2016). In comparison, intra-VTA bicuculline ($GABA_AR$ antagonist) had no effect on cocaine reinstatement. Stress-induced reinstatement depends on CRF actions in the VTA. CRF can exert opposing actions through the two receptor types, CRF-R1 and CRF-R2 (Williams et al. 2014). Within the VTA, CRF is known to enhance glutamate release (Wang et al. 2005), increase AMPAR/NMDAR ratio in DA neurons (Ungless et al. 2003), as well as promoting GABA release (Williams et al. 2014) and potentiating GIRK-mediated inhibitory currents in DA neurons upon D2R and $GABA_BR$ activation (Beckstead et al. 2009). The balance of these effects can be altered by cocaine self-administration, in that CRF-R2-mediated facilitation of GABA release becomes suppressed and the overall effect of CRF shifts to excitation (Williams et al. 2014). Due to the complexity of GABAergic signaling in the VTA, it is difficult to conclude whether the $GABA_BR$ signaling responsible for stress-induced cocaine reinstatement acts on DA neurons, local interneurons, or presynaptic terminals. Measurements of VTA neuron activity and DA release may help clarify the mechanism.

Chronic restraint stress facilitated the development of morphine CPP, in that stressed mice required lower doses and less conditioning sessions to develop a preference for morphine (Meng et al. 2014). Baclofen injection (1.25 or 2.5 mg/kg) 30 min before morphine conditioning abolished the emergence of stress-precipitated morphine CPP (Meng et al. 2014). However, it was not determined whether baclofen would also block the morphine CPP that develops over more pairings regardless of stress pre-exposure. In separate experiments, 2.5 mg/kg baclofen facilitated extinction of morphine CPP in stressed mice, and it prevented forced swim-induced reinstatement of morphine CPP in both stressed and non-stressed mice (Meng et al. 2014).

There are several instances where baclofen was used in conjunction with the $GABA_AR$ agonist muscimol to inhibit neural activity of a specific brain region in a drug addiction paradigm. Baclofen/muscimol injection into the NAc core or shell suppressed context-induced reinstatement of cocaine seeking (Fuchs et al. 2008). Baclofen/muscimol injection into the bed nucleus of the stria terminalis (BNST), a key convergence point of cues and stress, reduced cue and/or yohimbine stress-induced reinstatement of cocaine seeking (Buffalari and See 2011). The LHb encodes aversive and anxiogenic states, and when inactivated by baclofen/muscimol resulted in reduced cue-induced cocaine reinstatement in the presence of yohimbine stress (Gill et al. 2013). Finally, baclofen/muscimol inactivation of BLA or ventral hippocampus prevented stress-potentiated reacquisition of nicotine self-administration (Yu and Sharp 2015). Whether $GABA_BR$ modulation alone in these brain regions is sufficient to induce these behavioral effects remains to be investigated.

5.6 *Summary*

So far, preclinical research has revealed a general pattern that $GABA_BR$ agonists and PAMs are capable of (1) attenuating the initial reinforcing/stimulating effect of addictive drugs and blocking certain aspects of drug-induced plasticity; (2) reducing self-administration of drugs on various reinforcement schedules; (3) reducing drug seeking induced by cue, stress, or drug re-exposure; (4) attenuating withdrawal symptoms for opioids. The first three effects have been localized to the VTA, since intra-VTA application of $GABA_BR$ ligands usually recapitulates the effects seen with systemic application. Furthermore, $GABA_BR$-mediated inhibition of DA neurons and resulting decrease in DA release is the principle mechanism of these anti-reward effects. On the other hand, $GABA_BR$ signaling can also mediate some neural processes that contribute to drug addiction, especially in relation to stress and anxiety during drug abstinence. Compared with $GABA_BR$ agonists, $GABA_BR$ PAMs in most cases demonstrated comparable or better performance with less undesirable nonspecific effects. Novel $GABA_BR$ PAMs are continuously being discovered and can be tested in the context of drug addiction. Future studies should also investigate the pharmacokinetics and off-target activity of $GABA_BR$ PAMs, which are critical determinants of their clinical applicability. So far it is not known whether other allosteric binding sites exist on $GABA_BR$ or whether PAM binding affects modulation by auxiliary subunits (KCTDs) (Filip et al. 2015), which also interact with the $GABA_BR2$ subunit (Zuo et al. 2019; Zheng et al. 2019). Mapping of the $GABA_BR$ PAM binding site(s) will allow for in silico screening of new compounds based on homology models, which may help accelerate the drug discovery process.

6 Clinical Studies of $GABA_BR$ Modulators in Drug Addiction

For recent discussions on clinical trials using baclofen or other drugs that act indirectly through the $GABA_BR$ for the treatment of substance abuse, see (Tyacke et al. 2010; Brebner et al. 2002; Phillips and Reed 2014; Agabio and Colombo 2015). To date, baclofen is the only $GABA_BR$ ligand for which clinical data exist on drug addiction treatment, although human studies using the $GABA_BR$ PAM ADX 71441 developed by Addex Therapeutics (Kalinichev et al. 2017) for treating cocaine use disorder have recently been funded by the NIH. In 2018 baclofen was authorized for marketing as a treatment for alcohol use disorder in France although its effectiveness and safety have been questioned by a number of clinical studies (Braillon et al. 2020; Palpacuer et al. 2018; Chaignot et al. 2018). With the encouraging preclinical results discussed earlier, in the coming years we should expect to see more $GABA_BR$ PAMs in clinical studies.

Baclofen has been most extensively tested for cocaine use disorders. An open label study in Los Angeles on 10 patients found that 20 mg baclofen given three times per day (20 mg t.i.d.) decreased cocaine craving and use (Ling et al. 1998). A subsequent randomized placebo-controlled trial by the same group however found no significant effect of baclofen (20 mg t.i.d.) on cocaine craving although cocaine use was reduced as confirmed by urine testing (Shoptaw et al. 2003). On the other hand, in a few other studies cocaine craving in cocaine-dependent individuals was reported to be suppressed by baclofen (60 mg per day) (Kaplan et al. 2004; Haney et al. 2006). In addition, an fMRI study on cocaine-dependent individuals demonstrated a baclofen-induced suppression of neural activation by subliminal cocaine cues, but not sexual or aversive cues (Young et al. 2014). This provides a potential mechanism for baclofen's inhibition of cocaine craving. Interestingly, the subjective "high" of cocaine was consistently found to be unaffected by baclofen (Rotheram-Fuller et al. 2007; Haney et al. 2006; Lile et al. 2004; Ling et al. 1998). Discrepant results were reported concerning baclofen modulation of the cardiovascular effects of cocaine. While one study found a high dose of baclofen (60 mg) attenuated cocaine (50 mg)-induced increase in heart rate (Haney et al. 2006), another study showed that baclofen plus amantadine (30 mg and 100 mg t.i.d.) did not impact heart rate or blood pressure changes with 20 and 40 mg cocaine (Rotheram-Fuller et al. 2007). Yet another study reported only an increase in systolic blood pressure with 45 mg cocaine, which was not affected by up to 30 mg of baclofen (Lile et al. 2004).

A multi-center double-blind study involving 160 individuals with severe cocaine dependence tested the efficacy of an 8-week treatment with baclofen (60 mg/day) on cocaine use (Kahn et al. 2009). Unfortunately, there was no difference between baclofen and placebo groups in cocaine use (self-reported and urine test confirmed) or craving. While the severity of cocaine dependence may have contributed to the negative result, it has also been suggested that baclofen is more likely to help prevent relapse than to initiate abstinence (Phillips and Reed 2014; Tyacke et al. 2010; Brebner et al. 2002). Indeed, as discussed above, baclofen does not alter the subjective experience in taking cocaine, but rather reduces craving during abstinence. In addition, the 60 mg/day dose that was recommended for treating spasticity was well tolerated across aforementioned studies and may need to be gradually augmented to achieve better effects on addiction treatment. In line with this, a randomized double-blind trial of baclofen (20 mg t.i.d.) on methamphetamine dependent individuals revealed a significant treatment effect only among those who reported taking a higher percentage of baclofen (Heinzerling et al. 2006).

The acute effect of baclofen (20 mg) on cigarette smoking has been studied on smokers using a double-blind within-subject design (Cousins et al. 2001). In the 3-h ad libitum smoking period following drug administration, the number of cigarettes smoked and nicotine craving were not changed by baclofen. Baclofen did however decrease liking of cigarette smoking and produced a sedative/relaxing feeling. It would be of interest to see whether this acute change in subjective experience would have long-term impact if baclofen is repeatedly paired with smoking. A more recent double-blind placebo-controlled trial examined the effects of prolonged baclofen treatment (9 weeks) on cigarette smoking (Franklin et al. 2009). Here a higher dose of baclofen was used, i.e. 80 mg per day in 20 mg doses, which was titrated up over 12 days. Baclofen was significantly more effective in reducing the number of cigarettes smoked per day although a decline in smoking over the 9 weeks was seen in the placebo group as well. Importantly, baclofen did not produce more side effects than placebo, except mild sedation. The authors attributed this lack of side effects to the gradual increase of baclofen dose at the beginning of the study. These findings are very encouraging and suggest that baclofen could potentially be used for smoking relapse prevention. Future studies should also monitor cigarette consumption after cessation of baclofen treatment, as it has been reported that abrupt discontinuation of baclofen can produce withdrawal symptoms and relapse into cocaine use (Kaplan et al. 2004).

There are very few clinical studies so far evaluating baclofen's effect on opioid dependence and withdrawal, possibly due to the availability of methadone and buprenorphine as treatment options. One randomized double-blind clinical trial compared the effects of baclofen (40 mg/day) and alpha-2A adrenergic receptor agonist clonidine (0.8 mg/day) on opiate addicts undergoing detoxification, and showed that baclofen was more effective in treating both the physical and mental symptoms of withdrawal (Akhondzadeh et al. 2000). The side effects and retention rate for the two drugs were similar while clonidine generated more problems relating to hypotension (Ahmadi-Abhari et al. 2001). These studies suggest that baclofen may serve as a better detoxification medication than clonidine. However, the lack of placebo control in this trial prevented a direct evaluation of baclofen's intrinsic effects. Subsequently, the same group conducted a 12-week long double-blind placebo-controlled trial to evaluate the effect of 60 mg/day baclofen on opioid withdrawal (Assadi et al. 2003). It was found that retention rate was significantly higher for the baclofen group, and that baclofen treatment was significantly better than placebo in managing withdrawal and depressive symptoms. However, the rates of opioid-positive urine did not differ between the groups. One drawback of the present studies is the high dropout rate during the trial. In the future, larger clinical trials should be conducted to help validate these findings.

Baclofen generally does not produce subjective effects indicative of abuse potential in humans. However, in cannabis users trained to discriminate THC from placebo, baclofen alone (50 mg) was able to substitute for the THC discriminative stimulus (Lile et al. 2012). This suggests that baclofen may help relieve some symptoms associated with cannabis abstinence, such as craving and anxiety. Future studies should evaluate the potential of baclofen and other GABA$_B$R ligands in treating cannabis use disorders. Thus far, one study focusing on daily marijuana smokers found that baclofen (60, 90 mg/day) dose-dependently reduced craving for marijuana and tobacco during active smoking, yet it did not impact mood during abstinence and did not prevent relapse (Haney et al. 2010). Actually, baclofen led to slight worsening of cognitive performance and early wake-ups during marijuana abstinence.

In summary, clinical studies have demonstrated some preliminary success as well as limitations for baclofen in addiction treatments. Baclofen appears to help ameliorate craving and withdrawal symptoms, and can help prevent relapse especially at high doses. It may best serve as a compliment to existing forms of therapy and medication, instead of a stand-alone treatment. Whether baclofen is also effective in treating multidrug addiction will require more investigation. One limitation of baclofen is its relatively short pharmacokinetic profile, which necessitates multiple (3–4) doses per day. Typical dosage used is 60 mg per day given in three doses although this was based on the recommendation for spasticity and should be adjusted to achieve the best outcome for treating addiction. Many studies start with a lower dose and titrate it up over the course of a few days, and this seems to help reduce side effects. Another GABAergic drug vigabatrin induces a more prolonged increase in GABA levels due to its molecular mechanism as an irreversible inhibitor of GABA transaminase (Tyacke et al. 2010). However, the lack of $GABA_BR$ specificity gives rise to a wide range of side effects. With the development of newer $GABA_BR$ agonists and PAMs and the accumulating evidence that they are comparably or more effective than baclofen in suppressing addiction-like behavior while having less side effects in animal models, more efforts should be put in to push these drugs towards clinical research. In addition, the development of drugs that target signaling molecules downstream of $GABA_BRs$ may also provide options for treating addiction.

7 Conclusions

$GABA_BR$ signaling plays important roles in the reward circuit, and advancing techniques are enabling more in-depth dissection of the nuanced interactions between drugs of abuse and $GABA_BR$ signaling. On the other hand, pharmacological research, both preclinical and clinical, has revealed great potential for $GABA_BR$ agonists and PAMs as a supplemental treatment for drug addiction, mainly in suppressing craving, relapse, and withdrawal symptoms. Further research is needed to find the most efficacious, specific, and practical $GABA_BR$ agent for human use, as well as to understand the potential long-term side effects due to $GABA_BR$ modulation in various parts of the brain.

References

Agabio R, Colombo G (2015) GABAB receptor as therapeutic target for drug addiction: from baclofen to positive allosteric modulators. Psychiatr Pol 49(2):215–223

Ahmadi-Abhari SA, Akhondzadeh S, Assadi SM, Shabestari OL, Farzanehgan ZM, Kamlipour A (2001) Baclofen versus clonidine in the treatment of opiates withdrawal, side-effects aspect: a double-blind randomized controlled trial. J Clin Pharm Ther 26(1):67–71

Akhondzadeh S, Ahmadi-Abhari SA, Assadi SM, Shabestari OL, Kashani AR, Farzanehgan ZM (2000) Double-blind randomized controlled trial of baclofen vs. clonidine in the treatment of opiates withdrawal. J Clin Pharm Ther 25(5):347–353

Amantea D, Bowery NG (2004) Reduced inhibitory action of a GABAB receptor agonist on [3H]-dopamine release from rat ventral tegmental area in vitro after chronic nicotine administration. BMC Pharmacol 4:24

Amantea D, Tessari M, Bowery NG (2004) Reduced G-protein coupling to the GABAB receptor in the nucleus accumbens and the medial prefrontal cortex of the rat after chronic treatment with nicotine. Neurosci Lett 355(3):161–164

Arai S, Takuma K, Mizoguchi H, Ibi D, Nagai T, Kamei H, Kim HC, Yamada K (2009) GABAB receptor agonist baclofen improves methamphetamine-induced cognitive deficit in mice. Eur J Pharmacol 602(1):101–104

Arora D, Hearing M, Haluk DM, Mirkovic K, Fajardo-Serrano A, Wessendorf MW, Watanabe M, Lujan R, Wickman K (2011) Acute cocaine exposure weakens GABA(B) receptor-dependent G-protein-gated inwardly rectifying K+ signaling in dopamine neurons of the ventral tegmental area. J Neurosci 31(34):12251–12257

Assadi SM, Radgoodarzi R, Ahmadi-Abhari SA (2003) Baclofen for maintenance treatment of opioid dependence: a randomized double-blind placebo-controlled clinical trial [ISRCTN32121581]. BMC Psychiatry 3:16

Backes EN, Hemby SE (2008) Contribution of ventral tegmental GABA receptors to cocaine self-administration in rats. Neurochem Res 33(3):459–467

Baik JH (2013) Dopamine signaling in reward-related behaviors. Front Neural Circuits 7:152

Bartoletti M, Gubellini C, Ricci F, Gaiardi M (2004) The GABAB agonist baclofen blocks the expression of sensitisation to the locomotor stimulant effect of amphetamine. Behav Pharmacol 15(5–6):397–401

Bartoletti M, Gubellini C, Ricci F, Gaiardi M (2005) Baclofen blocks the development of sensitization to the locomotor stimulant effect of amphetamine. Behav Pharmacol 16(7):553–558

Beckstead MJ, Gantz SC, Ford CP, Stenzel-Poore MP, Phillips PE, Mark GP, Williams JT (2009) CRF enhancement of GIRK channel-mediated transmission in dopamine neurons. Neuropsychopharmacology 34(8):1926–1935

Bettler B, Tiao JY (2006) Molecular diversity, trafficking and subcellular localization of GABAB receptors. Pharmacol Ther 110(3):533–543

Bexis S, Ong J, White J (2001) Attenuation of morphine withdrawal signs by the GABA (B) receptor agonist baclofen. Life Sci 70(4):395–401

Bidaut-Russell M, Devane WA, Howlett AC (1990) Cannabinoid receptors and modulation of cyclic AMP accumulation in the rat brain. J Neurochem 55(1):21–26

Biermann B, Ivankova-Susankova K, Bradaia A, Abdel Aziz S, Besseyrias V, Kapfhammer JP, Missler M, Gassmann M, Bettler B (2010) The sushi domains of GABAB receptors function as axonal targeting signals. J Neurosci 30(4):1385–1394

Binet V, Brajon C, Le Corre L, Acher F, Pin JP, Prezeau L (2004) The heptahelical domain of GABA(B2) is activated directly by CGP7930, a positive allosteric modulator of the GABA (B) receptor. J Biol Chem 279(28):29085–29091

Blacktop JM, Vranjkovic O, Mayer M, van Hoof M, Baker DA, Mantsch JR (2016) Antagonism of GABA-B but not GABA-A receptors in the VTA prevents stress- and intra-VTA CRF-induced reinstatement of extinguished cocaine seeking in rats. Neuropharmacology 102:197–206

Bocklisch C, Pascoli V, Wong JC, House DR, Yvon C, de Roo M, Tan KR, Luscher C (2013) Cocaine disinhibits dopamine neurons by potentiation of GABA transmission in the ventral tegmental area. Science 341(6153):1521–1525

Bonci A, Williams JT (1997) Increased probability of GABA release during withdrawal from morphine. J Neurosci 17(2):796–803

Bowery NG (2006) GABAB receptor: a site of therapeutic benefit. Curr Opin Pharmacol 6 (1):37–43
Braillon A, Naudet F, Cristea IA, Lexchin J (2020) Baclofen and alcohol use disorders: breakthrough or great white elephant? Alcohol Alcohol 55(1):49–50
Brebner K, Froestl W, Andrews M, Phelan R, Roberts DC (1999) The GABA(B) agonist CGP 44532 decreases cocaine self-administration in rats: demonstration using a progressive ratio and a discrete trials procedure. Neuropharmacology 38(11):1797–1804
Brebner K, Phelan R, Roberts DC (2000a) Effect of baclofen on cocaine self-administration in rats reinforced under fixed-ratio 1 and progressive-ratio schedules. Psychopharmacology 148 (3):314–321
Brebner K, Phelan R, Roberts DC (2000b) Intra-VTA baclofen attenuates cocaine self-administration on a progressive ratio schedule of reinforcement. Pharmacol Biochem Behav 66(4):857–862
Brebner K, Childress AR, Roberts DC (2002) A potential role for GABA(B) agonists in the treatment of psychostimulant addiction. Alcohol Alcohol 37(5):478–484
Brebner K, Ahn S, Phillips AG (2005) Attenuation of d-amphetamine self-administration by baclofen in the rat: behavioral and neurochemical correlates. Psychopharmacology 177 (4):409–417
Brodie MS, Dunwiddie TV (1990) Cocaine effects in the ventral tegmental area: evidence for an indirect dopaminergic mechanism of action. Naunyn Schmiedeberg's Arch Pharmacol 342 (6):660–665
Buffalari DM, See RE (2011) Inactivation of the bed nucleus of the stria terminalis in an animal model of relapse: effects on conditioned cue-induced reinstatement and its enhancement by yohimbine. Psychopharmacology 213(1):19–27
Cameron DL, Williams JT (1993) Dopamine D1 receptors facilitate transmitter release. Nature 366 (6453):344–347
Cameron DL, Williams JT (1994) Cocaine inhibits GABA release in the VTA through endogenous 5-HT. J Neurosci 14(11 Pt 1):6763–6767
Campbell UC, Lac ST, Carroll ME (1999) Effects of baclofen on maintenance and reinstatement of intravenous cocaine self-administration in rats. Psychopharmacology 143(2):209–214
Campbell UC, Morgan AD, Carroll ME (2002) Sex differences in the effects of baclofen on the acquisition of intravenous cocaine self-administration in rats. Drug Alcohol Depend 66 (1):61–69
Cedillo LN, Miranda F (2013) Effects of co-administration of the GABAB receptor agonist baclofen and a positive allosteric modulator of the GABAB receptor, CGP7930, on the development and expression of amphetamine-induced locomotor sensitization in rats. Pharmacol Rep 65(5):1132–1143
Chaignot C, Zureik M, Rey G, Dray-Spira R, Coste J, Weill A (2018) Risk of hospitalisation and death related to baclofen for alcohol use disorders: comparison with nalmefene, acamprosate, and naltrexone in a cohort study of 165 334 patients between 2009 and 2015 in France. Pharmacoepidemiol Drug Saf 27(11):1239–1248
Chebib M, Johnston GA (1999) The 'ABC' of GABA receptors: a brief review. Clin Exp Pharmacol Physiol 26(11):937–940
Chen M, Zhao Y, Yang H, Luan W, Song J, Cui D, Dong Y, Lai B, Ma L, Zheng P (2015) Morphine disinhibits glutamatergic input to VTA dopamine neurons and promotes dopamine neuron excitation. elife 4
Ciccarelli A, Calza A, Panzanelli P, Concas A, Giustetto M, Sassoe-Pognetto M (2012) Organization of GABAergic synaptic circuits in the rat ventral tegmental area. PLoS One 7(10):e46250
Corre J, van Zessen R, Loureiro M, Patriarchi T, Tian L, Pascoli V, Luscher C (2018) Dopamine neurons projecting to medial shell of the nucleus accumbens drive heroin reinforcement. elife 7
Corrigall WA, Coen KM, Adamson KL, Chow BL, Zhang J (2000) Response of nicotine self-administration in the rat to manipulations of mu-opioid and gamma-aminobutyric acid receptors in the ventral tegmental area. Psychopharmacology 149(2):107–114

Cousins MS, Stamat HM, de Wit H (2001) Effects of a single dose of baclofen on self-reported subjective effects and tobacco smoking. Nicotine Tob Res 3(2):123–129
Cruz HG, Ivanova T, Lunn ML, Stoffel M, Slesinger PA, Luscher C (2004) Bi-directional effects of GABA(B) receptor agonists on the mesolimbic dopamine system. Nat Neurosci 7(2):153–159
Davies CH, Collingridge GL (1996) Regulation of EPSPs by the synaptic activation of GABAB autoreceptors in rat hippocampus. J Physiol 496(Pt 2):451–470
Davies CH, Starkey SJ, Pozza MF, Collingridge GL (1991) GABA autoreceptors regulate the induction of LTP. Nature 349(6310):609–611
de Beaurepaire R, Sinclair JMA, Heydtmann M, Addolorato G, Aubin HJ, Beraha EM, Caputo F, Chick JD, de La Selle P, Franchitto N, Garbutt JC, Haber PS, Jaury P, Lingford-Hughes AR, Morley KC, Muller CA, Owens L, Pastor A, Paterson LM, Pelissier F, Rolland B, Stafford A, Thompson A, van den Brink W, Leggio L, Agabio R (2018) The use of baclofen as a treatment for alcohol use disorder: a clinical practice perspective. Front Psych 9:708
de Miguel E, Vekovischeva O, Kuokkanen K, Vesajoki M, Paasikoski N, Kaskinoro J, Myllymaki M, Lainiola M, Janhunen SK, Hyytia P, Linden AM, Korpi ER (2018) GABAB receptor positive allosteric modulators with different efficacies affect neuroadaptation to and self-administration of alcohol and cocaine. Addict Biol
DePoy LM, Allen AG, Gourley SL (2016) Adolescent cocaine self-administration induces habit behavior in adulthood: sex differences and structural consequences. Trans Psychiatry 6(8):e875
Diaz SL, Barros VG, Antonelli MC, Rubio MC, Balerio GN (2006) Morphine withdrawal syndrome and its prevention with baclofen: autoradiographic study of mu-opioid receptors in prepubertal male and female mice. Synapse 60(2):132–140
Dobrovitsky V, Pimentel P, Duarte A, Froestl W, Stellar JR, Trzcinska M (2002) CGP 44532, a GABAB receptor agonist, is hedonically neutral and reduces cocaine-induced enhancement of reward. Neuropharmacology 42(5):626–632
Dupuis DS, Relkovic D, Lhuillier L, Mosbacher J, Kaupmann K (2006) Point mutations in the transmembrane region of GABAB2 facilitate activation by the positive modulator N,N'-dicyclopentyl-2-methylsulfanyl-5-nitro-pyrimidine-4,6-diamine (GS39783) in the absence of the GABAB1 subunit. Mol Pharmacol 70(6):2027–2036
Edwards NJ, Tejeda HA, Pignatelli M, Zhang S, McDevitt RA, Wu J, Bass CE, Bettler B, Morales M, Bonci A (2017) Circuit specificity in the inhibitory architecture of the VTA regulates cocaine-induced behavior. Nat Neurosci 20(3):438–448
Enna SJ (2001) A GABA(B) mystery: the search for pharmacologically distinct GABA (B) receptors. Mol Interv 1(4):208–218
Fadda P, Scherma M, Fresu A, Collu M, Fratta W (2003) Baclofen antagonizes nicotine-, cocaine-, and morphine-induced dopamine release in the nucleus accumbens of rat. Synapse 50(1):1–6
Faget L, Osakada F, Duan J, Ressler R, Johnson AB, Proudfoot JA, Yoo JH, Callaway EM, Hnasko TS (2016) Afferent inputs to neurotransmitter-defined cell types in the ventral tegmental area. Cell Rep 15(12):2796–2808
Filip M, Frankowska M (2007) Effects of GABA(B) receptor agents on cocaine priming, discrete contextual cue and food induced relapses. Eur J Pharmacol 571(2–3):166–173
Filip M, Frankowska M (2008) GABA(B) receptors in drug addiction. Pharmacol Rep 60 (6):755–770
Filip M, Frankowska M, Przegalinski E (2007) Effects of GABA(B) receptor antagonist, agonists and allosteric positive modulator on the cocaine-induced self-administration and drug discrimination. Eur J Pharmacol 574(2–3):148–157
Filip M, Frankowska M, Sadakierska-Chudy A, Suder A, Szumiec L, Mierzejewski P, Bienkowski P, Przegalinski E, Cryan JF (2015) GABAB receptors as a therapeutic strategy in substance use disorders: focus on positive allosteric modulators. Neuropharmacology 88:36–47
Floresco SB, West AR, Ash B, Moore H, Grace AA (2003) Afferent modulation of dopamine neuron firing differentially regulates tonic and phasic dopamine transmission. Nat Neurosci 6 (9):968–973

Francis TC, Gantz SC, Moussawi K, Bonci A (2019) Synaptic and intrinsic plasticity in the ventral tegmental area after chronic cocaine. Curr Opin Neurobiol 54:66–72

Franklin TR, Harper D, Kampman K, Kildea-McCrea S, Jens W, Lynch KG, O'Brien CP, Childress AR (2009) The GABA B agonist baclofen reduces cigarette consumption in a preliminary double-blind placebo-controlled smoking reduction study. Drug Alcohol Depend 103 (1–2):30–36

Frankowska M, Wydra K, Faron-Gorecka A, Zaniewska M, Kusmider M, Dziedzicka-Wasylewska M, Filip M (2008) Neuroadaptive changes in the rat brain GABA(B) receptors after withdrawal from cocaine self-administration. Eur J Pharmacol 599(1–3):58–64

Frankowska M, Nowak E, Filip M (2009) Effects of GABAB receptor agonists on cocaine hyperlocomotor and sensitizing effects in rats. Pharmacol Rep 61(6):1042–1049

Froestl W (2010) Chemistry and pharmacology of GABAB receptor ligands. Adv Pharmacol 58:19–62

Froger-Colleaux C, Castagne V (2016) Effects of baclofen and raclopride on reinstatement of cocaine self-administration in the rat. Eur J Pharmacol 777:147–155

Fu Z, Yang H, Xiao Y, Zhao G, Huang H (2012) The gamma-aminobutyric acid type B (GABAB) receptor agonist baclofen inhibits morphine sensitization by decreasing the dopamine level in rat nucleus accumbens. Behav Brain Funct 8:20

Fuchs RA, Ramirez DR, Bell GH (2008) Nucleus accumbens shell and core involvement in drug context-induced reinstatement of cocaine seeking in rats. Psychopharmacology 200(4):545–556

Gahwiler BH, Brown DA (1985) GABAB-receptor-activated K+ current in voltage-clamped CA3 pyramidal cells in hippocampal cultures. Proc Natl Acad Sci U S A 82(5):1558–1562

Galvez T, Parmentier ML, Joly C, Malitschek B, Kaupmann K, Kuhn R, Bittiger H, Froestl W, Bettler B, Pin JP (1999) Mutagenesis and modeling of the GABAB receptor extracellular domain support a venus flytrap mechanism for ligand binding. J Biol Chem 274 (19):13362–13369

Galvez T, Prezeau L, Milioti G, Franek M, Joly C, Froestl W, Bettler B, Bertrand HO, Blahos J, Pin JP (2000) Mapping the agonist-binding site of GABAB type 1 subunit sheds light on the activation process of GABAB receptors. J Biol Chem 275(52):41166–41174

Garcia-Gil L, de Miguel R, Romero J, Perez A, Ramos JA, Fernandez-Ruiz JJ (1999) Perinatal delta9-tetrahydrocannabinol exposure augmented the magnitude of motor inhibition caused by GABA(B), but not GABA(A), receptor agonists in adult rats. Neurotoxicol Teratol 21 (3):277–283

Gassmann M, Shaban H, Vigot R, Sansig G, Haller C, Barbieri S, Humeau Y, Schuler V, Muller M, Kinzel B, Klebs K, Schmutz M, Froestl W, Heid J, Kelly PH, Gentry C, Jaton AL, van der Putten H, Mombereau C, Lecourtier L, Mosbacher J, Cryan JF, Fritschy JM, Luthi A, Kaupmann K, Bettler B (2004) Redistribution of GABAB(1) protein and atypical GABAB responses in GABAB(2)-deficient mice. J Neurosci 24(27):6086–6097

Gill MJ, Ghee SM, Harper SM, See RE (2013) Inactivation of the lateral habenula reduces anxiogenic behavior and cocaine seeking under conditions of heightened stress. Pharmacol Biochem Behav 111:24–29

Giorgetti M, Hotsenpiller G, Froestl W, Wolf ME (2002) In vivo modulation of ventral tegmental area dopamine and glutamate efflux by local GABA(B) receptors is altered after repeated amphetamine treatment. Neuroscience 109(3):585–595

Guery S, Floersheim P, Kaupmann K, Froestl W (2007) Syntheses and optimization of new GS39783 analogues as positive allosteric modulators of GABA B receptors. Bioorg Med Chem Lett 17(22):6206–6211

Gysling K, Wang RY (1983) Morphine-induced activation of A10 dopamine neurons in the rat. Brain Res 277(1):119–127

Halbout B, Quarta D, Valerio E, Heidbreder CA, Hutcheson DM (2011) The GABA-B positive modulator GS39783 decreases psychostimulant conditioned-reinforcement and conditioned-reward. Addict Biol 16(3):416–427

Haney M, Hart CL, Foltin RW (2006) Effects of baclofen on cocaine self-administration: opioid- and nonopioid-dependent volunteers. Neuropsychopharmacology 31(8):1814–1821
Haney M, Hart CL, Vosburg SK, Comer SD, Reed SC, Cooper ZD, Foltin RW (2010) Effects of baclofen and mirtazapine on a laboratory model of marijuana withdrawal and relapse. Psychopharmacology 211(2):233–244
Heaney CF, Kinney JW (2016) Role of GABA(B) receptors in learning and memory and neurological disorders. Neurosci Biobehav Rev 63:1–28
Hearing M, Kotecki L, Marron Fernandez de Velasco E, Fajardo-Serrano A, Chung HJ, Lujan R, Wickman K (2013) Repeated cocaine weakens GABA(B)-Girk signaling in layer 5/6 pyramidal neurons in the prelimbic cortex. Neuron 80(1):159–170
Heinrichs SC, Leite-Morris KA, Carey RJ, Kaplan GB (2010) Baclofen enhances extinction of opiate conditioned place preference. Behav Brain Res 207(2):353–359
Heinzerling KG, Shoptaw S, Peck JA, Yang X, Liu J, Roll J, Ling W (2006) Randomized, placebo-controlled trial of baclofen and gabapentin for the treatment of methamphetamine dependence. Drug Alcohol Depend 85(3):177–184
Holtz NA, Carroll ME (2011) Baclofen has opposite effects on escalation of cocaine self-administration: increased intake in rats selectively bred for high (HiS) saccharin intake and decreased intake in those selected for low (LoS) saccharin intake. Pharmacol Biochem Behav 100(2):275–283
Hotsenpiller G, Wolf ME (2003) Baclofen attenuates conditioned locomotion to cues associated with cocaine administration and stabilizes extracellular glutamate levels in rat nucleus accumbens. Neuroscience 118(1):123–134
Howlett AC, Qualy JM, Khachatrian LL (1986) Involvement of Gi in the inhibition of adenylate cyclase by cannabimimetic drugs. Mol Pharmacol 29(3):307–313
Jacobson LH, Sweeney FF, Kaupmann K, O'Leary OF, Gassmann M, Bettler B, Cryan JF (2016) Differential roles of GABAB1 subunit isoforms on locomotor responses to acute and repeated administration of cocaine. Behav Brain Res 298(Pt B):12–16
Jacobson LH, Vlachou S, Slattery DA, Li X, Cryan JF (2018) The gamma-aminobutyric acid B receptor in depression and reward. Biol Psychiatry 83(11):963–976
Jayaram P, Steketee JD (2004) Effects of repeated cocaine on medial prefrontal cortical GABAB receptor modulation of neurotransmission in the mesocorticolimbic dopamine system. J Neurochem 90(4):839–847
Jhou TC, Fields HL, Baxter MG, Saper CB, Holland PC (2009a) The rostromedial tegmental nucleus (RMTg), a GABAergic afferent to midbrain dopamine neurons, encodes aversive stimuli and inhibits motor responses. Neuron 61(5):786–800
Jhou TC, Geisler S, Marinelli M, Degarmo BA, Zahm DS (2009b) The mesopontine rostromedial tegmental nucleus: a structure targeted by the lateral habenula that projects to the ventral tegmental area of Tsai and substantia nigra compacta. J Comp Neurol 513(6):566–596
Johnson SW, North RA (1992) Opioids excite dopamine neurons by hyperpolarization of local interneurons. J Neurosci 12(2):483–488
Kahn R, Biswas K, Childress AR, Shoptaw S, Fudala PJ, Gorgon L, Montoya I, Collins J, McSherry F, Li SH, Chiang N, Alathari H, Watson D, Liberto J, Beresford T, Stock C, Wallace C, Gruber V, Elkashef A (2009) Multi-center trial of baclofen for abstinence initiation in severe cocaine-dependent individuals. Drug Alcohol Depend 103(1–2):59–64
Kalinichev M, Girard F, Haddouk H, Rouillier M, Riguet E, Royer-Urios I, Mutel V, Lutjens R, Poli S (2017) The drug candidate, ADX71441, is a novel, potent and selective positive allosteric modulator of the GABAB receptor with a potential for treatment of anxiety, pain and spasticity. Neuropharmacology 114:34–47
Kalivas PW (1995) Interactions between dopamine and excitatory amino acids in behavioral sensitization to psychostimulants. Drug Alcohol Depend 37(2):95–100
Kalivas PW, Duffy P, Eberhardt H (1990) Modulation of A10 dopamine neurons by gamma-aminobutyric acid agonists. J Pharmacol Exp Ther 253(2):858–866

Kallo I, Molnar CS, Szoke S, Fekete C, Hrabovszky E, Liposits Z (2015) Area-specific analysis of the distribution of hypothalamic neurons projecting to the rat ventral tegmental area, with special reference to the GABAergic and glutamatergic efferents. Front Neuroanat 9:112

Kaplan GB, McRoberts RL 3rd, Smokler HJ (2004) Baclofen as adjunctive treatment for a patient with cocaine dependence and schizoaffective disorder. J Clin Psychopharmacol 24(5):574–575

Kaufling J, Veinante P, Pawlowski SA, Freund-Mercier MJ, Barrot M (2009) Afferents to the GABAergic tail of the ventral tegmental area in the rat. J Comp Neurol 513(6):597–621

Kaupmann K, Huggel K, Heid J, Flor PJ, Bischoff S, Mickel SJ, McMaster G, Angst C, Bittiger H, Froestl W, Bettler B (1997) Expression cloning of GABA(B) receptors uncovers similarity to metabotropic glutamate receptors. Nature 386(6622):239–246

Klitenick MA, DeWitte P, Kalivas PW (1992) Regulation of somatodendritic dopamine release in the ventral tegmental area by opioids and GABA: an in vivo microdialysis study. J Neurosci 12 (7):2623–2632

Kniazeff J, Galvez T, Labesse G, Pin JP (2002) No ligand binding in the GB2 subunit of the GABA (B) receptor is required for activation and allosteric interaction between the subunits. J Neurosci 22(17):7352–7361

Kumar K, Sharma S, Kumar P, Deshmukh R (2013) Therapeutic potential of GABA(B) receptor ligands in drug addiction, anxiety, depression and other CNS disorders. Pharmacol Biochem Behav 110:174–184

Kushner SA, Unterwald EM (2001) Chronic cocaine administration decreases the functional coupling of GABA(B) receptors in the rat ventral tegmental area as measured by baclofen-stimulated 35S-GTPgammaS binding. Life Sci 69(9):1093–1102

Labouebe G, Lomazzi M, Cruz HG, Creton C, Lujan R, Li M, Yanagawa Y, Obata K, Watanabe M, Wickman K, Boyer SB, Slesinger PA, Luscher C (2007) RGS2 modulates coupling between GABAB receptors and GIRK channels in dopamine neurons of the ventral tegmental area. Nat Neurosci 10(12):1559–1568

Lacey MG, Mercuri NB, North RA (1990) Actions of cocaine on rat dopaminergic neurones in vitro. Br J Pharmacol 99(4):731–735

Lalive AL, Lüscher C (2016) GABA B receptor functions in the mesolimbic dopamine system. In: GiancarloColombo (ed) GABAB receptors the receptors. Humana Press, Cham, pp 129–154

Laviolette SR, van der Kooy D (2001) GABA(A) receptors in the ventral tegmental area control bidirectional reward signalling between dopaminergic and non-dopaminergic neural motivational systems. Eur J Neurosci 13(5):1009–1015

Laviolette SR, Gallegos RA, Henriksen SJ, van der Kooy D (2004) Opiate state controls bi-directional reward signaling via GABAA receptors in the ventral tegmental area. Nat Neurosci 7(2):160–169

Le Foll B, Wertheim CE, Goldberg SR (2008) Effects of baclofen on conditioned rewarding and discriminative stimulus effects of nicotine in rats. Neurosci Lett 443(3):236–240

Lecca S, Melis M, Luchicchi A, Muntoni AL, Pistis M (2012) Inhibitory inputs from rostromedial tegmental neurons regulate spontaneous activity of midbrain dopamine cells and their responses to drugs of abuse. Neuropsychopharmacology 37(5):1164–1176

Leite-Morris KA, Fukudome EY, Shoeb MH, Kaplan GB (2004) GABA(B) receptor activation in the ventral tegmental area inhibits the acquisition and expression of opiate-induced motor sensitization. J Pharmacol Exp Ther 308(2):667–678

LeSage MG, Stafford D, Glowa JR (2000) Abuse liability of the anesthetic propofol: self-administration of propofol in rats under fixed-ratio schedules of drug delivery. Psychopharmacology 153(1):148–154

Lhuillier L, Mombereau C, Cryan JF, Kaupmann K (2007) GABA(B) receptor-positive modulation decreases selective molecular and behavioral effects of cocaine. Neuropsychopharmacology 32 (2):388–398

Li SM, Yin LL, Ren YH, Pan LS, Zheng JW (2001) GABA(B) receptor agonist baclofen attenuates the development and expression of d-methamphetamine-induced place preference in rats. Life Sci 70(3):349–356

Li J, Olinger AB, Dassow MS, Abel MS (2002a) GABA(B) receptor gene expression is not altered in cocaine-sensitized rats. J Neurosci Res 68(2):241–247

Li SP, Park MS, Bahk JY, Kim MO (2002b) Chronic nicotine and smoking exposure decreases GABA(B1) receptor expression in the rat hippocampus. Neurosci Lett 334(2):135–139

Li SP, Park MS, Kim JH, Kim MO (2004) Chronic nicotine and smoke treatment modulate dopaminergic activities in ventral tegmental area and nucleus accumbens and the gamma-aminobutyric acid type B receptor expression of the rat prefrontal cortex. J Neurosci Res 78 (6):868–879

Li X, Sturchler E, Kaczanowska K, Cameron M, Finn MG, Griffin P, McDonald P, Markou A (2017) KK-92A, a novel GABAB receptor positive allosteric modulator, attenuates nicotine self-administration and cue-induced nicotine seeking in rats. Psychopharmacology 234 (9–10):1633–1644

Lile JA, Stoops WW, Allen TS, Glaser PE, Hays LR, Rush CR (2004) Baclofen does not alter the reinforcing, subject-rated or cardiovascular effects of intranasal cocaine in humans. Psychopharmacology 171(4):441–449

Lile JA, Kelly TH, Hays LR (2012) Separate and combined effects of the GABA(B) agonist baclofen and Delta9-THC in humans discriminating Delta9-THC. Drug Alcohol Depend 126 (1–2):216–223

Ling W, Shoptaw S, Majewska D (1998) Baclofen as a cocaine anti-craving medication: a preliminary clinical study. Neuropsychopharmacology 18(5):403–404

Liu J, Maurel D, Etzol S, Brabet I, Ansanay H, Pin JP, Rondard P (2004) Molecular determinants involved in the allosteric control of agonist affinity in the GABAB receptor by the GABAB2 subunit. J Biol Chem 279(16):15824–15830

Lobina C, Carai MA, Froestl W, Mugnaini C, Pasquini S, Corelli F, Gessa GL, Colombo G (2011) Activation of the GABA(B) receptor prevents nicotine-induced locomotor stimulation in mice. Front Psych 2:76

Luscher C, Malenka RC (2011) Drug-evoked synaptic plasticity in addiction: from molecular changes to circuit remodeling. Neuron 69(4):650–663

Luscher C, Jan LY, Stoffel M, Malenka RC, Nicoll RA (1997) G protein-coupled inwardly rectifying K+ channels (GIRKs) mediate postsynaptic but not presynaptic transmitter actions in hippocampal neurons. Neuron 19(3):687–695

Macey DJ, Froestl W, Koob GF, Markou A (2001) Both GABA(B) receptor agonist and antagonists decreased brain stimulation reward in the rat. Neuropharmacology 40(5):676–685

Mahler SV, Vazey EM, Beckley JT, Keistler CR, McGlinchey EM, Kaufling J, Wilson SP, Deisseroth K, Woodward JJ, Aston-Jones G (2014) Designer receptors show role for ventral pallidum input to ventral tegmental area in cocaine seeking. Nat Neurosci 17(4):577–585

Malherbe P, Masciadri R, Norcross RD, Knoflach F, Kratzeisen C, Zenner MT, Kolb Y, Marcuz A, Huwyler J, Nakagawa T, Porter RH, Thomas AW, Wettstein JG, Sleight AJ, Spooren W, Prinssen EP (2008) Characterization of (R,S)-5,7-di-tert-butyl-3-hydroxy-3-trifluoromethyl-3H-benzofuran-2-one as a positive allosteric modulator of GABAB receptors. Br J Pharmacol 154(4):797–811

Mansvelder HD, Keath JR, McGehee DS (2002) Synaptic mechanisms underlie nicotine-induced excitability of brain reward areas. Neuron 33(6):905–919

Manzoni OJ, Williams JT (1999) Presynaptic regulation of glutamate release in the ventral tegmental area during morphine withdrawal. J Neurosci 19(15):6629–6636

Margeta-Mitrovic M, Mitrovic I, Riley RC, Jan LY, Basbaum AI (1999) Immunohistochemical localization of GABA(B) receptors in the rat central nervous system. J Comp Neurol 405 (3):299–321

Margolis EB, Hjelmstad GO, Fujita W, Fields HL (2014) Direct bidirectional mu-opioid control of midbrain dopamine neurons. J Neurosci 34(44):14707–14716

Matsui A, Williams JT (2011) Opioid-sensitive GABA inputs from rostromedial tegmental nucleus synapse onto midbrain dopamine neurons. J Neurosci 31(48):17729–17735

Matsui A, Jarvie BC, Robinson BG, Hentges ST, Williams JT (2014) Separate GABA afferents to dopamine neurons mediate acute action of opioids, development of tolerance, and expression of withdrawal. Neuron 82(6):1346–1356

Matthews RT, German DC (1984) Electrophysiological evidence for excitation of rat ventral tegmental area dopamine neurons by morphine. Neuroscience 11(3):617–625

Meng S, Quan W, Qi X, Su Z, Yang S (2014) Effect of baclofen on morphine-induced conditioned place preference, extinction, and stress-induced reinstatement in chronically stressed mice. Psychopharmacology 231(1):27–36

Mombereau C, Kaupmann K, Froestl W, Sansig G, van der Putten H, Cryan JF (2004) Genetic and pharmacological evidence of a role for GABA(B) receptors in the modulation of anxiety- and antidepressant-like behavior. Neuropsychopharmacology 29(6):1050–1062

Mombereau C, Lhuillier L, Kaupmann K, Cryan JF (2007) GABAB receptor-positive modulation-induced blockade of the rewarding properties of nicotine is associated with a reduction in nucleus accumbens DeltaFosB accumulation. J Pharmacol Exp Ther 321(1):172–177

Morrisett RA, Mott DD, Lewis DV, Swartzwelder HS, Wilson WA (1991) GABAB-receptor-mediated inhibition of the N-methyl-D-aspartate component of synaptic transmission in the rat hippocampus. J Neurosci 11(1):203–209

Munoz MB, Padgett CL, Rifkin R, Terunuma M, Wickman K, Contet C, Moss SJ, Slesinger PA (2016) A role for the GIRK3 subunit in methamphetamine-induced attenuation of GABAB receptor-activated GIRK currents in VTA dopamine neurons. J Neurosci 36(11):3106–3114

Negrete-Diaz JV, Shumilov K, Real MA, Medina-Luque J, Valderrama-Carvajal A, Flores G, Rodriguez-Moreno A, Rivera A (2019) Pharmacological activation of dopamine D4 receptor modulates morphine-induced changes in the expression of GAD65/67 and GABAB receptors in the basal ganglia. Neuropharmacology 152:22–29

Nestler EJ, Terwilliger RZ, Walker JR, Sevarino KA, Duman RS (1990) Chronic cocaine treatment decreases levels of the G protein subunits Gi alpha and Go alpha in discrete regions of rat brain. J Neurochem 55(3):1079–1082

Nieh EH, Matthews GA, Allsop SA, Presbrey KN, Leppla CA, Wichmann R, Neve R, Wildes CP, Tye KM (2015) Decoding neural circuits that control compulsive sucrose seeking. Cell 160 (3):528–541

Nieh EH, Vander Weele CM, Matthews GA, Presbrey KN, Wichmann R, Leppla CA, Izadmehr EM, Tye KM (2016) Inhibitory input from the lateral hypothalamus to the ventral tegmental area disinhibits dopamine neurons and promotes behavioral activation. Neuron 90(6):1286–1298

Odagaki Y, Koyama T (2001) Identification of galpha subtype(s) involved in gamma-aminobutyric acid(B) receptor-mediated high-affinity guanosine triphosphatase activity in rat cerebral cortical membranes. Neurosci Lett 297(2):137–141

Olds J, Milner P (1954) Positive reinforcement produced by electrical stimulation of septal area and other regions of rat brain. J Comp Physiol Psychol 47(6):419–427

Olpe HR, Steinmann MW, Ferrat T, Pozza MF, Greiner K, Brugger F, Froestl W, Mickel SJ, Bittiger H (1993) The actions of orally active GABAB receptor antagonists on GABAergic transmission in vivo and in vitro. Eur J Pharmacol 233(2–3):179–186

Overton PG, Richards CD, Berry MS, Clark D (1999) Long-term potentiation at excitatory amino acid synapses on midbrain dopamine neurons. Neuroreport 10(2):221–226

Padgett CL, Lalive AL, Tan KR, Terunuma M, Munoz MB, Pangalos MN, Martinez-Hernandez J, Watanabe M, Moss SJ, Lujan R, Luscher C, Slesinger PA (2012) Methamphetamine-evoked depression of GABA(B) receptor signaling in GABA neurons of the VTA. Neuron 73 (5):978–989

Paladini CA, Tepper JM (1999) GABA(A) and GABA(B) antagonists differentially affect the firing pattern of substantia nigra dopaminergic neurons in vivo. Synapse 32(3):165–176

Palpacuer C, Duprez R, Huneau A, Locher C, Boussageon R, Laviolle B, Naudet F (2018) Pharmacologically controlled drinking in the treatment of alcohol dependence or alcohol use disorders: a systematic review with direct and network meta-analyses on nalmefene, naltrexone, acamprosate, baclofen and topiramate. Addiction 113(2):220–237

Paterson NE, Froestl W, Markou A (2004) The GABAB receptor agonists baclofen and CGP44532 decreased nicotine self-administration in the rat. Psychopharmacology 172(2):179–186

Paterson NE, Bruijnzeel AW, Kenny PJ, Wright CD, Froestl W, Markou A (2005a) Prolonged nicotine exposure does not alter GABA(B) receptor-mediated regulation of brain reward function. Neuropharmacology 49(7):953–962

Paterson NE, Froestl W, Markou A (2005b) Repeated administration of the GABAB receptor agonist CGP44532 decreased nicotine self-administration, and acute administration decreased cue-induced reinstatement of nicotine-seeking in rats. Neuropsychopharmacology 30 (1):119–128

Paterson NE, Vlachou S, Guery S, Kaupmann K, Froestl W, Markou A (2008) Positive modulation of GABA(B) receptors decreased nicotine self-administration and counteracted nicotine-induced enhancement of brain reward function in rats. J Pharmacol Exp Ther 326(1):306–314

Pedron VT, Varani AP, Balerio GN (2016) Baclofen prevents the elevated plus maze behavior and BDNF expression during naloxone precipitated morphine withdrawal in male and female mice. Synapse 70(5):187–197

Perdona E, Costantini VJ, Tessari M, Martinelli P, Carignani C, Valerio E, Mok MH, Zonzini L, Visentini F, Gianotti M, Gordon L, Rocheville M, Corsi M, Capelli AM (2011) In vitro and in vivo characterization of the novel GABAB receptor positive allosteric modulator, 2-{1-[2-(4-chlorophenyl)-5-methylpyrazolo[1,5-a]pyrimidin-7-yl]-2-piperidinyl}ethanol (CMPPE). Neuropharmacology 61(5–6):957–966

Perrotti LI, Bolanos CA, Choi KH, Russo SJ, Edwards S, Ulery PG, Wallace DL, Self DW, Nestler EJ, Barrot M (2005) DeltaFosB accumulates in a GABAergic cell population in the posterior tail of the ventral tegmental area after psychostimulant treatment. Eur J Neurosci 21(10):2817–2824

Phillips TJ, Reed C (2014) Targeting GABAB receptors for anti-abuse drug discovery. Expert Opin Drug Discov 9(11):1307–1317

Pitman KA, Puil E, Borgland SL (2014) GABA(B) modulation of dopamine release in the nucleus accumbens core. Eur J Neurosci 40(10):3472–3480

Polter AM, Barcomb K, Tsuda AC, Kauer JA (2018) Synaptic function and plasticity in identified inhibitory inputs onto VTA dopamine neurons. Eur J Neurosci 47(10):1208–1218

Prosser HM, Gill CH, Hirst WD, Grau E, Robbins M, Calver A, Soffin EM, Farmer CE, Lanneau C, Gray J, Schenck E, Warmerdam BS, Clapham C, Reavill C, Rogers DC, Stean T, Upton N, Humphreys K, Randall A, Geppert M, Davies CH, Pangalos MN (2001) Epileptogenesis and enhanced prepulse inhibition in GABA(B1)-deficient mice. Mol Cell Neurosci 17 (6):1059–1070

Ramshini E, Alaei H, Reisi P, Alaei S, Shahidani S (2013) The role of GABAB receptors in morphine self-administration. Int J Prev Med 4(2):158–164

Ranaldi R (2014) Dopamine and reward seeking: the role of ventral tegmental area. Rev Neurosci 25(5):621–630

Ranaldi R, Poeggel K (2002) Baclofen decreases methamphetamine self-administration in rats. Neuroreport 13(9):1107–1110

Riahi E, Mirzaii-Dizgah I, Karimian SM, Sadeghipour HR, Dehpour AR (2009) Attenuation of morphine withdrawal signs by a GABAB receptor agonist in the locus coeruleus of rats. Behav Brain Res 196(1):11–14

Rifkin RA, Huyghe D, Li X, Parakala M, Aisenberg E, Moss SJ, Slesinger PA (2018) GIRK currents in VTA dopamine neurons control the sensitivity of mice to cocaine-induced locomotor sensitization. Proc Natl Acad Sci U S A 115(40):E9479–E9488

Roberts DC, Andrews MM (1997) Baclofen suppression of cocaine self-administration: demonstration using a discrete trials procedure. Psychopharmacology 131(3):271–277

Roberts DC, Andrews MM, Vickers GJ (1996) Baclofen attenuates the reinforcing effects of cocaine in rats. Neuropsychopharmacology 15(4):417–423

Rotheram-Fuller E, De La Garza R 2nd, Mahoney JJ 3rd, Shoptaw S, Newton TF (2007) Subjective and cardiovascular effects of cocaine during treatment with amantadine and baclofen in combination. Psychiatry Res 152(2–3):205–210

Saal D, Dong Y, Bonci A, Malenka RC (2003) Drugs of abuse and stress trigger a common synaptic adaptation in dopamine neurons. Neuron 37(4):577–582

Saigusa T, Aono Y, Sekino R, Uchida T, Takada K, Oi Y, Koshikawa N, Cools AR (2012) In vivo neurochemical evidence that newly synthesised GABA activates GABA(B), but not GABA(A), receptors on dopaminergic nerve endings in the nucleus accumbens of freely moving rats. Neuropharmacology 62(2):907–913

Schuler V, Luscher C, Blanchet C, Klix N, Sansig G, Klebs K, Schmutz M, Heid J, Gentry C, Urban L, Fox A, Spooren W, Jaton AL, Vigouret J, Pozza M, Kelly PH, Mosbacher J, Froestl W, Kaslin E, Korn R, Bischoff S, Kaupmann K, van der Putten H, Bettler B (2001) Epilepsy, hyperalgesia, impaired memory, and loss of pre- and postsynaptic GABA (B) responses in mice lacking GABA(B(1)). Neuron 31(1):47–58

Schwenk J, Metz M, Zolles G, Turecek R, Fritzius T, Bildl W, Tarusawa E, Kulik A, Unger A, Ivankova K, Seddik R, Tiao JY, Rajalu M, Trojanova J, Rohde V, Gassmann M, Schulte U, Fakler B, Bettler B (2010) Native GABA(B) receptors are heteromultimers with a family of auxiliary subunits. Nature 465(7295):231–235

Seddik R, Jungblut SP, Silander OK, Rajalu M, Fritzius T, Besseyrias V, Jacquier V, Fakler B, Gassmann M, Bettler B (2012) Opposite effects of KCTD subunit domains on GABA (B) receptor-mediated desensitization. J Biol Chem 287(47):39869–39877

Self DW (2004) Regulation of drug-taking and -seeking behaviors by neuroadaptations in the mesolimbic dopamine system. Neuropharmacology 47(Suppl 1):242–255

Selley DE, Nestler EJ, Breivogel CS, Childers SR (1997) Opioid receptor-coupled G-proteins in rat locus coeruleus membranes: decrease in activity after chronic morphine treatment. Brain Res 746(1–2):10–18

Selley DE, Cassidy MP, Martin BR, Sim-Selley LJ (2004) Long-term administration of Delta9-tetrahydrocannabinol desensitizes CB1-, adenosine A1-, and GABAB-mediated inhibition of adenylyl cyclase in mouse cerebellum. Mol Pharmacol 66(5):1275–1284

Sharpe AL, Varela E, Bettinger L, Beckstead MJ (2014) Methamphetamine self-administration in mice decreases GIRK channel-mediated currents in midbrain dopamine neurons. Int J Neuropsychopharmacol 18(5)

Sharpe MJ, Marchant NJ, Whitaker LR, Richie CT, Zhang YJ, Campbell EJ, Koivula PP, Necarsulmer JC, Mejias-Aponte C, Morales M, Pickel J, Smith JC, Niv Y, Shaham Y, Harvey BK, Schoenbaum G (2017) Lateral hypothalamic GABAergic neurons encode reward predictions that are relayed to the ventral tegmental area to regulate learning. Curr Biol 27 (14):2089–2100.e5

Shoaib M, Swanner LS, Beyer CE, Goldberg SR, Schindler CW (1998) The GABAB agonist baclofen modifies cocaine self-administration in rats. Behav Pharmacol 9(3):195–206

Shoji S, Simms D, McDaniel WC, Gallagher JP (1997) Chronic cocaine enhances gamma-aminobutyric acid and glutamate release by altering presynaptic and not postsynaptic gamma-aminobutyric acidB receptors within the rat dorsolateral septal nucleus. J Pharmacol Exp Ther 280(1):129–137

Shoptaw S, Yang X, Rotheram-Fuller EJ, Hsieh YC, Kintaudi PC, Charuvastra VC, Ling W (2003) Randomized placebo-controlled trial of baclofen for cocaine dependence: preliminary effects for individuals with chronic patterns of cocaine use. J Clin Psychiatry 64(12):1440–1448

Sim LJ, Hampson RE, Deadwyler SA, Childers SR (1996) Effects of chronic treatment with delta9-tetrahydrocannabinol on cannabinoid-stimulated [35S]GTPgammaS autoradiography in rat brain. J Neurosci 16(24):8057–8066

Sinha R (2008) Chronic stress, drug use, and vulnerability to addiction. Ann N Y Acad Sci 1141:105–130

Slattery DA, Markou A, Froestl W, Cryan JF (2005) The GABAB receptor-positive modulator GS39783 and the GABAB receptor agonist baclofen attenuate the reward-facilitating effects of cocaine: intracranial self-stimulation studies in the rat. Neuropsychopharmacology 30 (11):2065–2072

Smith MA, Yancey DL, Morgan D, Liu Y, Froestl W, Roberts DC (2004) Effects of positive allosteric modulators of the GABAB receptor on cocaine self-administration in rats. Psychopharmacology 173(1–2):105–111
Smith KS, Tindell AJ, Aldridge JW, Berridge KC (2009) Ventral pallidum roles in reward and motivation. Behav Brain Res 196(2):155–167
Soden ME, Chung AS, Cuevas B, Resnick JM, Awatramani R, Zweifel LS (2020) Anatomic resolution of neurotransmitter-specific projections to the VTA reveals diversity of GABAergic inputs. Nat Neurosci
Steketee JD, Beyer CE (2005) Injections of baclofen into the ventral medial prefrontal cortex block the initiation, but not the expression, of cocaine sensitization in rats. Psychopharmacology 180 (2):352–358
Sturchler E, Li X, de Lourdes Ladino M, Kaczanowska K, Cameron M, Griffin PR, Finn MG, Markou A, McDonald P (2017) GABAB receptor allosteric modulators exhibit pathway-dependent and species-selective activity. Pharmacol Res Perspect 5(2):e00288
Sugita S, Johnson SW, North RA (1992) Synaptic inputs to GABAA and GABAB receptors originate from discrete afferent neurons. Neurosci Lett 134(2):207–211
Takahashi T, Kajikawa Y, Tsujimoto T (1998) G-protein-coupled modulation of presynaptic calcium currents and transmitter release by a GABAB receptor. J Neurosci 18(9):3138–3146
Tan KR, Brown M, Labouebe G, Yvon C, Creton C, Fritschy JM, Rudolph U, Luscher C (2010) Neural bases for addictive properties of benzodiazepines. Nature 463(7282):769–774
Tan D, Nuno-Perez A, Mameli M, Meye FJ (2018) Cocaine withdrawal reduces GABAB R transmission at entopeduncular nucleus – lateral habenula synapses. Eur J Neurosci 50 (3):2124–2133
Taylor SR, Badurek S, Dileone RJ, Nashmi R, Minichiello L, Picciotto MR (2014) GABAergic and glutamatergic efferents of the mouse ventral tegmental area. J Comp Neurol 522(14):3308–3334
Thompson SM, Gahwiler BH (1992) Comparison of the actions of baclofen at pre- and postsynaptic receptors in the rat hippocampus in vitro. J Physiol 451:329–345
Tsuji M, Nakagawa Y, Ishibashi Y, Yoshii T, Takashima T, Shimada M, Suzuki T (1996) Activation of ventral tegmental GABAB receptors inhibits morphine-induced place preference in rats. Eur J Pharmacol 313(3):169–173
Turecek R, Schwenk J, Fritzius T, Ivankova K, Zolles G, Adelfinger L, Jacquier V, Besseyrias V, Gassmann M, Schulte U, Fakler B, Bettler B (2014) Auxiliary GABAB receptor subunits uncouple G protein betagamma subunits from effector channels to induce desensitization. Neuron 82(5):1032–1044
Tyacke RJ, Lingford-Hughes A, Reed LJ, Nutt DJ (2010) GABAB receptors in addiction and its treatment. In: GABAB receptor pharmacology – a tribute to Norman bowery advances in pharmacology, pp 373–396
Ulrich D, Bettler B (2007) GABA(B) receptors: synaptic functions and mechanisms of diversity. Curr Opin Neurobiol 17(3):298–303
Ungless MA, Whistler JL, Malenka RC, Bonci A (2001) Single cocaine exposure in vivo induces long-term potentiation in dopamine neurons. Nature 411(6837):583–587
Ungless MA, Singh V, Crowder TL, Yaka R, Ron D, Bonci A (2003) Corticotropin-releasing factor requires CRF binding protein to potentiate NMDA receptors via CRF receptor 2 in dopamine neurons. Neuron 39(3):401–407
Urwyler S (2011) Allosteric modulation of family C G-protein-coupled receptors: from molecular insights to therapeutic perspectives. Pharmacol Rev 63(1):59–126
Urwyler S, Mosbacher J, Lingenhoehl K, Heid J, Hofstetter K, Froestl W, Bettler B, Kaupmann K (2001) Positive allosteric modulation of native and recombinant gamma-aminobutyric acid (B) receptors by 2,6-Di-tert-butyl-4-(3-hydroxy-2,2-dimethyl-propyl)-phenol (CGP7930) and its aldehyde analog CGP13501. Mol Pharmacol 60(5):963–971
Urwyler S, Pozza MF, Lingenhoehl K, Mosbacher J, Lampert C, Froestl W, Koller M, Kaupmann K (2003) N,N'-Dicyclopentyl-2-methylsulfanyl-5-nitro-pyrimidine-4,6-diamine (GS39783) and

structurally related compounds: novel allosteric enhancers of gamma-aminobutyric acidB receptor function. J Pharmacol Exp Ther 307(1):322–330

Van Etten ML, Anthony JC (1999) Comparative epidemiology of initial drug opportunities and transitions to first use: marijuana, cocaine, hallucinogens and heroin. Drug Alcohol Depend 54 (2):117–125

Varani AP, Balerio GN (2012) GABA(B) receptors involvement in the effects induced by nicotine on anxiety-related behaviour in mice. Pharmacol Res 65(5):507–513

Varani AP, Moutinho LM, Bettler B, Balerio GN (2012) Acute behavioural responses to nicotine and nicotine withdrawal syndrome are modified in GABAB1 knockout mice. Neuropharmacology 63(5):863–872

Varani AP, Pedron VT, Bettler B, Balerio GN (2014) Involvement of GABAB receptors in biochemical alterations induced by anxiety-related responses to nicotine in mice: genetic and pharmacological approaches. Neuropharmacology 81:31–41

Varani AP, Pedron VT, Machado LM, Antonelli MC, Bettler B, Balerio GN (2015) Lack of GABAB receptors modifies behavioural and biochemical alterations induced by precipitated nicotine withdrawal. Neuropharmacology 90:90–101

Varani AP, Pedron VT, Aon AJ, Hocht C, Acosta GB, Bettler B, Balerio GN (2018) Nicotine-induced molecular alterations are modulated by GABAB receptor activity. Addict Biol 23 (1):230–246

Vengeliene V, Takahashi TT, Dravolina OA, Belozertseva I, Zvartau E, Bespalov AY, Spanagel R (2018) Efficacy and side effects of baclofen and the novel GABAB receptor positive allosteric modulator CMPPE in animal models for alcohol and cocaine addiction. Psychopharmacology 235(7):1955–1965

Vigot R, Barbieri S, Brauner-Osborne H, Turecek R, Shigemoto R, Zhang YP, Lujan R, Jacobson LH, Biermann B, Fritschy JM, Vacher CM, Muller M, Sansig G, Guetg N, Cryan JF, Kaupmann K, Gassmann M, Oertner TG, Bettler B (2006) Differential compartmentalization and distinct functions of $GABA_B$ receptor variants. Neuron 50(4):589–601

Vlachou S, Markou A (2010) GABAB receptors in reward processes. In: GABAB receptor pharmacology – a tribute to Norman bowery advances in pharmacology, pp 315–371

Vlachou S, Guery S, Froestl W, Banerjee D, Benedict J, Finn MG, Markou A (2011) Repeated administration of the GABAB receptor positive modulator BHF177 decreased nicotine self-administration, and acute administration decreased cue-induced reinstatement of nicotine seeking in rats. Psychopharmacology 215(1):117–128

Voigt RM, Herrold AA, Napier TC (2011a) Baclofen facilitates the extinction of methamphetamine-induced conditioned place preference in rats. Behav Neurosci 125 (2):261–267

Voigt RM, Herrold AA, Riddle JL, Napier TC (2011b) Administration of GABA(B) receptor positive allosteric modulators inhibit the expression of previously established methamphetamine-induced conditioned place preference. Behav Brain Res 216(1):419–423

Wanat MJ, Willuhn I, Clark JJ, Phillips PE (2009) Phasic dopamine release in appetitive behaviors and drug addiction. Curr Drug Abuse Rev 2(2):195–213

Wang B, Shaham Y, Zitzman D, Azari S, Wise RA, You ZB (2005) Cocaine experience establishes control of midbrain glutamate and dopamine by corticotropin-releasing factor: a role in stress-induced relapse to drug seeking. J Neurosci 25(22):5389–5396

Weerts EM, Froestl W, Griffiths RR (2005) Effects of GABAergic modulators on food and cocaine self-administration in baboons. Drug Alcohol Depend 80(3):369–376

Weerts EM, Froestl W, Kaminski BJ, Griffiths RR (2007) Attenuation of cocaine-seeking by GABA B receptor agonists baclofen and CGP44532 but not the GABA reuptake inhibitor tiagabine in baboons. Drug Alcohol Depend 89(2–3):206–213

Westerink BH, Kwint HF, deVries JB (1996) The pharmacology of mesolimbic dopamine neurons: a dual-probe microdialysis study in the ventral tegmental area and nucleus accumbens of the rat brain. J Neurosci 16(8):2605–2611

Williams CL, Buchta WC, Riegel AC (2014) CRF-R2 and the heterosynaptic regulation of VTA glutamate during reinstatement of cocaine seeking. J Neurosci 34(31):10402–10414
Willick ML, Kokkinidis L (1995) The effects of ventral tegmental administration of GABAA, GABAB and NMDA receptor agonists on medial forebrain bundle self-stimulation. Behav Brain Res 70(1):31–36
Xi ZX, Stein EA (1998) Nucleus accumbens dopamine release modulation by mesolimbic GABAA receptors-an in vivo electrochemical study. Brain Res 798(1–2):156–165
Xi ZX, Stein EA (1999) Baclofen inhibits heroin self-administration behavior and mesolimbic dopamine release. J Pharmacol Exp Ther 290(3):1369–1374
Xi ZX, Ramamoorthy S, Shen H, Lake R, Samuvel DJ, Kalivas PW (2003) GABA transmission in the nucleus accumbens is altered after withdrawal from repeated cocaine. J Neurosci 23 (8):3498–3505
Xia Y, Driscoll JR, Wilbrecht L, Margolis EB, Fields HL, Hjelmstad GO (2011) Nucleus accumbens medium spiny neurons target non-dopaminergic neurons in the ventral tegmental area. J Neurosci 31(21):7811–7816
Yang B, Wang BF, Lai MJ, Zhang FQ, Yang XW, Zhou WH, Lian QQ (2011) Differential involvement of GABAA and GABAB receptors in propofol self-administration in rats. Acta Pharmacol Sin 32(12):1460–1465
Yang H, de Jong JW, Tak Y, Peck J, Bateup HS, Lammel S (2018) Nucleus accumbens subnuclei regulate motivated behavior via direct inhibition and disinhibition of VTA dopamine subpopulations. Neuron 97(2):434–449.e4
Yoon SS, Lee BH, Kim HS, Choi KH, Yun J, Jang EY, Shim I, Kim JA, Kim MR, Yang CH (2007) Potential roles of GABA receptors in morphine self-administration in rats. Neurosci Lett 428 (1):33–37
Young KA, Franklin TR, Roberts DC, Jagannathan K, Suh JJ, Wetherill RR, Wang Z, Kampman KM, O'Brien CP, Childress AR (2014) Nipping cue reactivity in the bud: baclofen prevents limbic activation elicited by subliminal drug cues. J Neurosci 34(14):5038–5043
Yu G, Sharp BM (2015) Basolateral amygdala and ventral hippocampus in stress-induced amplification of nicotine self-administration during reacquisition in rat. Psychopharmacology 232 (15):2741–2749
Zhang XF, Hu XT, White FJ, Wolf ME (1997) Increased responsiveness of ventral tegmental area dopamine neurons to glutamate after repeated administration of cocaine or amphetamine is transient and selectively involves AMPA receptors. J Pharmacol Exp Ther 281(2):699–706
Zhang K, Tarazi FI, Campbell A, Baldessarini RJ (2000) GABA(B) receptors: altered coupling to G-proteins in rats sensitized to amphetamine. Neuroscience 101(1):5–10
Zhao-Shea R, Liu L, Soll LG, Improgo MR, Meyers EE, McIntosh JM, Grady SR, Marks MJ, Gardner PD, Tapper AR (2011) Nicotine-mediated activation of dopaminergic neurons in distinct regions of the ventral tegmental area. Neuropsychopharmacology 36(5):1021–1032
Zheng S, Abreu N, Levitz J, Kruse AC (2019) Structural basis for KCTD-mediated rapid desensitization of GABAB signalling. Nature 567(7746):127–131
Zuo H, Glaaser I, Zhao Y, Kurinov I, Mosyak L, Wang H, Liu J, Park J, Frangaj A, Sturchler E, Zhou M, McDonald P, Geng Y, Slesinger PA, Fan QR (2019) Structural basis for auxiliary subunit KCTD16 regulation of the GABAB receptor. Proc Natl Acad Sci U S A 116 (17):8370–8379

$GABA_B$ Receptors and Alcohol Use Disorders: Preclinical Studies

August F. Holtyn and Elise M. Weerts

Contents

Abstract Preclinical research over the past several decades has demonstrated a role for the γ-aminobutyric $acid_B$ ($GABA_B$) receptor in alcohol use disorder (AUD). This chapter offers an examination of preclinical evidence on the role of the $GABA_B$ receptor on alcohol-related behaviors with a particular focus on the $GABA_B$ receptor agonist baclofen, for which effects have been most extensively studied, and positive allosteric modulators (PAMs) of the $GABA_B$ receptor. Studies employing rodent and non-human primate models have shown that activation of the $GABA_B$ receptor can reduce (1) stimulating and rewarding effects of alcohol; (2) signs of alcohol withdrawal in rats made physically dependent on alcohol; (3) acquisition and maintenance of alcohol drinking under a two-bottle alcohol versus water choice procedure; (4) alcohol intake under oral operant self-administration procedures; (5) motivational properties of alcohol measured using extinction and progressive ratio procedures;

A. F. Holtyn (✉) and E. M. Weerts
Johns Hopkins University School of Medicine, Baltimore, MD, USA
e-mail: aholtyn1@jhmi.edu; eweerts@jhmi.edu

Curr Topics Behav Neurosci (2022) 52: 157–194
https://doi.org/10.1007/7854_2020_178
Published Online: 18 August 2020

(6) the increase in alcohol intake after a period of alcohol abstinence (the alcohol deprivation effect or ADE); and (7) the ability of alcohol cues and stress to reinstate alcohol seeking when alcohol is no longer available. Baclofen and $GABA_B$ PAMs reduce the abovementioned behaviors across different preclinical models, which provides strong evidence for a significant role of the $GABA_B$ receptor in alcohol-related behaviors and supports development of medications targeting $GABA_B$ receptors for the treatment of AUD. This chapter highlights the value of examining mechanisms of alcohol-related behaviors across multiple animal models to increase the confidence in identification of new therapeutic targets.

Keywords Alcohol · Animal models · Baclofen · GABA · Positive allosteric modulators

1 $GABA_B$ Receptors and Alcohol Use Disorders: Preclinical Studies

Alcohol use disorder (AUD) can be a chronic, relapsing condition with many individuals returning to heavy drinking after detoxification and abstinence. Advances in our understanding of the behavioral, neurobiological, genetic, and environmental mechanisms that perpetuate heavy alcohol use will shed light on the most promising methods to promote sustained reductions in alcohol use. Central to the advancement of our understanding of these mechanisms, and their interactions, is the use of preclinical animal models that capture key aspects of AUD. While it is not possible to model all aspects of the human disorder in animals, models have been developed to study different features related to AUD. Examining these different features separately affords more precise experimental control of alcohol exposure and allows for control over the multiple, confounding influences involved in the development and maintenance of AUD.

In considering animal models, it is important to remember that models are constantly evolving as we learn more about AUD and its diagnosis. For example, the fourth edition of the *Diagnostic and Statistical Manual of Mental Disorders* (DSM-IV) had two distinct disorders, alcohol abuse and alcohol dependence, with specific diagnostic criteria for each. Diagnosis of alcohol dependence differed from the historical use of the term "dependence," which referred to a physiological adaption to chronic alcohol consumption in which alcohol withdrawal syndrome occurred when alcohol use was discontinued. This differentiation is important as physical dependence is a function of dose and duration of drug exposure, and occurs outside of substance use disorders (e.g., medications for pain, anxiety, or depression). The fifth edition of the DSM (DSM-5) now has a single disorder called AUD with mild, moderate, and severe subclassifications based on the number of diagnostic criteria met. AUD includes a range of drinking patterns and behavioral components, and spans a continuum of severity. Thus, the use of multiple animal models that capture different features of AUD and encompass this continuum is important.

The present chapter offers an examination of preclinical evidence on the role of the $GABA_B$ receptor in alcohol-related behaviors with a particular focus on $GABA_B$ receptor agonists and positive modulators, for which effects have been most extensively studied. This includes baclofen, a $GABA_B$ receptor agonist that has been studied in rodents and non-human primates, and positive allosteric modulators of $GABA_B$ receptors ($GABA_B$ PAMs) – CGP7930, GS39783, BHF177, rac-BHFF, ADX71441, CMPPE, COR659, and ORM-27669 – studied primarily in rodent models. $GABA_B$ PAMs bind to sites of the $GABA_B$ receptor that are distinct from the binding site of endogenous GABA or direct agonists, such as baclofen. $GABA_B$ PAMs have no or modest intrinsic agonist activity; they activate the $GABA_B$ receptor system only when and where endogenous GABA has been released and can potentiate the action of endogenous GABA (or agonists; Urwyler 2016). The following sections review preclinical evidence featuring baclofen (see Table 1 for study details) and $GABA_B$ PAMs (see Table 2 for study details) and are organized by the procedures utilized.

2 Locomotor Activity

Acutely, alcohol produces a biphasic effect on spontaneous locomotor behavior in rodents, where lower doses stimulate activity and very high doses suppress activity and produce a loss of the righting reflex (i.e., the rodent no longer flips onto its feet when placed on its back) (Pohorecky 1977). Alcohol-induced stimulation of locomotor activity has been used to model the stimulating and euphorigenic effects of alcohol in humans. It has been proposed that a drug that can reduce alcohol-induced stimulation of locomotor activity in rodents may also reduce the euphorigenic effects of alcohol in humans (Wise and Bozarth 1987). In a seminal study, acute intraperitoneal (i.p.) administration of 5 mg/kg baclofen prior to alcohol administration prevented alcohol-induced stimulation of locomotor activity in mice (Cott et al. 1976). This finding was replicated in subsequent studies showing that acute administration of baclofen prior to alcohol prevented increases in locomotor activity induced by low-to-moderate doses of alcohol in several different rat and mouse strains (Arias et al. 2009; Boehm II et al. 2002; Broadbent and Harless 1999; Chester and Cunningham 1999; Holstein et al. 2009a; Humeniuk et al. 1993; Quintanilla et al. 2008; Shen et al. 1998). Acute administration of the $GABA_B$ PAM GS39783 (30 mg/kg, i.p.) also has been shown to suppress locomotor activity induced by acute administration of 2 g/kg alcohol in mice (Kruse et al. 2012).

Table 1 Description of studies examining effects of the $GABA_B$ receptor agonist, baclofen, on alcohol-related behaviors

Reference	Dose and route	Treatment duration	Alcohol-related behavior	Experimental procedure	Alcohol dose	Animal	Effect	Outcome
Cott et al. (1976)	5 mg/kg, i.p.	Acute	Locomotor activity	Locomotor activity induced by acute alcohol	2.4 g/kg, i.p.	Female NMRI mice	Decrease	Locomotor activity (activity counts)
Quintanilla et al. (2008)	R(+)-baclofen: 1, 2, and 3 mg/kg, i.p.	Acute	Locomotor activity	Locomotor activity induced by acute alcohol	0.5 g/kg, i.p.	Male and female UChB rats	Dose-dependent decrease	Locomotor activity (number of square crossings)
Arias et al. (2009)	1, 1.5, and 2.5 mg/kg, s.c.	Acute	Locomotor activity	Locomotor activity induced by acute alcohol	2.5 g/kg, i.g.	Male and female preweanling Sprague-Dawley rats	Decrease at 1 and 2.5 mg/kg	Locomotor activity (number of beam breaks)
Humeniuk et al. (1993)	Baclofen: 0.25, 0.5, and 1 mg/kg, i.p.; (+)-baclofen: 0.5, 1, and 5 mg/kg, i.p.; (−)-baclofen: 0.25, 0.5, 0.75, and 1 mg/kg, i.p.	Acute	Locomotor activity	Locomotor activity induced by acute alcohol	1.75 g/kg, i.p.	Female BALB/c mice	Baclofen: dose-dependent decrease; (+)-baclofen: no change; (−)-baclofen: decrease at 0.75 and 1 mg/kg	Locomotor activity (activity counts)
Shen et al. (1998)	0.625, 1.25, 2.5, and 5 mg/kg, i.p.	Acute	Locomotor activity	Locomotor activity induced by acute alcohol	2 g/kg, i.p. (FAST mice); 1.5 g/kg, i.p. (C57BL/6J, and DBA/2J mice)	Male FAST, C57BL/6J, and DBA/2J mice	Dose-dependent decrease (FAST and DBA/2J mice); no alcohol-induced locomotor activity (C57BL/6J mice)	Locomotor activity (distance traveled)

Broadbent and Harless (1999)	5, 6.25, and 7.5 mg/kg, i.p.	Repeated (injections given before each activity trial)	Locomotor activity	Locomotor activity induced by repeated alcohol (4 activity trials at 48-h intervals)	2 g/kg, i.p.	Male DBA/2J mice	Dose-dependent decrease	Locomotor activity (activity counts)
Boehm II et al. (2002)	0.69, 1.39, and 2.77 μg, i.c.v.	Acute	Locomotor activity	Locomotor activity induced by acute alcohol	2 g/kg, i.p.	Male and female FAST mice	Decrease at 1.39 and 2.77 μg	Locomotor activity (distance traveled)
Boehm II et al. (2002)	0.01 and 0.02 μg, intra-VTA	Acute	Locomotor activity	Locomotor activity induced by acute alcohol	2 g/kg, i.p.	Male FAST and SLOW mice	Dose-dependent decrease (anterior VTA) in FAST mice	Locomotor activity (distance traveled)
Holstein et al. (2009b)	0.625, 1.25, 2.5, and 5 mg/kg, i. p.	Acute	Locomotor activity	Locomotor activity induced by acute alcohol	1 g/kg, i.p.	Male FAST mice	Decrease at 5 mg/kg (FAST-1 mice); no change (FAST-2 mice)	Locomotor activity (distance traveled)
Chester and Cunningham (1999)	2.5, 5, and 7.5 mg/kg, i.p.	Repeated (before each alcohol conditioning session)	Conditioned place preference (CPP) and locomotor activity	Eight conditioning sessions, four with alcohol and four with saline	2 g/kg, i.p.	Male DBA/2J mice	No change in CPP; dose-dependent decrease in locomotor activity	Time on grid floor paired with alcohol and locomotor activity (activity counts)
Bechtholt and Cunningham (2005)	25 and 50 ng, intra-VTA	Acute	Conditioned place preference (CPP)	Eight conditioning sessions, four with alcohol and four with saline	2 g/kg, i.p.	Male DBA/2J mice	Decrease	Time on grid floor paired with alcohol
Colombo et al. (2000)	10, 20, and 40 mg/kg, i.p.	Acute	Alcohol withdrawal	Withdrawal from intragastric alcohol gavage	20% w/v	Male Wistar rats	Dose-dependent decrease	Withdrawal severity score

(continued)

Table 1 (continued)

Reference	Dose and route	Treatment duration	Alcohol-related behavior	Experimental procedure	Alcohol dose	Animal	Effect	Outcome
Humeniuk et al. (1994)	5, 10, and 20 mg/kg, i.p.	Acute	Alcohol withdrawal	Withdrawal from alcohol-containing diet	6% v/v	Female C57BL/6 mice	No change in hypothermia, tremor, and tail arch; increase in convulsive behavior at 10 and 20 mg/kg	Hypothermia, tremor, tail arch, and convulsions
File et al. (1991)	1.25 and 2.5 mg/kg, i.p.	Acute	Alcohol withdrawal	Withdrawal from alcohol-containing diet	10% v/v	Male Lister rats	Decreased anxiogenic withdrawal responses and aggression; decreased tremor at 2.5 mg/kg	Anxiogenic withdrawal responses (time spent in social interaction and in open arms of plus maze), aggression (number of aggressive acts), and tremor
Knapp et al. (2007)	1.25 and 2.5 mg/kg, i.p.	Acute	Alcohol withdrawal	Withdrawal from alcohol-containing diet	7% w/v	Sprague-Dawley rats	Attenuation of decreased social interaction	Time spent in social interaction
Knapp et al. (2007)	1.25 mg/kg, i.p.	Acute (during the third withdrawal)	Alcohol withdrawal	Withdrawal from cycled alcohol-containing diet	4.5% w/v	Sprague-Dawley rats	Attenuation of decreased social interaction	Time spent in social interaction

Knapp et al. (2007)	1.25, 2.5, and 5 mg/kg, i.p.	Repeated (during the first and second withdrawal)	Alcohol withdrawal	Withdrawal from cycled alcohol-containing diet	4.5% w/v	Sprague-Dawley rats	Attenuation of decreased social interaction at 2.5 and 5 mg/kg dose	Time spent in social interaction
Knapp et al. (2007)	1.25, 2.5, and 5 mg/kg, i.p.	Repeated (once daily for 2 days prior to stress sessions)	Alcohol withdrawal	Restraint stress and withdrawal from alcohol-containing diet	4.5% w/v	Sprague-Dawley rats	Attenuation of decreased social interaction at 2.5 and 5 mg/kg doses	Time spent in social interaction
Knapp et al. (2007)	5 mg/kg, i.p.	Repeated (2 injections 5 days apart plus one injection prior to a stress session)	Alcohol withdrawal	Restraint stress and withdrawal from cycled alcohol-containing diet	4.5% w/v	Sprague-Dawley rats	Attenuation of decreased social interaction	Time spent in social interaction
Colombo et al. (2002)	1 and 3 mg/kg, i.p.	Repeated (once daily for 10 consecutive days)	Acquisition of alcohol drinking	Two-bottle alcohol versus water choice (unlimited access)	10% v/v	Male sP rats	Decrease at 3 mg/kg	Alcohol intake (g/kg)
Colombo et al. (2005)	1 mg/kg baclofen, i.p. with and without 0.5 mg/kg naltrexone, i.p.	Repeated (once daily for 10 consecutive days)	Acquisition of alcohol drinking	Two-bottle alcohol versus water choice (unlimited access)	10% v/v	Male sP rats	Decrease at 1 mg/kg baclofen + 0.5 mg/kg naltrexone	Alcohol intake (g/kg)
Smith et al. (1992)	10 mg/kg, i.p.	Repeated (once daily for 10 days; every other day)	Acquisition of alcohol drinking	Two-bottle alcohol versus water choice (limited access; every other day)	2, 4, 6, 8, and 10% v/v (increasing across days)	Male Long-Evans rats	Increase	Alcohol intake (g/kg)

(continued)

Table 1 (continued)

Reference	Dose and route	Treatment duration	Alcohol-related behavior	Experimental procedure	Alcohol dose	Animal	Effect	Outcome
Quintanilla et al. (2008)	R(+)-baclofen: 1, 2, and 3 mg/kg, i.p.	Acute	Alcohol drinking	Two-bottle alcohol versus water choice (limited access)	10% v/v	Female UChB rats	Dose-dependent decrease	Alcohol intake (g/kg)
Quintanilla et al. (2008)	R(+)-baclofen: 1, 2, and 3 mg/kg, i.p.	Repeated (once daily for 4 consecutive days)	Alcohol drinking	Two-bottle alcohol versus water choice (unlimited access)	10% v/v	Male and female UChB rats	Dose-dependent decrease	Alcohol intake (g/kg)
Kemppainen et al. (2012)	3, 10, and 30 ng, microinjections into the VP	Acute	Alcohol drinking	Two-bottle alcohol versus water choice (limited access; every other day)	10% v/v	Male AA rats	Dose-dependent decrease	Alcohol intake (g/kg)
Daoust et al. (1987)	3 mg/kg, i.p.	Repeated (once daily for 14 consecutive days)	Alcohol drinking	Three-bottle alcohol versus water choice (unlimited access)	12% v/v	Male Long-Evans rats	Decrease	Alcohol intake (g/kg)
Boas et al. (2012)	1.25, 2.5, and 5 mg/kg, i.p.	Repeated (once daily for 2 consecutive days)	Alcohol drinking	Three-bottle alcohol versus water choice (unlimited access)	5% and 10% v/v	Male Swiss mice	Decrease at 1.25 mg/kg in "light drinkers"	Alcohol intake (g/kg)

Stromberg (2004)	2.5, 5, and 7.5 mg/kg, i.p. with and without 1.0 mg/kg naltrexone, i.p.	Repeated (once daily for 2 consecutive days alone followed by once daily for 2 consecutive days with naltrexone)	Alcohol drinking	Two-bottle alcohol versus water choice (limited access)	8% v/v	Male Wistar rats	Decrease at 5 mg/kg and at all doses in combination with naltrexone	Alcohol intake (g/kg)
Stromberg (2004)	2.5, 5, and 7.5 mg/kg, i.p.	Repeated (once daily for 4 consecutive days)	Alcohol drinking	Two-bottle alcohol versus water choice (limited access)	8% v/v	Male Wistar rats	Dose-dependent decrease	Alcohol intake (g/kg)
Kasten et al. (2015)	R(+)-baclofen: 10 mg/kg, i.p.; S(−)-baclofen: 10 mg/kg, i.p.	Acute	Alcohol drinking	Two-bottle alcohol versus water choice (unlimited access)	10% v/v	Male and female HAP1 mice	R(+)-baclofen: decrease; S(−)-baclofen: increase	Alcohol intake (g/kg)
Peters et al. (2012)	2.5 mg/kg, i.p.	Acute	Alcohol drinking	Two-bottle alcohol versus water choice (unlimited access)	8% v/v	Male C57BL/6N mice	Decrease in socially stressed and non-socially stressed mice	Alcohol consumed (% of baseline)
Smith et al. (1999)	10 mg/kg, i.p.	Repeated (once daily for 5 days, every other day)	Alcohol drinking	Two-bottle alcohol versus water choice (limited access; every other day)	10% v/v	Male Long-Evans rats	Increase	Alcohol intake (g/kg and volume consumed), preference for alcohol (preference ratio)

(continued)

Table 1 (continued)

Reference	Dose and route	Treatment duration	Alcohol-related behavior	Experimental procedure	Alcohol dose	Animal	Effect	Outcome
Colombo et al. (2000)	2.5, 5, and 10 mg/kg, i.p.	Repeated (once daily for 14 consecutive days)	Alcohol drinking	Two-bottle alcohol versus water choice (unlimited access)	10% v/v	Male sP rats	Dose-dependent decrease	Alcohol intake (g/kg)
Colombo et al. (2004)	0.5 and 1 mg/kg, i.p.	Acute	Alcohol drinking	Morphine and WIN 55,212-2 induced alcohol drinking: two-bottle alcohol versus water choice (unlimited access)	10% v/v	Male sP rats	Dose-dependent decrease	Alcohol intake (g/kg)
Kasten et al. (2015)	R(+)-baclofen: 1, 3, and 10 mg/kg, i.p.; S(−)-baclofen: 1, 3, 10, and 30 mg/kg, i.p.	Acute	Binge-like drinking	Drinking in the dark	20% v/v	Male C57BL/6J mice	R(+)-baclofen: decrease at 10 mg/kg; S(−)-baclofen: increase at 10 mg/kg	Alcohol intake (g/kg)
Moore et al. (2009)	0.01 and 0.02 μg, intra-VTA	Acute	Binge-like drinking	Drinking in the dark	20% v/v	Male C57BL/6J mice	Decrease (anterior VTA)	Alcohol intake (g/kg)
Crabbe et al. (2017)	R(+)-baclofen: 5 and 10 mg/kg, i.p.	Acute	Binge-like drinking	Drinking in the dark	20% v/v	Male and female HDID mice	Dose-dependent decrease	Alcohol intake (g/kg)
Tanchuck et al. (2011)	2.5 or 5 mg/kg, i.p.	Acute	Binge-like drinking	Scheduled high alcohol consumption (SHAC)	5% v/v	Male WSC mice	Dose-dependent decrease	Alcohol intake (g/kg)

Colombo et al. (2003a)	1, 1.7, and 3 mg/kg, i.p.	Acute	Relapse-like drinking	Alcohol deprivation effect (7 days of alcohol deprivation after two-bottle alcohol versus water choice; unlimited access)	10% v/v	Male sP rats	Dose-dependent decrease of ADE	Alcohol intake (g/kg)
Colombo et al. (2006)	1 mg/kg, i.p.	Acute	Relapse-like drinking	Alcohol deprivation effect (14 days of alcohol deprivation after four-bottle alcohol versus water choice; (unlimited access)	10%, 20%, and 30% v/v	Male sP rats	Decrease of ADE	Alcohol intake (g/kg)
Vengeliene et al. (2018)	1 and 3 mg/kg, i.p.	Repeated (five injections across alcohol deprivation and return to alcohol access)	Relapse-like drinking	Alcohol deprivation effect (repeated deprivation periods after four-bottle alcohol versus water choice; (unlimited access)	5%, 10%, and 20% v/v	Male Wistar rats	Decrease at 3 mg/kg	Alcohol intake (g/kg)
Dean et al. (2012)	1 mg/kg baclofen, s.c. with and without 3 mg/kg loperamide, i.g.	Acute	Operant self-administration	FR 2	10% v/v (in 0.04% saccharine)	Male Wistar rats	Decrease with and without loperamide (potentiation by loperamide)	Alcohol self-administration (percent change from baseline)

(continued)

Table 1 (continued)

Reference	Dose and route	Treatment duration	Alcohol-related behavior	Experimental procedure	Alcohol dose	Animal	Effect	Outcome
Anstrom et al. (2003)	1.8, 3.2, and 5.6 mg/kg, i.p.	Acute	Operant self-administration	FR 1	10% v/v	Male Long-Evans rats	Dose-dependent decrease	Alcohol intake (g/kg)
Janak and Michael Gill (2003)	0.3, 1, and 3 mg/kg, i.p.	Acute	Operant self-administration	FR 1	10% v/v	Male Long-Evans rats	Decrease by 1 and 3 mg/kg	Lever presses
Walker and Koob (2007)	0.5, 1, 2, and 4 mg/kg, i.p.	Acute	Operant self-administration	FR 1	10% w/v	Dependent and non-dependent male Wistar rats	Dose-dependent decrease	Number of reinforcers and alcohol intake (g/kg)
Lorrai et al. (2016)	Baclofen: 3 mg/kg, i.p.; R(+)-baclofen: 0.75, 1.5, and 3 mg/kg, i.p.; S(−)-baclofen: 6, 12, and 24 mg/kg; i.p.	Acute	Operant self-administration	Concurrent schedule of alcohol and water reinforcement (FR 4 alcohol lever, FR 1 water lever)	15% v/v	Male sP rats	Baclofen: decrease at 3 mg/kg; R(+)-baclofen: decrease at 1.5 and 3 mg/kg; S(−)-baclofen: no change	Lever presses and alcohol intake (g/kg)
Besheer et al. (2004)	1, 3, 10, and 17 mg/kg, i.p.	Acute	Operant self-administration	Concurrent schedule of alcohol and water reinforcement (FR 1 alcohol lever, FR 1 water lever)	10% v/v	Male C57BL/6J mice	Decrease at 10 and 17 mg/kg	Lever presses on the alcohol lever
Liang et al. (2006)	2 and 3 mg/kg, i.p.	Acute	Operant self-administration	FR 3	10% v/v	Male P rats	Decrease at 3 mg/kg	Lever presses

Maccioni et al. (2005)	1.7 and 3 mg/kg, i.p.	Acute	Operant self-administration	FR 4	15% v/v	Male sP rats	Decrease at 3 mg/kg	Lever presses, alcohol intake (g/kg), and latency to first response
Maccioni et al. (2018)	0.03, 0.1, and 0.3 μg; intra-VTA	Acute	Operant self-administration	FR 4	15% v/v	Male sP rats	Decrease at 0.1 and 0.3 μg (posterior VTA)	Lever presses and alcohol intake (g/kg)
Orrù et al. (2012)	1.25, 2.5, and 5 mg/kg, i.p.	Acute	Operant self-administration	FR 3	9% v/v (alcoholic beer)	Male C57BL/6J mice	Decrease at 5 mg/kg	Lever presses and alcohol intake (g/kg)
Williams et al. (2016)	0.3, 1, and 3 mg/kg; i.p.	Acute	Operant self-administration	Stress (yohimbine)-induced self-administration	10% w/v	Male Wistar rats	Decrease at 1 and 3 mg/kg	Number of reinforcers and alcohol intake (g/kg)
Williams et al. (2016)	0.3, 1, and 3 mg/kg, i.p.	Acute	Operant self-administration	FR 1	10% w/v	Male Wistar rats	Decrease at 3 mg/kg	Number of reinforcers and alcohol intake (g/kg)
Maccioni et al. (2012)	1, 1.7, and 3 mg/kg, i.p.	Acute	Operant self-administration	FR 4	15% v/v	Male P, sP, and AA rats	Decrease at 1.7 and 3 mg/kg (P rats) and decrease at 3 mg/kg (sP and AA rats)	Lever presses and alcohol intake (g/kg)
Maccioni et al. (2012)	1, 1.7, and 3 mg/kg, i.p.	Acute	Operant self-administration (motivation to consume alcohol)	PR	15% v/v	Male P, sP, and AA rats	Decrease at 1.7 and 3 mg/kg (P and sP rats) and decrease at 3 mg/kg (AA rats)	PR breakpoint and cumulative responses

(continued)

Table 1 (continued)

Reference	Dose and route	Treatment duration	Alcohol-related behavior	Experimental procedure	Alcohol dose	Animal	Effect	Outcome
Maccioni et al. (2008b)	1 and 3 mg/kg, i.p.	Acute	Operant self-administration (motivation to consume alcohol)	PR	15% v/v	Male sP rats	Dose-dependent decrease	PR breakpoint and cumulative responses
Walker and Koob (2007)	2 mg/kg, i.p.	Acute	Operant self-administration (motivation to consume alcohol)	PR	10% w/v	Dependent and non-dependent male Wistar rats	Decrease	PR breakpoint and cumulative responses
Colombo et al. (2003b)	1, 2, and 3 mg/kg, i.p.	Acute	Motivation to consume alcohol	Within-session extinction	10% v/v	Male sP rats	Dose-dependent decrease	Extinction responses
Duke et al. (2014)	0.3, 0.56, 1, 1.8, and 2.4 mg/kg; i.m.	Acute	Motivation to consume alcohol	Within-session extinction	4% w/v	Male baboons	Decrease at 1.8 and 2.4 mg/kg	Extinction responses
Maccioni et al. (2008a)	3 mg/kg, i.p.	Acute	Reinstatement of alcohol seeking	Cue-induced alcohol seeking	15% v/v	Male sP rats	Decrease	Lever presses, latency to first response, and response rate
Vengeliene et al. (2018)	1 and 3 mg/kg, i.p.	Acute	Reinstatement of alcohol seeking	Cue-induced alcohol seeking	10% v/v	Male Wistar rats	Dose-dependent decrease	Nose pokes

Tanchuck et al. (2011)	2.5 or 5 mg/kg, i.p.	Acute	Operant seeking and self-administration	Sipper	10% v/v	Male C57BL/6J mice	Seeking: decrease at 5 mg/kg; drinking: no change	Seeking: lever presses; drinking: alcohol intake (g/kg)
Czachowski et al. (2006)	0.3, 1, and 3 mg/kg, i.p.	Repeated (once daily for 3 consecutive days)	Operant seeking and self-administration	Sipper	10% v/v	Male Long-Evans rats	Seeking: decrease at 3 mg/kg; drinking: increase at 1 mg/kg	Seeking: lever presses and latency to first lever press; drinking: alcohol intake (g/kg)
Duke et al. (2014)	0.1, 0.3, 0.56, 1.0, 1.8, and 2.4 mg/kg; i.m.	Acute	Operant seeking and self-administration	Three-component chained schedule	4% w/v	Male baboons	Seeking: no change; drinking: decrease at 1.8 and 2.4 mg/kg	Seeking: lever presses and latency to first lever press; drinking: alcohol intake (g/kg)
Holtyn et al. (2017)	0.1, 0.32, 0.56, 1.0, and 1.8 mg/kg; i.m.	Repeated (once daily for 10 consecutive days across alcohol deprivation and return to alcohol access)	Operant seeking and self-administration	Three-component chained schedule	4% w/v	Male baboons	Seeking and drinking: no change	Seeking: lever presses and latency to first lever press; drinking: alcohol intake (g/kg) and number of drinks in drinking bout

(continued)

Table 2 (continued)

Reference	Drug	Dose and route	Treatment duration	Alcohol-related behavior	Experimental procedure	Alcohol dose	Animal	Effect	Outcome
Orru et al. (2005)	GS39783	6.25, 12.5, and 25 mg/kg, i.g.	Repeated (once daily for 5 consecutive days)	Acquisition of alcohol drinking	Two-bottle alcohol versus water choice (unlimited access)	10% v/v	Male sP rats	Dose-dependent decrease	Alcohol intake (g/kg)
Orru et al. (2005)	CGP7930	25, 50, and 100 mg/kg, i.g.	Repeated (once daily for 5 consecutive days)	Acquisition of alcohol drinking	Two-bottle alcohol versus water choice (unlimited access)	10% v/v	Male sP rats	Dose-dependent decrease	Alcohol intake (g/kg)
Orru et al. (2005)	GS39783	50 and 100 mg/kg, i.g.	Repeated (once daily for 5 consecutive days)	Alcohol drinking	Two-bottle alcohol versus water choice (unlimited access)	10% v/v	Male sP rats	Decrease at 100 mg/kg	Alcohol intake (g/kg)
Loi et al. (2013)	Rac-BHFF	50, 100, and 200 mg/kg, i.g.	Repeated (once daily for 7 consecutive days)	Alcohol drinking	Two-bottle alcohol versus water choice (unlimited access)	10% v/v	Male sP rats	Dose-dependent decrease	Alcohol intake (g/kg)
Orru et al. (2005)	CGP7930	50 and 100 mg/kg, i.g.	Repeated (once daily for 5 consecutive days)	Alcohol drinking	Two-bottle alcohol versus water choice (unlimited access)	10% v/v	Male sP rats	Decrease at 100 mg/kg	Alcohol intake (g/kg)

Hwa et al. (2014)	ADX71441	3, 10, 17 mg/kg, i.g.	Acute	Alcohol drinking	Two-bottle alcohol versus water choice (repeated cycles of limited access)	20% w/v	Male C57BL/6J mice	Decrease at 10 and 17 mg/kg	Alcohol intake (g/kg)
Hwa et al. (2014)	ADX71441	3, 10, 30 mg/kg, i.g.	Acute	Binge-like drinking	Drinking in the dark	20% w/v	Male C57BL/6J mice	Dose-dependent decrease	Alcohol intake (g/kg)
Colombo et al. (2015)	GS39783	25, 50, and 100 mg/kg, i.g.	Acute	Binge-like drinking	Four-bottle alcohol versus water choice (limited and unpredictable access)	10%, 20%, and 30% v/v	Male sP rats	Decrease	Alcohol intake (g/kg)
de Miguel et al. (2018)	Rac-BHFF	30 mg/kg, i.p.	Acute	Binge-like drinking	Drinking in the dark	20% v/v	Male C57BL/6J mice	Decrease	Alcohol intake (g/kg)
de Miguel et al. (2018)	ORM-27669	100 mg/kg, i.p.	Acute	Binge-like drinking	Drinking in the dark	20% v/v	Male C57BL/6J mice	Decrease	Alcohol intake (g/kg)
Linsenbardt and Boehm (2014)	GS39783	30 mg/kg, i.p.	Acute	Binge-like drinking	Drinking in the dark	20% v/v	Male C57BL/6J mice	Decrease	Alcohol intake (g/kg)
Vengeliene et al. (2018)	CMPPE	10 and 30 mg/kg, i.p.	Repeated (five injections across alcohol deprivation and return to alcohol access)	Relapse-like drinking	Alcohol deprivation effect (repeated deprivation periods after four-bottle alcohol versus water choice; unlimited access)	5%, 10%, and 20% v/v	Male Wistar rats	Decrease of ADE	Alcohol intake (g/kg)

(continued)

Table 2 (continued)

Reference	Drug	Dose and route	Treatment duration	Alcohol-related behavior	Experimental procedure	Alcohol dose	Animal	Effect	Outcome
Augier et al. (2017)	ADX71441	1, 3, 10, and 30 mg/kg, i.p.	Acute	Operant self-administration	FR 2–3	20% v/v	Dependent and non-dependent male Wistar rats	Dose-dependent decrease (higher potency in alcohol-dependent rats)	Lever presses and reinforcers earned
Lorrai et al. (2019)	GS39783	25, 50, and 100 mg/kg, i.g.	Acute	Operant self-administration	FR 4	15% v/v	Male and female sP rats	Dose-dependent decrease	Lever presses and alcohol intake (g/kg)
Maccioni et al. (2017)	GS39783	2.5, 5, and 10 mg/kg, i.p.	Acute	Operant self-administration	FR 4	15% v/v	Male sP rats	Decrease at 5 and 10 mg/kg	Lever presses and alcohol intake (g/kg)
Maccioni et al. (2007)	GS39783	25, 50, and 100 mg/kg, i.g.	Acute	Operant self-administration	FR 4	15% v/v	Male sP rats	Dose-dependent decrease	Lever presses and alcohol intake (g/kg)
Maccioni et al. (2015)	GS39783	50 mg/kg, i.g.	Repeated (once daily for 10 consecutive days)	Operant self-administration	FR 4	15% v/v	Male sP rats	Decrease	Lever presses and alcohol intake (g/kg)
Maccioni et al. (2015)	GS39783 and baclofen	5 mg/kg, i.g. (GS39783) and 1 mg/kg, i.p. (baclofen)	Acute	Operant self-administration	FR 4	15% v/v	Male sP rats	Decrease	Lever presses and alcohol intake (g/kg)

Maccioni et al. (2012)	GS39783	25, 50, and 100 mg/kg, i.g.	Acute	Operant self-administration	FR 4	15% v/v	Male P, sP, and AA rats	Decrease at all three doses (P and sP rats) and decrease at 50 and 100 mg/kg (AA rats)	Lever presses and alcohol intake (g/kg)
Maccioni et al. (2010b)	Rac-BHFF	50, 100, and 200 mg/kg, i.g.	Acute	Operant self-administration	FR 4	15% v/v	Male sP rats	Dose-dependent decrease	Lever presses and alcohol intake (g/kg)
Maccioni et al. (2015)	Rac-BHFF	50 mg/kg, i.g.	Repeated (once daily for 5 consecutive days)	Operant self-administration	FR 4	15% v/v	Male sP rats	Decrease	Lever presses and alcohol intake (g/kg)
Maccioni et al. (2015)	Rac-BHFF and baclofen	5 mg/kg, i.g. (rac-BHFF) and 1 mg/kg, i.p. (baclofen)	Acute	Operant self-administration	FR 4	15% v/v	Male sP rats	Decrease	Lever presses and alcohol intake (g/kg)
Liang et al. (2006)	CGP7930	10 and 20 mg/kg, i.p.	Acute	Operant self-administration	FR 3	10% v/v	Male P rats	Decrease at 20 mg/kg	Lever presses
Liang et al. (2006)	CGP7930 and baclofen	10 mg/kg (CGP7930) and 2 mg/kg (baclofen), i.p.	Acute	Operant self-administration	FR 3	10% v/v	Male P rats	Decrease	Lever presses
Maccioni et al. (2018)	CGP7930	5, 10, and 20 μg; intra-VTA	Acute	Operant self-administration	FR 4	15% v/v	Male sP rats	Decrease at 10 and 20 μg	Lever presses and alcohol intake (g/kg)
Maccioni et al. (2009)	BHF177	12.5, 25, and 50 mg/kg, i.g.	Acute	Operant self-administration	FR 4	15% v/v	Male sP rats	Decrease at 25 and 50 mg/kg	Lever presses and alcohol intake (g/kg)

(continued)

Table 2 (continued)

Reference	Drug	Dose and route	Treatment duration	Alcohol-related behavior	Experimental procedure	Alcohol dose	Animal	Effect	Outcome
Orrù et al. (2012)	BHF177	3.75, 7.5, 15, and 30 mg/kg, i.p.	Acute	Operant self-administration	FR 3	9% v/v (alcoholic beer)	Male C57BL/6J mice	Decrease at 15 and 30 mg/kg	Lever presses and alcohol intake (g/kg)
Maccioni et al. (2019a)	COR659	2.5, 5, and 10 mg/kg, i.p.	Repeated (once daily for 10 consecutive days)	Operant self-administration	FR 4	15% v/v	Male sP rats	Decrease at 5 mg/kg and 10 mg/kg	Lever presses and alcohol intake (g/kg)
Maccioni et al. (2017)	COR659	2.5, 5, and 10 mg/kg, i.p.	Acute	Operant self-administration	FR 4	15% v/v	Male sP rats	Decrease at 5 and 10 mg/kg	Lever presses and alcohol intake (g/kg)
Maccioni et al. (2019b)	CMPPE	2.5, 5, and 10 mg/kg, i.p.	Acute	Operant self-administration	FR 5	15% v/v	Female sP rats	Decrease at 5 and 10 mg/kg	Lever presses and alcohol intake (g/kg)
Maccioni et al. (2017)	GS39783	2.5, 5, and 10 mg/kg, i.p.	Acute	Operant self-administration (motivation to consume alcohol)	PR	15% v/v	Male sP rats	No significant change	PR breakpoint and cumulative responses
Maccioni et al. (2008b)	GS39783	25, 50, and 100 mg/kg, i.g.	Acute	Operant self-administration (motivation to consume alcohol)	PR	15% v/v	Male sP rats	Dose-dependent decrease	PR breakpoint and cumulative responses
Augier et al. (2017)	ADX71441	3 and 10 mg/kg, i.p.	Acute	Operant self-administration (motivation to consume alcohol)	PR	20% v/v	Male Wistar rats	Dose-dependent decrease	PR breakpoint

Maccioni et al. (2012)	GS39783	25, 50, and 100 mg/kg, i.g.	Acute	Operant self-administration	PR	15% v/v	Male P, sP, and AA rats	Decrease at all three doses (P and sP rats) and no significant change (AA rats)	PR breakpoint and cumulative responses
Maccioni et al. (2009)	BHF177	12.5, 25, and 50 mg/kg, i.g.	Acute	Operant self-administration (motivation to consume alcohol)	PR	15% v/v	Male sP rats	Decrease at 50 mg/kg	PR breakpoint and cumulative responses
Maccioni et al. (2017)	COR659	2.5, 5, and 10 mg/kg, i.p.	Acute	Operant self-administration (motivation to consume alcohol)	PR	15% v/v	Male sP rats	Decrease at 5 and 10 mg/kg	PR breakpoint and cumulative responses
Maccioni et al. (2019b)	CMPPE	2.5, 5, and 10 mg/kg, i.p.	Acute	Operant self-administration (motivation to consume alcohol)	PR	15% v/v	Female sP rats	Decrease at 5 and 10 mg/kg	PR breakpoint and cumulative responses
Maccioni et al. (2010a)	GS39783	25, 50, and 100 mg/kg, i.g.	Acute	Operant self-administration	Sipper	15% v/v	Male sP rats	Seeking: decrease at 100 mg/kg in response requirement achieved and no significant change in latency; drinking: decrease at all three doses	Seeking: response requirement achieved and latency to first response; drinking: number of licks and alcohol intake (g/kg)

(continued)

Table 2 (continued)

Reference	Drug	Dose and route	Treatment duration	Alcohol-related behavior	Experimental procedure	Alcohol dose	Animal	Effect	Outcome
Maccioni et al. (2019a)	COR659	2.5, 5, and 10 mg/kg, i.p.	Repeated (once daily for 10 consecutive days)	Reinstatement of alcohol seeking	Cue-induced alcohol seeking	15% v/v	Male sP rats	Dose-dependent decrease	Lever presses
Maccioni et al. (2019b)	CMPPE	2.5, 5, and 10 mg/kg, i.p.	Acute	Reinstatement of alcohol seeking	Cue-induced alcohol seeking	15% v/v	Female sP rats	Decrease at 5 and 10 mg/kg	Lever presses
Augier et al. (2017)	ADX71441	3 and 10 mg/kg, i.p.	Acute	Reinstatement of alcohol seeking	Cue- and stress (footshock)-induced alcohol seeking	20% v/v	Male Wistar rats	Dose-dependent decrease	Lever presses
Vengeliene et al. (2018)	CMPPE	10 and 30 mg/kg, i.p.	Acute	Reinstatement of alcohol seeking	Cue-induced alcohol seeking	10% v/v	Male Wistar rats	Dose-dependent decrease	Nose pokes

Drug administration routes: *i.p.* intraperitoneal, *i.g.* intragastric, *VTA* ventral tegmental area; rat strains, *sP* Sardinian alcohol-preferring, *AA* Alko alcohol, *P* Indiana alcohol-preferring; experimental procedures: *ADE* alcohol deprivation effect, *FR* fixed ratio, *PR* progressive ratio

3 Place Conditioning

Place conditioning can be used to study the rewarding and aversive effects of drugs. The place conditioning apparatus used for rodents is typically a 2- or 3-sided chambered compartment with dividers between sides; each side of the box has distinct environmental cues (e.g., floor texture, color). Animals are treated with a drug/dose and placed in one side of the compartment (CS+); vehicle administrations are paired with the other side of the compartment (CS−). After repeated drug-CS+ or vehicle-CS− pairings, the divider is removed and drug-free animals are tested for side preferences. A conditioned place preference (CPP) assesses the rewarding effects of drug administration by measuring increased approach and contact behaviors to the location containing the distinct environmental cues previously paired with the drug and is measured as increased time spent in the drug-paired compartment during the drug-free test.

The effects of activation of the $GABA_B$ receptor on CPP have been mixed. An early study reported that the repeated administration of baclofen (before each conditioning session) did not affect the acquisition of CPP associated with 2 g/kg alcohol in mice (Chester and Cunningham 1999). In a subsequent study, acute microinjection of baclofen into the ventral tegmental area (VTA) prior to preference testing decreased CPP associated with 2 g/kg alcohol in mice (Bechtholt and Cunningham 2005). One study has investigated effects of $GABA_B$ PAMs on CPP using the positive modulators ORM-27669 and rac-BHFF. In that study, the repeated administration of ORM-27669 or rac-BHFF (before each conditioning session) did not significantly change the acquisition of CPP associated with 0.5 g/kg alcohol in mice. The injection-free test session was conducted 48 h after the last conditioning session, which may have contributed to the lack of effects of ORM-27669 and rac-BHFF (de Miguel et al. 2018). Effects of baclofen and $GABA_B$ PAMs on CPP have not been examined in rats.

4 Alcohol Withdrawal Syndrome

Heavy drinkers who abstain from alcohol use may experience alcohol withdrawal, which can include mild to moderate symptoms such as tremors, irritability, or anxiety to more severe symptoms such as delirium tremens, hallucinations, and seizures (Saitz 1998). When animals are exposed to a period of chronic alcohol consumption or administration, followed by a period of abstinence in which access to alcohol is withheld, animals often show signs of withdrawal (e.g., tremors, anxiety-like behavior, seizures, irritability/aggression) that can resemble those observed in humans experiencing alcohol withdrawal (Becker 2000). In rats made physically dependent on alcohol via repeated intragastric alcohol administration, baclofen has been shown to reduce tremors and seizures (Colombo et al. 2000). Baclofen was also effective in the reduction of anxiety-like behaviors and tremors in rats made physically dependent on alcohol by prolonged exposure to an alcohol-containing diet (File et al. 1991; Knapp et al. 2007). However, in mice made physically dependent by exposure to an alcohol-containing diet, baclofen did not affect tremors or tail arch, and induced convulsive behavior (Humeniuk et al. 1994).

5 Alcohol Drinking

5.1 Acquisition and Maintenance of Alcohol Drinking

A common method to assess alcohol drinking is to give rodents access to alcohol and water concurrently in two separate bottles in their home cages and measure consumption of both fluids. This two-bottle "alcohol versus water" choice regimen allows for measurement of the amount of alcohol that animals will consume voluntarily when alcohol and water are both available. Access to alcohol can be unlimited (i.e., 24 h/day) or limited (i.e., <24 h/day or access on alternating days). Outbred rodents (e.g., Swiss mice, Long-Evans rats, Wister rats) rarely consume sufficient amounts of alcohol to achieve a blood alcohol level (BAL) of 0.08% or more (Leeman et al. 2010). High alcohol intake and BALs exceeding 0.08% can be obtained in inbred mouse (e.g., C57BL/6N; high-alcohol-preferring, HAP mice) and rat (e.g., Sardinian alcohol-preferring, sP; University of Chile bibulous, UChB; Alko Alcohol, AA rats) lines selectively bred for high alcohol drinking, preference, or blood alcohol levels (Crabbe 2008), and such lines have been used extensively to investigate novel pharmacotherapies, often in comparison to non-preferring or low drinking strains.

Several studies have used the two (or more)-bottle "alcohol versus water" choice regimen to investigate effects of baclofen on the acquisition and maintenance of alcohol drinking. Water availability allows for evaluation of whether a change in alcohol intake after administration of a test drug produced a specific effect on alcohol. Some studies (Colombo et al. 2002, 2005) have shown that the repeated administration of baclofen during initial access to alcohol decreased acquisition of alcohol drinking in Sardinian alcohol-preferring (sP) rats given unlimited access to alcohol. During initial access to alcohol, sP rats will typically escalate alcohol intake over several sessions eventually reaching asymptote (Colombo et al. 2002, 2005). However, baclofen increased acquisition of alcohol drinking in another study using outbred Long-Evans rats given limited alcohol access (Smith et al. 1992). In most studies, acute or repeated administration of baclofen reduced ongoing alcohol drinking in several different rat (UChB, AA, sP, Long-Evans, and Wistar rats) and mouse (Swiss, C57BL/6N, and HAP mice) strains (Boas et al. 2012; Colombo et al. 2000, 2004; Daoust et al. 1987; Kasten et al. 2015; Kemppainen et al. 2012; Peters et al. 2012; Quintanilla et al. 2008; Stromberg 2004). In the studies in which baclofen reduced the acquisition or maintenance of alcohol drinking, baclofen either did not change water intake or increased water intake so that total fluid intake remained unchanged, indicating a specific reductive effect on alcohol intake (Boas et al. 2012; Colombo et al. 2000, 2002, 2004, 2005; Kasten et al. 2015; Kemppainen et al. 2012; Peters et al. 2012; Quintanilla et al. 2008; Stromberg 2004).

A few studies have used the two-bottle "alcohol versus water" choice regimen to investigate effects of GABA$_B$ PAMs on the acquisition and maintenance of alcohol drinking. In those studies, repeated treatment with GS39783 and CGP7930 during initial access to alcohol decreased acquisition of alcohol drinking in alcohol-preferring sP rats (Orru et al. 2005). Repeated administration of GS39783, CGP7930, and rac-BHFF also reduced ongoing alcohol drinking in sP rats; these effects were accompanied by increases in daily water intake, leaving daily total fluid intake unchanged (Loi et al. 2013; Orru et al. 2005). Exposure to repeated cycles of alcohol access and deprivation (i.e., chronic intermittent alcohol access) can generate high levels of voluntary alcohol drinking in rodents (Crabbe et al. 2011; Hwa et al. 2011). Acute administration of the GABA$_B$ PAM ADX71441 reduced alcohol intake, without altering water intake, in alcohol-preferring C57BL/6J mice repeatedly given intermittent alcohol access (Hwa et al. 2014).

5.2 *Binge-Like Drinking*

Binge drinking (drinking "too much, too fast") is defined by the National Institute on Alcohol Abuse and Alcoholism (NIAAA) as alcohol consumption sufficient to achieve a blood alcohol level of 80 mg/dL (0.08%) or more within a 2–3 h drinking period; this corresponds to consumption of about 0.8–1 g/kg. There are two mouse models of binge-like drinking in which alcohol is voluntarily consumed – "drinking in the dark" (DID) (Rhodes et al. 2005) and "scheduled high alcohol consumption" (SHAC) (Finn et al. 2005) – that have been used to investigate effects of baclofen on alcohol consumption. Under the DID procedure, brief periods of alcohol access (2–4 h/day) are provided at fixed times during the early period of the dark phase of the light/dark cycle. Under the SHAC procedure, water-restricted mice are given daily access to water for periods of fluid availability ranging from 4 to 10 h; every third day, access to alcohol is provided during the initial 30 min of fluid availability followed by access to water during the remainder of the fluid-availability period. Acute administration of baclofen reduced binge-like drinking in both models in several different mouse strains, including C57BL/6J, high DID (HDID), and withdrawal seizure control (WSC) mice (Crabbe et al. 2017; Kasten et al. 2015; Moore et al. 2009; Tanchuck et al. 2011).

A few studies have used the DID procedure to investigate effects of GABA$_B$ PAMs on binge-like drinking. In those studies, acute administration of GS39783, ADX71441, rac-BHFF, and ORM-27669 reduced binge-like drinking in C57BL/6J mice (de Miguel et al. 2018; Hwa et al. 2014; Linsenbardt and Boehm 2014). Acutely administered GS39783 also has been shown to reduce binge-like drinking in sP rats exposed to limited and unpredictable alcohol access during the dark phase of the light/dark cycle. Water and food intake were unchanged (Colombo et al. 2015).

Primate models have been developed that generate high levels of voluntary alcohol intake and can produce binge-like drinking patterns (Baker et al. 2014, 2017; Kaminski et al. 2008; Katner et al. 2004, 2007; Weed et al. 2008; Weerts et al. 2006). However, these models have not been well utilized to investigate effects of baclofen or $GABA_B$ PAMs. Only two studies have examined effects of baclofen using these primate models (Duke et al. 2014; Holtyn et al. 2017; see the "appetitive-consummatory procedures" section for a description of these studies).

5.3 *Relapse-Like Drinking*

When animals are exposed to a period of alcohol drinking, followed by a period of abstinence in which access to alcohol is withheld (i.e., deprivation), animals often show a transient but marked increase in alcohol-drinking behavior upon return to alcohol access. This phenomenon, termed the "alcohol deprivation effect" or ADE, was first described in rats by Sinclair and Senter under the two-bottle "alcohol versus water" choice regimen (Sinclair and Senter 1968). Since this initial finding, the ADE has been observed in mice (Khisti et al. 2006; Melendez et al. 2006) and non-human primates (Kornet et al. 1990; Weerts et al. 2006) under both two (or more)-bottle "alcohol versus water" choice procedures (Spanagel et al. 1996; Wolffgramm and Heyne 1995) and operant self-administration procedures (Heyser et al. 1997; Hölter et al. 1997). The magnitude of the ADE is a function of the duration of alcohol abstinence in mice, rats, and non-human primates (Middaugh et al. 2000; Rodd-Henricks et al. 2000; Weerts et al. 2006), and in rodents, is augmented with repeated alcohol deprivations (Oster et al. 2006; Rodd et al. 2003). It does not appear that physical dependence is a primary factor in expression of the ADE; animals that show the ADE do not show signs of a withdrawal syndrome upon alcohol abstinence (Heyser et al. 1997), and when withdrawal symptoms are present, the ADE still occurs after symptoms have dissipated (Cicero et al. 1971; Waller et al. 1982).

Four rodent studies have shown treatment with baclofen or the $GABA_B$ PAM CMPPE to reduce ADE under the two-bottle "alcohol versus water" choice regimen. In two studies, acute administration of baclofen upon return to alcohol access after 7 (Colombo et al. 2003a) or 14 (Colombo et al. 2006) days of alcohol abstinence reduced ADE. Baclofen effects were specific to alcohol intake, as food and water intake and spontaneous locomotor activity were unchanged. In the third study, rats were exposed to repeated alcohol deprivations interspersed with long periods of alcohol access. Repeated administration of baclofen (3 mg/kg) or CMPPE (10 and 30 mg/kg) during the final alcohol deprivation and return to alcohol access decreased ADE. The reduction in ADE was accompanied by a decrease in spontaneous locomotor activity for baclofen (suggestive of nonspecific effects), but not CMPPE (Vengeliene et al. 2018).

6 Operant Alcohol Self-Administration

Operant self-administration paradigms are commonly used to study voluntary alcohol intake in laboratory animals. Under this procedure, an animal performs a response, such as pressing a lever, to obtain access to alcohol, and reinforcement is demonstrated when self-administration of alcohol is greater when compared to self-administration of the vehicle (what alcohol is mixed in). Baclofen and $GABA_B$ PAM effects on alcohol self-administration have been investigated using a number of different self-administration procedures (described below), including the fixed and progressive ratio procedures, appetitive-consummatory procedures ("sipper" and chained schedule of alcohol reinforcement procedures), and extinction and reinstatement of alcohol seeking procedures.

6.1 Fixed Ratio (FR) Procedure

Under the prototypical fixed ratio (FR) procedure, a specific number of responses are required to obtain access to alcohol. The response requirement to gain access to alcohol remains unchanged within and across sessions. Several studies using the FR procedure have shown that acute and repeated treatment with baclofen reduced both the number of lever responses for alcohol and amount of self-administered alcohol in several different rat and mouse strains (Anstrom et al. 2003; Besheer et al. 2004; Dean et al. 2012; Janak and Michael Gill 2003; Liang et al. 2006; Lorrai et al. 2016; Maccioni et al. 2005, 2012, 2018; Orrù et al. 2012; Walker and Koob 2007; Williams et al. 2016). Similar findings have been reported with $GABA_B$ PAMs: acute treatment with CGP7930, GS39783, rac-BHFF, BHF177, ADX71441, COR659, and CMPPE (Augier et al. 2017; Liang et al. 2006; Lorrai et al. 2019; Maccioni et al. 2007, 2009, 2010b, 2012, 2017, 2018, 2019b; Orrù et al. 2012) and repeated treatment with GS39783, rac-BHFF, and COR659 (Maccioni et al. 2015, 2019a) reduced both the number of lever responses for alcohol and amount of self-administered alcohol in rodents. Some of the $GABA_B$ PAMs (BHF177, rac-BHFF, and CMPPE) had effects that were selective for alcohol as self-administration of alternative reinforcers (e.g., sucrose or saccharin solutions) were unchanged, while others (COR659 and ADX71441) had non-selective effects (Augier et al. 2017; Maccioni et al. 2017).

Some studies also have investigated effects of combining low doses of $GABA_B$ PAMs and baclofen. In the initial study of CGP7930 (Liang et al. 2006), the combination of low doses of CGP7930 (10 mg/kg, i.p.) and baclofen (2 mg/kg, i. p.) decreased alcohol self-administration in alcohol-preferring Indiana P rats. In the same study, 10 mg/kg CGP7930 administered alone and 2 mg/kg baclofen administered alone were ineffective at reducing self-administration. This finding was extended in subsequent studies showing that "ineffective" doses of GS39783 (5 mg/kg, i.g.) or rac-BHFF (5 mg/kg, i.g.) and baclofen (1 mg/kg, i.p.) reduced both the number of lever responses for alcohol and amount of self-administered alcohol in rats (Maccioni et al. 2015). Both drug combinations (GS39783 + baclofen and rac-BHFF + baclofen) were selective for alcohol (i.e., sucrose self-administration in a control group did not change after administration of the drug combinations).

6.2 Progressive Ratio (PR) Procedure

Under the within-session progressive ratio (PR) procedure, the number of responses required to obtain access to alcohol is progressively increased after each delivery of alcohol. The "breakpoint," or the lowest ratio not completed, is taken as a measure of the motivational properties of alcohol. Acute administration of baclofen reduced the breakpoint for alcohol and the cumulative (overall) number of responses in several rat strains (Maccioni et al. 2008b, 2012; Walker and Koob 2007). However, treatment with baclofen also tended to reduce self-administration of sucrose or saccharin solutions comparable to the self-administration of alcohol (Anstrom et al. 2003; Colombo et al. 2003b; Janak and Michael Gill 2003; Maccioni et al. 2005, 2008b, 2012). Acute administration of the $GABA_B$ PAMs GS39783, ADX71441, BHF177, COR659, and CMPPE reduced the breakpoint for alcohol in several rat strains, including Wistar, sP, and P rats (Augier et al. 2017; Maccioni et al. 2008b, 2009, 2012, 2017, 2019b). Treatment with GS39783 and BHF177 did not alter breakpoints for a nondrug, sucrose solution, whereas COR659 reduced breakpoints for sucrose.

6.3 Extinction and Reinstatement of Alcohol Seeking Procedures

In the extinction and reinstatement procedures, alcohol seeking is observed under conditions in which access to alcohol has been discontinued. Under the extinction procedure, the cues presented are those that previously had been directly paired with alcohol self-administration. Animals trained to self-administer alcohol will eventually stop responding if access to alcohol is discontinued. The persistence of responding (e.g., the highest number of responses) provides a quantifiable measure of the extent to which stimuli previously associated with alcohol maintain responding in its absence. Acute administration of baclofen decreased lever pressing during extinction of alcohol seeking in rats (Colombo et al. 2003b) and non-human primates (Duke et al. 2014). The ability of a pharmacotherapy to facilitate extinction may be important to the development of medications targeting alcohol cravings and urges to drink.

The reinstatement paradigm measures the ability of an alcohol-associated stimulus (cue-induced reinstatement), a priming dose of alcohol or another drug (drug-induced reinstatement), or a stressor such as a footshock (stress-induced reinstatement) to reinstate responding (i.e., alcohol seeking) following extinction (i.e., in the absence of alcohol). These procedures are designed to model conditions that can trigger craving and relapse. Two studies in rats showed that acute administration of baclofen prior to presentation of alcohol-associated stimuli reduced reinstatement of alcohol seeking (Maccioni et al. 2008b; Vengeliene et al. 2018). Various GABA$_B$ PAMs have shown efficacy in reducing alcohol- and stress-induced reinstatement of alcohol seeking in sP and Wistar rats. Acute pretreatment with ADX71441 and CMPEE prior to presentation of alcohol-associated stimuli or a stressor (footshock) reduced reinstatement of alcohol seeking (Augier et al. 2017; Maccioni et al. 2019b; Vengeliene et al. 2018), and repeated treatment with COR659 similarly reduced cue-induced reinstatement of alcohol seeking (Maccioni et al. 2019a).

6.4 *Appetitive-Consummatory Procedures*

The "sipper" procedure and the chained schedule of alcohol reinforcement (CSR) procedure have been developed to separately examine seeking and consumption under conditions of ongoing alcohol availability. In these procedures, seeking is defined as responding that produces the opportunity to consume alcohol. Under the typical sipper procedure, developed by Samson et al. (1998), a specific number of responses (e.g., 30 lever presses) were required to gain access to alcohol. Lever responses in this phase were defined as appetitive alcohol seeking. Once the response requirement was completed, alcohol was freely available for a fixed duration (e.g., 20 min), which encompassed the consummatory phase. In two studies that used the sipper procedure, baclofen reduced alcohol seeking but did not reduce alcohol intake in male C57BL/6J mice and male Long-Evans rats (Czachowski et al. 2006; Tanchuck et al. 2011). One additional study used the "sipper" procedure to examine effects of the GABA$_B$ PAM GS39783 on alcohol seeking and consumption (Maccioni et al. 2010a). In that study, acute treatment with GS39783 decreased both alcohol seeking and consumption in male sP rats. Selectively of effects of GS39783 on alcohol seeking and consumption were not investigated (Maccioni et al. 2010a).

The CSR, developed by Holtyn et al. (2014), Kaminski et al. (2008), and Weerts et al. (2006), consisted of three sequential components – each contained different schedule requirements – that modeled different phases of alcohol anticipation, seeking, and consumption. Fulfilling the schedule requirement in each successive component was necessary to progress to the next component with alcohol available for self-administration only in the final component. The CSR can include fixed ratio and within-session or across-session progressive ratio manipulations. Under the CSR, baboons consume significant amounts of alcohol ($\sim$1.0 g/kg per day) to reach BALs exceeding 0.08% (i.e., binge drinking) and maintain this level of consumption 7 days per week for prolonged periods (Holtyn et al. 2014; Kaminski et al. 2008). In the two studies that used the CSR procedure, administration of baclofen during active daily drinking decreased alcohol self-administration behaviors and total consumption, but did not alter seeking responses (Duke et al. 2014; Holtyn et al. 2017). In contrast, when baclofen treatment was initiated during abstinence, baclofen did not significantly alter alcohol self-administration upon return to alcohol access conditions (Holtyn et al. 2017). Baclofen did, however, facilitate extinction of responses previously reinforced by alcohol or by a non-alcohol reinforcer (Tang, a calorically equivalent, orange-flavored beverage) under within-session extinction conditions, particularly during early extinction (Duke et al. 2014). Non-selective effects of baclofen on the non-alcohol reinforcer (i.e., Tang) and transient side effects (i.e., vomiting, decreased food intake, and lethargy) also were reported at the highest doses tested.

7 Discussion

The present chapter describes and highlights preclinical research on the role of the $GABA_B$ receptor on alcohol-related behaviors. The therapeutic effects of the $GABA_B$ receptor agonist, baclofen, in preclinical animal models of alcohol reinforcement, motivation, and self-administration are well established. Despite the promising results in preclinical models, the use of baclofen in clinical practice may be limited by side effects, such as sedation and motor incoordination (Garbutt 2018). Such effects were also suggested by some preclinical research. For example, doses of baclofen that reduced alcohol self-administration potentiated the sedative effects of alcohol, even at non-sedative doses of alcohol, and also reduced locomotor activity (Besheer et al. 2004), or reduced self-administration of a non-alcoholic reinforcer such as an orange-flavored sweet beverage (Duke et al. 2014; Holtyn et al. 2017). Positive allosteric modulation of the $GABA_B$ receptor produced an effect on alcohol self-administration and consumption similar to that produced by baclofen. A potential advantage of $GABA_B$ PAMs is that they may possess a wider therapeutic window compared to full agonists, such as baclofen. In general, the $GABA_B$ PAMs have been shown to reduce alcohol self-administration at doses lower than those inducing sedation and motor-incoordination.

To examine the selectivity of effects, preclinical studies have compared self-administration and consumption of alcohol versus alternative, non-alcohol reinforcers (e.g., sucrose or saccharin solutions) following treatment with baclofen and $GABA_B$ PAMs. The selectivity of the reducing effect of baclofen on alcohol self-administration and consumption has been shown to be mixed, as treatment with baclofen reduced consumption and self-administration of alternative, non-alcohol reinforcers in some studies (Anstrom et al. 2003; Czachowski et al. 2006; Janak and Michael Gill 2003; Maccioni et al. 2005, 2008b; Tanchuck et al. 2011). The majority of studies examining $GABA_B$ PAMs have shown they selectively reduced alcohol consumption and did not reduce consumption of other non-alcohol reinforcers (Maccioni and Colombo 2019). Both baclofen and $GABA_B$ PAMs may be useful treatments to promote alcohol abstinence and reduce drinking if it occurs, although nonspecific effects of baclofen and some $GABA_B$ PAMs must be considered.

The use of preclinical animal models is critical to understanding the development of AUD and the consequences of chronic alcohol exposure. Furthermore, these models provide valuable tools for the concurrent testing of potential pharmacotherapies for AUD in a variety of paradigms. It is our contention that this concurrent testing will increase the probability of developing a wider spectrum of efficacious pharmacotherapies that can benefit a greater majority of individuals with AUD and help identify the best candidates for advancement for in-human testing. The preclinical studies reviewed in the present chapter have greatly increased our understanding of the role of the $GABA_B$ receptor in the control of alcohol-related behaviors. The demonstration that baclofen and $GABA_B$ PAMs modify alcohol-related behaviors across multiple animal models allows greater confidence in the generality of the findings.

References

Anstrom KK, Cromwell HC, Markowski T, Woodward DJ (2003) Effect of baclofen on alcohol and sucrose self-administration in rats. Alcohol Clin Exp Res 27(6):900–908

Arias C, Mlewski EC, Molina JC, Spear NE (2009) Naloxone and baclofen attenuate ethanol's locomotor-activating effects in preweanling Sprague-Dawley rats. Behav Neurosci 123(1):172

Augier E, Dulman RS, Damadzic R, Pilling A, Hamilton JP, Heilig M (2017) The GABA B positive allosteric modulator ADX71441 attenuates alcohol self-administration and relapse to alcohol seeking in rats. Neuropsychopharmacology 42(9):1789

Baker EJ, Farro J, Gonzales S, Helms C, Grant KA (2014) Chronic alcohol self-administration in monkeys shows long-term quantity/frequency categorical stability. Alcohol Clin Exp Res 38 (11):2835–2843

Baker EJ, Walter NA, Salo A, Rivas Perea P, Moore S, Gonzales S, Grant KA (2017) Identifying future drinkers: behavioral analysis of monkeys initiating drinking to intoxication is predictive of future drinking classification. Alcohol Clin Exp Res 41(3):626–636

Bechtholt AJ, Cunningham CL (2005) Ethanol-induced conditioned place preference is expressed through a ventral tegmental area dependent mechanism. Behav Neurosci 119(1):213–223

Becker HC (2000) Animal models of alcohol withdrawal. Alcohol Res Health 24(2):105–114

Besheer J, Lepoutre V, Hodge CW (2004) GABA(B) receptor agonists reduce operant ethanol self-administration and enhance ethanol sedation in C57BL/6J mice. Psychopharmacology 174 (3):358–366

Boas GRV, Zamboni CG, Peretti MC, Correia D, Rueda AVL, Camarini R, Brunialti-Godard AL, Boerngen-Lacerda R (2012) GABAB receptor agonist only reduces ethanol drinking in light-drinking mice. Pharmacol Biochem Behav 102(2):233–240

Boehm S II, Piercy M, Bergstrom H, Phillips T (2002) Ventral tegmental area region governs GABAB receptor modulation of ethanol-stimulated activity in mice. Neuroscience 115 (1):185–200

Broadbent J, Harless WE (1999) Differential effects of GABAA and GABAB agonists on sensitization to the locomotor stimulant effects of ethanol in DBA/2 J mice. Psychopharmacology 141(2):197–205

Chester JA, Cunningham CL (1999) Baclofen alters ethanol-stimulated activity but not conditioned place preference or taste aversion in mice. Pharmacol Biochem Behav 63(2):325–331

Cicero TJ, Snider SR, Perez VJ, Swanson LW (1971) Physical dependence on and tolerance to alcohol in the rat. Physiol Behav 6(2):191–198

Colombo G, Agabio R, Carai MA, Lobina C, Pani M, Reali R, Addolorato G, Gessa GL (2000) Ability of baclofen in reducing alcohol intake and withdrawal severity: I--preclinical evidence. Alcohol Clin Exp Res 24(1):58–66

Colombo G, Serra S, Brunetti G, Atzori G, Pani M, Vacca G, Addolorato G, Froestl W, Carai MA, Gessa GL (2002) The GABAB receptor agonists baclofen and CGP 44532 prevent acquisition of alcohol drinking behaviour in alcohol-preferring rats. Alcohol Alcohol 37(5):499–503

Colombo G, Serra S, Brunetti G, Vacca G, Carai MA, Gessa GL (2003a) Suppression by baclofen of alcohol deprivation effect in Sardinian alcohol-preferring (sP) rats. Drug Alcohol Depend 70 (1):105–108

Colombo G, Vacca G, Serra S, Brunetti G, Carai MA, Gessa GL (2003b) Baclofen suppresses motivation to consume alcohol in rats. Psychopharmacology 167(3):221–224

Colombo G, Serra S, Vacca G, Gessa GL, Carai MA (2004) Suppression by baclofen of the stimulation of alcohol intake induced by morphine and WIN 55,212-2 in alcohol-preferring rats. Eur J Pharmacol 492(2–3):189–193

Colombo G, Serra S, Vacca G, Carai MA, Gessa GL (2005) Effect of the combination of naltrexone and baclofen, on acquisition of alcohol drinking behavior in alcohol-preferring rats. Drug Alcohol Depend 77(1):87–91

Colombo G, Serra S, Vacca G, Carai MA, Gessa GL (2006) Baclofen-induced suppression of alcohol deprivation effect in Sardinian alcohol-preferring (sP) rats exposed to different alcohol concentrations. Eur J Pharmacol 550(1–3):123–126

Colombo G, Lobina C, Maccioni P, Carai MA, Lorrai I, Zaru A, Contini A, Mugnaini C, Corelli F, Gessa GL (2015) Anxiety-like behaviors at the end of the nocturnal period in sP rats with a "history" of unpredictable, limited access to alcohol. Alcohol 49(7):707–712

Cott J, Carlsson A, Engel J, Lindqvist M (1976) Suppression of ethanol-induced locomotor stimulation by GABA-like drugs. Naunyn Schmiedeberg's Arch Pharmacol 295(3):203–209

Crabbe JC (2008) Neurogenetic studies of alcohol addiction. Philos Trans R Soc B Biol Sci 363 (1507):3201–3211

Crabbe JC, Harris RA, Koob GF (2011) Preclinical studies of alcohol binge drinking. Ann N Y Acad Sci 1216:24

Crabbe JC, Ozburn AR, Metten P, Barkley-Levenson A, Schlumbohm JP, Spence SE, Hack WR, Huang LC (2017) High drinking in the dark (HDID) mice are sensitive to the effects of some clinically relevant drugs to reduce binge-like drinking. Pharmacol Biochem Behav 160:55–62

Czachowski CL, Legg BH, Stansfield KH (2006) Ethanol and sucrose seeking and consumption following repeated administration of the GABA(B) agonist baclofen in rats. Alcohol Clin Exp Res 30(5):812–818

Daoust M, Saligaut C, Lhuintre J, Moore N, Flipo J, Boismare F (1987) GABA transmission, but not benzodiazepine receptor stimulation, modulates ethanol intake by rats. Alcohol 4 (6):469–472

de Miguel E, Vekovischeva O, Kuokkanen K, Vesajoki M, Paasikoski N, Kaskinoro J, Myllymäki M, Lainiola M, Janhunen SK, Hyytiä P (2018) GABAB receptor positive allosteric modulators with different efficacies affect neuroadaptation to and self-administration of alcohol and cocaine. Addict Biol 24(6):1191–1203

Dean RL, Eyerman D, Todtenkopf MS, Turncliff RZ, Bidlack JM, Deaver DR (2012) Effects of oral loperamide on efficacy of naltrexone, baclofen and AM-251 in blocking ethanol self-administration in rats. Pharmacol Biochem Behav 100(3):530–537

Duke AN, Kaminski BJ, Weerts EM (2014) Baclofen effects on alcohol seeking, self-administration and extinction of seeking responses in a within-session design in baboons. Addict Biol 19(1):16–26

File SE, Zharkovsky A, Gulati K (1991) Effects of baclofen and nitrendipine on ethanol withdrawal responses in the rat. Neuropharmacology 30(2):183–190

Finn DA, Belknap JK, Cronise K, Yoneyama N, Murillo A, Crabbe JC (2005) A procedure to produce high alcohol intake in mice. Psychopharmacology 178(4):471–480

Garbutt J (2018) Use of baclofen for alcohol use disorders in the United States. Front Psych 9:448

Heyser CJ, Schulteis G, Koob GF (1997) Increased ethanol self-administration after a period of imposed ethanol deprivation in rats trained in a limited access paradigm. Alcohol Clin Exp Res 21(5):784–791

Holstein SE, Dobbs L, Phillips TJ (2009a) Attenuation of the stimulant response to ethanol is associated with enhanced ataxia for a GABA, but not a GABA, receptor agonist. Alcohol Clin Exp Res 33(1):108–120

Holstein SE, Dobbs L, Phillips TJ (2009b) Attenuation of the stimulant response to ethanol is associated with enhanced ataxia for a GABAA, but not a GABAB, receptor agonist. Alcohol Clin Exp Res 33(1):108–120

Hölter SM, Landgraf R, Zieglgänsberger W, Spanagel R (1997) Time course of acamprosate action on operant ethanol self-administration after ethanol deprivation. Alcohol Clin Exp Res 21 (5):862–868

Holtyn AF, Kaminski BJ, Wand GS, Weerts EM (2014) Differences in extinction of cue-maintained conditioned responses associated with self-administration: alcohol versus a nonalcoholic reinforcer. Alcohol Clin Exp Res 38(10):2639–2646

Holtyn AF, Kaminski BJ, Weerts EM (2017) Baclofen and naltrexone effects on alcohol self-administration: comparison of treatment initiated during abstinence or ongoing alcohol access in baboons. Drug Alcohol Depend 179:47–54

Humeniuk RE, White JM, Ong J (1993) The role of GABAB receptors in mediating the stimulatory effects of ethanol in mice. Psychopharmacology 111(2):219–224

Humeniuk RE, White JM, Ong J (1994) The effects of GABAB ligands on alcohol withdrawal in mice. Pharmacol Biochem Behav 49(3):561–566

Hwa LS, Chu A, Levinson SA, Kayyali TM, DeBold JF, Miczek KA (2011) Persistent escalation of alcohol drinking in C57BL/6J mice with intermittent access to 20% ethanol. Alcohol Clin Exp Res 35(11):1938–1947

Hwa LS, Kalinichev M, Haddouk H, Poli S, Miczek KA (2014) Reduction of excessive alcohol drinking by a novel GABA B receptor positive allosteric modulator ADX71441 in mice. Psychopharmacology 231(2):333–343

Janak PH, Michael Gill T (2003) Comparison of the effects of allopregnanolone with direct GABAergic agonists on ethanol self-administration with and without concurrently available sucrose. Alcohol 30(1):1–7

Kaminski BJ, Goodwin AK, Wand G, Weerts EM (2008) Dissociation of alcohol-seeking and consumption under a chained schedule of oral alcohol reinforcement in baboons. Alcohol Clin Exp Res 32(6):1014–1022

Kasten CR, Blasingame SN, Boehm SL II (2015) Bidirectional enantioselective effects of the GABAB receptor agonist baclofen in two mouse models of excessive ethanol consumption. Alcohol 49(1):37–46

Katner SN, Flynn CT, Von Huben SN, Kirsten AJ, Davis SA, Lay CC, Cole M, Roberts AJ, Fox HS, Taffe MA (2004) Controlled and behaviorally relevant levels of oral ethanol intake in rhesus macaques using a flavorant-fade procedure. Alcohol Clin Exp Res 28(6):873–883

Katner SN, Von Huben SN, Davis SA, Lay CC, Crean RD, Roberts AJ, Fox HS, Taffe MA (2007) Robust and stable drinking behavior following long-term oral alcohol intake in rhesus macaques. Drug Alcohol Depend 91(2–3):236–243

Kemppainen H, Raivio N, Kiianmaa K (2012) Role for ventral pallidal GABAergic mechanisms in the regulation of ethanol self-administration. Psychopharmacology 223(2):211–221

Khisti RT, Wolstenholme J, Shelton KL, Miles MF (2006) Characterization of the ethanol-deprivation effect in substrains of C57BL/6 mice. Alcohol 40(2):119–126

Knapp DJ, Overstreet DH, Breese GR (2007) Baclofen blocks expression and sensitization of anxiety-like behavior in an animal model of repeated stress and ethanol withdrawal. Alcohol Clin Exp Res 31(4):582–595

Kornet M, Goosen C, Van Ree JM (1990) The effect of interrupted alcohol supply on spontaneous alcohol consumption by rhesus monkeys. Alcohol Alcohol 25(4):407–412

Shen EH, Dorow J, Harland R, Burkhart-Kasch S, Phillips TJ (1998) Seizure sensitivity and GABAergic modulation of ethanol sensitivity in selectively bred FAST and SLOW mouse lines. J Pharmacol Exp Ther 287(2):606–615

Sinclair J, Senter R (1968) Development of an alcohol-deprivation effect in rats. Q J Stud Alcohol 29(4):863–867

Smith BR, Robidoux J, Amit Z (1992) GABAergic involvement in the acquisition of voluntary ethanol intake in laboratory rats. Alcohol Alcohol 27(3):227–231

Smith BR, Boyle AE, Amit Z (1999) The effects of the GABA(B) agonist baclofen on the temporal and structural characteristics of ethanol intake. Alcohol 17(3):231–240

Spanagel R, Hölter SM, Allingham K, Landgraf R, Zieglgänsberger W (1996) Acamprosate and alcohol: I. effects on alcohol intake following alcohol deprivation in the rat. Eur J Pharmacol 305(1–3):39–44

Stromberg MF (2004) The effect of baclofen alone and in combination with naltrexone on ethanol consumption in the rat. Pharmacol Biochem Behav 78(4):743–750

Tanchuck MA, Yoneyama N, Ford MM, Fretwell AM, Finn DA (2011) Assessment of GABA-B, metabotropic glutamate, and opioid receptor involvement in an animal model of binge drinking. Alcohol 45(1):33–44. https://doi.org/10.1016/j.alcohol.2010.07.009

Urwyler S (2016) Allosteric modulators: the new generation of GABA B receptor ligands. In: GABAB Receptor. Springer, Berlin, pp 357–375

Vengeliene V, Takahashi TT, Dravolina OA, Belozertseva I, Zvartau E, Bespalov AY, Spanagel R (2018) Efficacy and side effects of baclofen and the novel GABA B receptor positive allosteric modulator CMPPE in animal models for alcohol and cocaine addiction. Psychopharmacology 235(7):1955–1965

Walker BM, Koob GF (2007) The γ-aminobutyric acid-B receptor agonist baclofen attenuates responding for ethanol in ethanol-dependent rats. Alcohol Clin Exp Res 31(1):11–18

Waller M, McBride W, Lumeng L, Li T-K (1982) Induction of dependence on ethanol by free-choice drinking in alcohol-preferring rats. Pharmacol Biochem Behav 16(3):501–507

Weed MR, Wilcox KM, Ator NA, Hienz RD (2008) Consistent, high-level ethanol consumption in pig-tailed macaques via a multiple-session, limited-intake, oral self-dosing procedure. Alcohol Clin Exp Res 32(6):942–951

Weerts EM, Goodwin AK, Kaminski BJ, Hienz RD (2006) Environmental cues, alcohol seeking, and consumption in baboons: effects of response requirement and duration of alcohol abstinence. Alcohol Clin Exp Res 30(12):2026–2036

Williams KL, Nickel MM, Bielak JT (2016) Baclofen blocks yohimbine-induced increases in ethanol-reinforced responding in rats. Pharmacol Biochem Behav 144:20–25

Wise RA, Bozarth MA (1987) A psychomotor stimulant theory of addiction. Psychol Rev 94 (4):469

Wolffgramm J, Heyne A (1995) From controlled drug intake to loss of control: the irreversible development of drug addiction in the rat. Behav Brain Res 70(1):77–94

GABA$_B$ Receptors and Alcohol Use Disorders: Clinical Studies

Warren B. Logge, Kirsten C. Morley, and Paul S. Haber

Contents

Abstract Harmful alcohol use and alcohol use disorders (AUD) result in major health and community burden worldwide, yet treatment options are limited. Novel pharmacotherapies are urgently required, and treatments involving GABA$_B$ receptors have been used in treating alcohol-related disorders. This chapter will review the clinical evidence of GABA$_B$ pharmacotherapies, such as baclofen and γ-hydroxybutyric acid. This includes the use of these treatments in individuals experiencing alcohol withdrawal symptoms and outlining the outcomes of studies

W. B. Logge (✉) and P. S. Haber
Discipline of Addiction Medicine, Central Clinical School, Faculty of Medicine and Health, University of Sydney, Camperdown, NSW, Australia

Drug Health Services, Royal Prince Alfred Hospital, Camperdown, NSW, Australia
e-mail: warren.logge@sydney.edu.au

K. C. Morley
Discipline of Addiction Medicine, Central Clinical School, Faculty of Medicine and Health, University of Sydney, Camperdown, NSW, Australia

Curr Topics Behav Neurosci (2022) 52: 195–212
https://doi.org/10.1007/7854_2020_182
Published Online: 13 February 2021

of alcohol relapse prevention relapse including case studies, comparative studies and randomised controlled trials. Laboratory research investigating biobehavioural effects of baclofen will also be summarised and polymorphisms associated with baclofen treatment, and safety concerns of $GABA_B$ treatments will be addressed. In summary, pharmacological treatments targeting $GABA_B$ receptors such as baclofen may be modestly effective in the management of alcohol use disorder, but safety concerns limit the widespread applicability of the currently available agents.

Keywords Alcohol use disorder · Alcohol withdrawal · Baclofen · Neuroimaging · Pharmacotherapy · Psychophysiology · Randomised controlled trial · Treatment

1 Introduction

Harmful alcohol use is a pervasive worldwide issue linked to significant health and community burden, accounting for 5.3% of all deaths and a causal factor in over 200 disease and injury conditions (World Health Organization 2018a). Globally, an estimated 6.2% of men and 1.2% of women have alcohol use disorders (AUDs) (World Health Organization 2018a). Diagnostic criteria for AUD have been developed and progressively validated and are included in the International Classification of Diseases (World Health Organization 2018b) and the Diagnostic and Statistical Manual of the American Psychiatric Association (American Psychiatric Association 2013). The cardinal features of AUD are loss of control over alcohol use and continuing use despite evident harms. Other important diagnostic criteria result from neuroadaptation leading to increased tolerance of the effects of alcohol and a characteristic withdrawal syndrome upon cessation of alcohol consumption (American Psychiatric Association 2013). Alcohol-related liver disease is a particularly important consequence of alcohol use (Thursz et al. 2018). Liver disease is the leading cause of alcohol-related mortality (Asrani et al. 2019), and, conversely, alcohol is the leading cause of advanced liver disease worldwide (Rehm et al. 2013). Liver disease complicates the pharmacological management of alcohol use disorder (AUD) because drug metabolism may be impaired in the presence of liver failure.

While alcohol use disorders are leading causes of preventable death, treatment options are still limited. Currently, there are few pharmacological treatments specifically indicated for alcohol dependence in Europe, the USA and Australia, including acamprosate, naltrexone, nalmefene and disulfiram. These agents have been extensively evaluated in a large number of double-blind, randomised controlled trials (RCTs), a study design that randomly assigns participants to receive either a placebo (or active placebo) or treatment, and considered the "gold-standard" research design for evaluating treatment efficacy. RCTs of these agents have generally demonstrated modest effect sizes for use in treatment of AUD (Johnson 2008), and treatment uptake in the community is low (Morley et al. 2016). There is thus an urgent need for the development of novel agents for the treatment of alcohol use disorder that are both more effective and more appealing to the target population.

2 Overview of Pharmacotherapies Modulating the GABAB Receptor

A number of current therapeutic drugs influence $GABA_A$ transmission including pregabalin, gabapentin and the benzodiazepines, but few act principally on $GABA_B$ receptors, the focus of this review. Accordingly, we restrict this review of $GABA_B$ receptor treatments for alcohol use disorders to agents that directly act upon the $GABA_B$ receptor such as baclofen and γ-hydroxybutyric acid.

2.1 *Baclofen*

Baclofen is a selective $GABA_B$ receptor agonist that is emerging as a potential treatment for alcohol dependence. Baclofen has minimal liver metabolism (~10–15%), and few reported hepatic side effects (Addolorato et al. 2007), making this drug of particular interest amongst those with alcohol-related liver disease. Chronic alcohol use results in a downregulation of GABA receptor activity and disinhibition of the dopaminergic pathway. $GABA_B$ receptors are expressed on dopamine, and GABA neurons and preclinical studies have demonstrated that baclofen can diminish self-administration of alcohol, maintenance and reinstatement of alcohol-drinking behaviour (Agabio and Colombo 2014), possibly due to inhibition of the mesolimbic reward system.

There has been expanded utilisation of baclofen in the treatment of alcohol dependence, particularly in Europe. Prescriptions for baclofen significantly increased between 2007 and 2013 with a large proportion being initiated in primary care (Chaignot et al. 2015). The provision of a temporary recommendation for baclofen to be prescribed to alcohol-dependent patients was granted in France in 2014, and further approval was granted in October 2018 for use of baclofen in alcohol-dependent patients up to a dose of 80 mg/day (Rolland et al. 2019). Nonetheless, there is ongoing controversy in the field, and, among the few high-quality trials in the literature, the results are mixed. The superiority of baclofen over placebo is yet to be established (Agabio et al. 2018), and baclofen has been recommended as a second-line treatment for alcohol use disorders for individuals who do not respond to other treatments, though it can be an effective first-line treatment for those with contraindications to currently approved medications.

2.2 γ-Hydroxybutyric Acid

Sodium oxybate is the sodium salt of γ-hydroxybutyric acid (GHB). It has been applied as a treatment for AUD since the early 1990s, particularly in European countries such as Italy and Austria (Keating 2014). GHB is a short-chain fatty acid structurally similar to GABA and is a weak agonist at the $GABA_B$ receptor (Sewell and Petrakis 2010). It may be converted to GABA leading to direct activation of $GABA_A$ and $GABA_B$ receptors, leading to sedative and anxiolytic manifestations that also resemble alcohol-like effects (Bay et al. 2014). GHB is also a potent agonist at the excitatory GHB receptor (Cash et al. 1999). It is likely that GABAergic activity mediates the therapeutic effect of GHB on withdrawal and relapse in AUD (Sewell and Petrakis 2010). As alcohol withdrawal results from reduced GABAergic activity in the central nervous system, exogenous GHB may ameliorate alcohol withdrawal through both its conversion to GABA and indirect activation of $GABA_A$ receptors. GHB's efficacy in relapse prevention may be as an effective agonist treatment, as it affects the $GABA_A$ receptor in a similar fashion to alcohol. This effect is analogous to the use of methadone for treatment of opioid dependence.

GHB can have a biphasic response upon dopamine release according to dose concentration, which can increase the risk of overdose when GHB is used as a treatment for AUD. At low concentrations of GHB, dopamine is released following GHB receptor stimulation. However, dopamine release is inhibited by higher doses of GHB that stimulate $GABA_B$ receptors (Caputo et al. 2009). This therefore produces a biphasic response, whereby lower doses of GHB cause euphoria, but higher doses may lead to deep sedation and potentially fatal overdose (Keating 2014). Consequently, abuse and toxicity limit the potential for therapeutic use of GHB.

One systematic review evaluating the use of GHB in the management of alcohol withdrawal syndrome and also for relapse prevention (Leone et al. 2010) was inconclusive regarding GHB's efficacy over placebo or other pharmacological treatments specifically for relapse prevention in AUD, due to an insufficient number of RCTs. Recently, an expert group of European alcohol researchers and clinicians evaluated the data for GHB in the treatment of AUD, and secondary analyses indicated GHB is effective in alcohol-dependent patients with very high-risk drinking (van den Brink et al. 2018). Nonetheless, these analyses were post hoc, and only a small number of patients were included with a wide study duration range (3–12 months).

3 Treatment for Alcohol Withdrawal Syndrome

Heavy drinkers that suddenly reduce their alcohol consumption or abstain altogether may consequently experience symptoms of alcohol withdrawal. These symptoms may range from mild to moderate effects, including tremors, anxiety, irritability and agitation, to more severe effects including hallucinations, delirium tremens and seizures (Fiellin et al. 2002). Alcohol withdrawal syndrome (AWS), which comprises a cluster of these symptoms occurring in individuals with alcohol use disorder, can occur after decreased chronic or heavy alcohol consumption or from cessation. Resultant experienced physiological manifestations can be mild to moderate (e.g. tremor, paroxysmal sweats, fever), but more severe presentations of symptoms such as seizures and delirium can have serious outcomes, including death (Connor et al. 2016). As several of these symptoms involve disruption of several neurotransmitter systems, including inhibitory systems such as GABA, treatments for AWS aim to target these systems to reduce these symptoms.

While benzodiazepines remain the preferred treatment for AWS, their use is limited by abuse liability, tolerance, lack of efficacy in severe cases and complications related to sedation (Haber et al. 2009). Other pharmacotherapies for treating alcohol withdrawal symptoms affecting $GABA_B$ receptors have been investigated, such as baclofen, and the evidence of baclofen for treating AWS will be covered briefly. Baclofen was initially used for treatment of AWS in alcohol-dependent patients in two case report studies (a detailed examination of one or a small set of patients administered a treatment) by the same research group (Addolorato et al. 2002b, 2003). In the first study (Addolorato et al. 2002b), a relatively low dose of baclofen (10 mg day t.i.d.) was administered to five patients, who demonstrated a rapid reduction of severity of AWS symptoms, as assessed using the Clinical Institute Withdrawal Assessment for Alcohol-revised scale (CIWA-Ar) (Sullivan et al. 1989). These patients maintained abstinence with continued baclofen treatment for a subsequent 30 days, with no reported major side effects. Another single case study of a patient (Addolorato et al. 2003) presenting with severe AWS who was administered the comparatively high initial dose of baclofen (75 mg/day) demonstrated significantly decreased AWS severity within 1 h of initial administration, and after stabilisation the patient was abstinent for 30 days with a resultant lower baclofen dose (30 mg/day) with no reported side effects.

Comparative studies that compare the target treatment versus an active "gold-standard" treatment have evaluated baclofen versus benzodiazepines – considered the gold-standard comparator for AWS treatment. Addolorato et al. (2006) found that patients receiving either oral baclofen (30 mg/day for 10 days) or diazepam (0.55–0.75 mg/kg/day for 6 consecutive days, with 25% tapered dose daily for days 7–10), experienced similar reductions in AWS severity, reflected by decreased CIWA-Ar scores. Reddy et al. (2014) compared progressive reduction in baclofen doses (30 mg/day t.i.d. reducing to 10 mg/day) versus chlordiazepoxide (75 mg/day reducing to 25 mg/day) over 9 days and found both treatments decreased CIWA-Ar scores, but baclofen was the less effective treatment. Recently, Gulati et al. (2019) found comparable efficacy for baclofen (10 mg/day t.i.d.) and lorazepam (8–12 mg/day t.i.d.) in an open-label study. However, in open-label study designs, the treatment received is known to patients and researchers; this may lead to biased outcomes as compared to blinded designs such as RCTs.

There have been two RCTs assessing the efficacy of baclofen for treating AWS compared to placebo, including usual care for symptom-triggered benzodiazepine treatment. In the first study (Lyon et al. 2011), patients with moderate AWS symptoms were randomised for a complete 3-day course of oral baclofen ($n = 18$, 30 mg/day t.i.d.) or placebo ($n = 13$), and baclofen treatment was associated with significantly reduced lorazepam use for management of AWS. In a larger study by Heppe et al. (2019) of 101 patients admitted for AWS, the same baclofen dose regimen ($n = 50$, 30 mg/day t.i.d.) compared to placebo ($n = 51$) was administered for 72 h. Baclofen did not significantly reduce progression of moderate of severe AWS in these patients, although there was a trend of baclofen reducing benzodiazepine use.

Taken together, while there is some evidence that baclofen may be effective in reducing alcohol withdrawal symptoms, it does not have clear superiority over benzodiazepines as a preferred pharmacotherapy for AWS. Two recent literature reviews similarly concluded that the evidence was insufficient to justify routine use of baclofen for AWS (Cooney et al. 2019; Liu and Wang 2019).

4 Relapse Prevention in Alcohol Use Disorders

Prevention of relapse is a major goal of pharmacotherapy of severe alcohol use disorders (Freyer et al. 2016). Relapse involves resumption of heavy alcohol consumption after a prolonged period of abstinence, and vulnerability to relapse is clinically commonly associated with intense cravings desires to drink that are often provoked by drinking-related cues (Becker 2008). It is a major obstacle to treatment efforts, and patterns of fluctuating remission and relapse are common in individuals in AUD treatment. Evidence from clinical studies of baclofen has been reported here. Baclofen has been extensively evaluated as a potential pharmacotherapy for relapse prevention in severe alcohol use disorder, with several case studies, open-label studies and RCTs. Additionally, there have been multiple meta-analyses evaluating the efficacy of baclofen (Bschor et al. 2018; Lesouef et al. 2014; Pierce et al. 2018; Rose and Jones 2018). Despite this body of work, the evidence is still equivocal regarding baclofen's superiority over placebo for relapse prevention, as presented in a recent consensus statement by leading researchers investigating baclofen (Agabio et al. 2018). Low-to-moderate baclofen doses are defined here as $\leq$60 mg/day, with high doses >60 mg/day (Pierce et al. 2018) – the latter including very high doses of more than 300 mg/day (de Beaurepaire 2014).

4.1 Baclofen: Low-to-Moderate Doses

With regard to low-to-moderate dosing regimens, the initial RCT investigating baclofen as a treatment of alcohol dependence was conducted by Addolorato et al. (2002a), which demonstrated significantly higher rates of abstinence and reduced alcohol intake after 4 weeks of baclofen (30 mg/day t.i.d.) treatment relative to placebo in 39 Italian outpatients. In a subsequent double-blinded RCT (where both

participants and researchers are blinded to treatment conditions) by the same research group (Addolorato et al. 2011), a beneficial dose-response effect of baclofen (30 mg/day versus 60 mg/day) was found, through secondary analyses, in reducing the number of drinks per drinking day after 12 weeks compared to placebo, with no reported significant differences for several alcohol dependence measures. An Australian multisite double-blinded RCT by Morley et al. (2018a) of 104 outpatients, with or without liver disease, demonstrated a greater time to lapse and relapse for baclofen-treated patients (30 mg/day or 75 mg/day) and a significant treatment effect for days abstinent compared to placebo.

Conversely, findings from a double-blinded RCT in the USA (Garbutt et al. 2010) found no beneficial effect of baclofen (30 mg/day) versus placebo in 80 alcohol-dependent patients randomised to 12 weeks of treatment. No significant effects were seen for any reported drinking outcomes, although baclofen significantly reduced subjective craving across the trial and reduced symptoms of anxiety. Similarly, Ponizovsky et al. (2015) also did not observe any beneficial treatment effect of baclofen during a 12-week RCT with a moderate dose (50 mg/day) in 64 alcohol-dependent patients. Factors that may account for the disparity in these studies' findings (Leggio et al. 2010) have included the relatively lower severity of AUD in the USA study, coupled with a high placebo response for both studies (Garbutt et al. 2010; Ponizovsky et al. 2015), as compared to studies demonstrating significant baclofen treatment effects. Indeed, a recent meta-analysis showed that baclofen was more effective for high drinking levels (Pierce et al. 2018), and similar effect was found in the Australian baclofen trial (Rombouts et al. 2019), suggesting that baclofen may have a significant beneficial effect in severe AUD cases only.

4.2 *Baclofen: High-Doses*

The use of high-dose baclofen was advocated by an influential anecdotal report published by French cardiologist Olivier Ameisen (Ameisen 2004). This led to widespread use of this medication, particularly in France. The first double-blinded RCT of high-dose baclofen (Müller et al. 2015) was conducted in a sample of German alcohol-dependent patients ($N = 56$) administered individually titrated baclofen doses (mean dose of 180 mg/day) versus placebo. There was a strong positive treatment effect on abstinence rates, with a number needed to treat of 2.3. Beraha et al. (2016) randomised 151 patients to high-dose baclofen (up to 150 mg/day), low dose (30 mg/day) or placebo and found no beneficial treatment effect of baclofen for time to first relapse or rates of abstinence. However, in post hoc analyses, this group did report that higher doses were associated with a longer first relapse. A long-term maintenance RCT of 180 mg/day in 320 patients by Reynaud et al. (2017) found only a trend for a positive baclofen effect versus placebo at 6-month follow-up in reducing alcohol consumption, although they reported a significant effect of baclofen to reduce craving.

4.3 Specific Populations

The effectiveness of baclofen in treating AUD may be reliant on specific patient characteristics increasing the probability of a favourable applicability of baclofen – that is, a personalised approach to pharmacotherapy in these patients. One potential reason of baclofen's differential effects on different populations may be related to greater depletion of brain GABA+ levels in specific populations, such as those with ALD, which may predicate a more effective baclofen treatment response (Morley et al. 2018b). Here, we briefly cover treatment effects of baclofen investigated in individuals with AUD and comorbid issues, including liver disease and mood disorders such as anxiety.

4.3.1 Alcohol-Related Liver Disease

Liver disease associated with alcohol is a major consequence of chronic alcohol consumption with a dose-response relationship, which can be exacerbated by conditions such as hepatitis C, with chronic infections of the hepatitis C virus leading to both acute and long-term liver damage (Goldberg et al. 2017). Current approved treatments for alcohol use disorders (e.g. disulfiram, naltrexone, nalmefene) are not suitable for patients with liver disease as they are contraindicated due to metabolism by the liver, whereas baclofen is processed primarily through the kidneys with low liver metabolism (about 15%) (Davidoff 1985). Baclofen is thus an ideal candidate for treating alcohol use disorder in patients with liver disease, and its effectiveness has been investigated in this subgroup. An RCT by Addolorato et al. (2007) demonstrated clear treatment effects of baclofen in reducing drinking outcomes specifically in Italian patients with cirrhosis ($N = 84$) in a 12-week trial, whereby patients receiving baclofen (30 mg/day t.i.d) had higher rates of abstinence and significantly reduced craving compared to placebo. This is contrasted by findings of RCT of USA military veterans (Hauser et al. 2017) with comorbid chronic hepatitis C and AUD, which found that a low dose of baclofen (30 mg/day) had no effect on days abstinent or reducing alcohol use during a 12-week trial compared to placebo. However, this sample was largely male (96.3% of sample) and had relatively low levels of baseline consumption. Most recently, Morley et al. (2018a) have confirmed the effectiveness of baclofen for treatment of AUD with high drinking levels and comorbid liver disease, and this study is the first replication of the influential findings of Addolorato et al. (2007). Furthermore, using magnetic resonance spectroscopy imaging techniques, lower cortical levels of GABA were revealed in alcohol-dependent patients with liver disease, and this suggests that a $GABA_B$ agonist may restore GABA transmission in this setting (Morley et al. 2018b).

4.3.2 Comorbid Anxiety

Morley et al. (2014) have previously reported no beneficial effect of baclofen (30 mg/day, 60 mg/day) in an RCT in 42 alcohol-dependent patients in Australia. No treatment effect was found for time to lapse or time to relapse, although a beneficial treatment effect of baclofen to reduce alcohol consumption was found in post hoc analyses in patients with a comorbid anxiety disorder. Further work by the same group (Rombouts et al. 2019) analysing a subsequent larger trial (Morley et al. 2018a) failed to observe that anxiety was a significant predictor of baclofen to reduce alcohol consumption. Garbutt et al. (2010) found that baclofen reduced anxiety levels during treatment of AUD, although this sample was not diagnosed for comorbid anxiety and did not find evidence of reduced drinking.

5 Meta-Analyses Assessing Baclofen's Efficacy

The efficacy of baclofen in treating relapse in AUD is therefore unclear. This has been further indicated through numerous meta-analyses conducted evaluating the efficacy of baclofen, with multiple meta-analyses published in 2018 alone. The inconsistency of findings between the individual trials can be partially explained by differing criteria used for study inclusion, such as the subpopulations assessed, and the reported outcomes. An earlier meta-analysis focused on RCTs using low baclofen doses of 30 mg/day (Lesouef et al. 2014) and reported that baclofen treatment had a significant effect on rates of abstinence but was limited by only including studies administering a relatively low baclofen dose. More recently, one meta-analysis determined that baclofen was associated with higher abstinence rates in AUD samples versus placebo, but not with other drinking outcomes of increased abstinence days or reduced heavy drinking days (Rose and Jones 2018). Alternatively, the largest meta-analysis identified a modest positive effect superior to placebo when considering noncomplicated AUD samples. Yet, baclofen's overall superior clinical utility was not clearly established, partly due to the significant heterogeneity among the RCTs (Bschor et al. 2018). Moreover, another meta-analysis examining alcohol-dependent samples indicates a dose-specific effect with efficacy in low doses (<60 mg/day) for achieving abstinence, but no benefit in higher doses (Pierce et al. 2018). Finally, benefit for those with high drinking levels was identified by Pierce et al. (2018), an important finding that helps explain numerous inconsistencies in the literature. In summary, it is apparent that there are mixed findings from double-blinded RCTs evaluating baclofen as a pharmacotherapy for AUD, and a consensus regarding superiority compared to placebo for treatment of AUD has not been reached.

de Beaurepaire R (2014) The use of very high-doses of baclofen for the treatment of alcohol-dependence: a case series. Front Psych 5:143–143
Dore GM, Lo K, Juckes L, Bezyan S, Latt N (2011) Clinical experience with baclofen in the management of alcohol-dependent patients with psychiatric comorbidity: a selected case series. Alcohol Alcohol 46:714–720
Enoch M-A, Zhou Z, Kimura M, Mash DC, Yuan Q, Goldman D (2012) GABAergic gene expression in postmortem hippocampus from alcoholics and cocaine addicts; corresponding findings in alcohol-Naïve P and NP rats. PLoS One 7:e29369
Enoch M-A, Hodgkinson CA, Shen P-H et al (2016) GABBR1 and SLC6A1, two genes involved in modulation of GABA synaptic transmission, influence risk for alcoholism: results from three ethnically diverse populations. Alcohol Clin Exp Res 40:93–101
Evans SM, Bisaga A (2009) Acute interaction of baclofen in combination with alcohol in heavy social drinkers. Alcohol Clin Exp Res 33:19–30
Farokhnia M, Schwandt ML, Lee MR et al (2017) Biobehavioral effects of baclofen in anxious alcohol-dependent individuals: a randomized, double-blind, placebo-controlled, laboratory study. Transl Psychiatry 7:e1108
Farokhnia M, Deschaine SL, Sadighi A et al (2018a) A deeper insight into how GABA-B receptor agonism via baclofen may affect alcohol seeking and consumption: lessons learned from a human laboratory investigation. Mol Psychiatry. https://doi.org/10.1038/s41380-018-0287-y
Farokhnia M, Sheskier MB, Lee MR et al (2018b) Neuroendocrine response to GABA-B receptor agonism in alcohol-dependent individuals: results from a combined outpatient and human laboratory experiment. Neuropharmacology 137:230–239
Fiellin DA, O'Connor PG, Holmboe ES, Horwitz RI (2002) Risk for delirium tremens in patients with alcohol withdrawal syndrome. Subst Abus 23:83–94
Freyer CH, Morley KC, Haber PS (2016) Alcohol use disorders in Australia. Intern Med J 46:1259–1268
Garbutt JC, Kampov-Polevoy AB, Gallop R, Kalka-Juhl L, Flannery BA (2010) Efficacy and safety of baclofen for alcohol dependence: a randomized, double-blind, placebo-controlled trial. Alcohol Clin Exp Res 34:1849–1857
Goldberg D, Ditah IC, Saeian K et al (2017) Changes in the prevalence of hepatitis C virus infection, nonalcoholic steatohepatitis, and alcoholic liver disease among patients with cirrhosis or liver failure on the waitlist for liver transplantation. Gastroenterology 152:1090–99.e1
Gulati P, Chavan B, Sidana A (2019) Comparative efficacy of baclofen and lorazepam in the treatment of alcohol withdrawal syndrome. Indian J Psychiatry 61:60–64
Haber P, Lintzeris N, Proude E, Lopatko O (2009) Guidelines for the treatment of alcohol problems. Department of Health and Ageing, Canberra
Hauser P, Fuller B, Ho SB, Thuras P, Kern S, Dieperink E (2017) The safety and efficacy of baclofen to reduce alcohol use in veterans with chronic hepatitis C: a randomized controlled trial. Addiction 112:1173–1183
Heppe DB, Keniston A, Bendelow T et al (2019) Reducing the severity of alcohol withdrawal with oral baclofen: a randomized controlled trial. Addict Res Theory 27:220–225
Hetland A, Carr DB (2014) Medications and impaired driving. Ann Pharmacother 48:494–506
Holla B, Karthik S, Biswal J et al (2018) Brain functional magnetic resonance imaging cue-reactivity can predict baclofen response in alcohol use disorders. Clin Psychopharmacol Neurosci 16:290–301
Jamshidi N, Morley KC, Cairns R, Dawson A, Haber PS (2018) A review of baclofen overdoses in Australia: calls to a poisons information centre and a case series. Alcohol Alcohol 54:73–78
Jasinska AJ, Stein EA, Kaiser J, Naumer MJ, Yalachkov Y (2014) Factors modulating neural reactivity to drug cues in addiction: a survey of human neuroimaging studies. Neurosci Biobehav Rev 38:1–16
Johnson BA (2008) Update on neuropharmacological treatments for alcoholism: scientific basis and clinical findings. Biochem Pharmacol 75:34–56

Keating GM (2014) Sodium oxybate: a review of its use in alcohol withdrawal syndrome and in the maintenance of abstinence in alcohol dependence. Clin Drug Investig 34:63–80

Lee C, Mayfield RD, Harris RA (2014) Altered gamma-aminobutyric acid type B receptor subunit 1 splicing in alcoholics. Biol Psychiatry 75:765–773

Leggio L, Garbutt J, Addolorato G (2010) Effectiveness and safety of baclofen in the treatment of alcohol dependent patients. CNS Neurol Disord Drug Targets 9:33–44

Leggio L, Zywiak WH, McGeary JE et al (2013) A human laboratory pilot study with baclofen in alcoholic individuals. Pharmacol Biochem Behav 103:784–791

Leone MA, Vigna-Taglianti F, Avanzi G, Brambilla R, Faggiano F (2010) Gamma-hydroxybutyrate (GHB) for treatment of alcohol withdrawal and prevention of relapses. Cochrane Database Syst Rev 2:CD006266

Lesouef N, Bellet F, Mounier G, Beyens M-N (2014) Efficacy of baclofen on abstinence and craving in alcohol-dependent patients: a meta-analysis of randomized controlled trials. Therapie 69:427–435

Liu J, Wang LN (2019) Baclofen for alcohol withdrawal. Cochrane Database Syst Rev 8: CD008502

Logge WB, Baillie AJ, Haber PS, Morley KC (2019a) Baclofen modulates cardiovascular responses to appetitive cues in treatment-seeking alcohol use disorder individuals. Hum Psychopharmacol Clin Exp 35(2):e2722

Logge WB, Morris RW, Baillie AJ, Haber PS, Morley KC (2019b) Baclofen attenuates fMRI alcohol cue reactivity in treatment-seeking alcohol dependent individuals. Psychopharmacology. https://doi.org/10.1007/s00213-019-05192-5

Lyon JE, Khan RA, Gessert CE, Larson PM, Renier CM (2011) Treating alcohol withdrawal with oral baclofen: a randomized, double-blind, placebo-controlled trial. J Hosp Med 6:469–474

Maccioni P, Colombo G (2019) Potential of GABAB receptor positive allosteric modulators in the treatment of alcohol use disorder. CNS Drugs 33:107–123

Morley KC, Baillie A, Leung S, Addolorato G, Leggio L, Haber PS (2014) Baclofen for the treatment of alcohol dependence and possible role of comorbid anxiety. Alcohol Alcohol 49:654–660

Morley KC, Logge W, Pearson S-A, Baillie A, Haber PS (2016) National trends in alcohol pharmacotherapy: findings from an Australian claims database. Drug Alcohol Depend 166:254–257

Morley KC, Baillie A, Fraser I et al (2018a) Baclofen in the treatment of alcohol dependence with or without liver disease: multisite, randomised, double-blind, placebo-controlled trial. Br J Psychiatry 212:362–369

Morley KC, Lagopoulos J, Logge W, Baillie A, Adams C, Haber PS (2018b) Brain GABA levels are reduced in alcoholic liver disease: a proton magnetic resonance spectroscopy study. Addict Biol 25(1):e1270

Morley KC, Luquin N, Baillie A et al (2018c) Moderation of baclofen response by a GABAB receptor polymorphism: results from the BacALD randomized controlled trial. Addiction 113:2205–2213

Müller CA, Geisel O, Pelz P et al (2015) High-dose baclofen for the treatment of alcohol dependence (BACLAD study): a randomized, placebo-controlled trial. Eur Neuropsychopharmacol 25:1167–1177

Nasti JJ, Brakoulias V (2011) Chronic baclofen abuse and withdrawal delirium. Aust N Z J Psychiatry 45:86–87

Pierce M, Sutterland A, Beraha EM, Morley KC, van den Brink W (2018) Efficacy, tolerability, and safety of low-dose and high-dose baclofen in the treatment of alcohol dependence: a systematic review and meta-analysis. Eur Neuropsychopharmacol 28:795–806

Ponizovsky AM, Rosca P, Aronovich E, Weizman A, Grinshpoon A (2015) Baclofen as add-on to standard psychosocial treatment for alcohol dependence: a randomized, double-blind, placebo-controlled trial with 1 year follow-up. J Subst Abus Treat 52:24–30 released GABA in response to peripheral inflammatory pain (Martins et al. 2015). Knock-down of $GABA_{B1a}$ as well as local injection of the $GABA_B$ receptor antagonist CGP 35348 into the DRt reduced inflammatory pain in the second phase of the formalin test, whereas injection of baclofen enhanced nociceptive behaviour (Martins et al. 2015). As $GABA_B$ receptors are predominantly expressed in enkephalinergic DRt neurons (Martins et al. 2015), local release of GABA upon noxious stimulation might activate presynaptic $GABA_B$ receptors and inhibit the release of endogenous opioids, abrogating tonic inhibition of projection neurons descending to the spinal cord.

3 $GABA_B$ Receptors and Pathological Pain

In unfortunate circumstances, nociceptive pain can become pathological, i.e. pain persists although the pain-causing insult has been removed. Chronic pain is a major health problem as it is widespread and pharmacologically difficult to address and massively affects life quality. Chronic pain is associated with profound plastic changes in peripheral and central nociceptive pathways, resulting in enhanced pain sensation to painful stimuli (hyperalgesia) or even in pain upon innocuous stimuli (allodynia) such as light touch (Yam et al. 2018). Reduced inhibitory control profoundly contributes to pain sensitization and the development of chronic pain (Zeilhofer et al. 2012a; Gwak and Hulsebosch 2011).

3.1 $GABA_B$ Receptors and Inflammatory Pain

Inflammation is an endogenous response of the body to clean up damaged tissue and promote its repair. Tissue damage triggers the production of a variety of inflammatory mediators that activate nociceptors. Prolonged inflammation results in peripheral as well as central sensitization. Peripheral sensitization is caused by excessive activation of nociceptors. In contrast, central sensitization results from plastic changes leading to enhanced excitation of neurons in the dorsal horn and the brain (Gangadharan and Kuner 2013; Basbaum et al. 2009).

Systemic application of baclofen diminished nocifensive behaviours in animal models of acute and chronic inflammatory pain (Smith et al. 1994; Patel et al. 2001; Shafizadeh et al. 1997). Thus, in principle, $GABA_B$ receptors at all anatomical levels of the pain pathway could contribute to the analgesic effects of baclofen. Indeed, there is evidence that baclofen acts at peripheral nociceptors as well as at spinal and supraspinal levels to relieve inflammatory pain. For instance, direct injection of baclofen into an inflamed rat paw produced an antinociceptive effect upon prostaglandin E2-induced hyperalgesia (Reis and Duarte 2006; Whitehead et al. 2012). The antinociceptive effect of baclofen was prevented by pre-injecting K^+ channel blockers, indicating the involvement of $GABA_B$ receptor-activated K^+ channels in the peripheral action of baclofen (Reis and Duarte 2006). Intrathecal injection of baclofen also reduced inflammatory pain responses in rodents, demonstrating its activity at spinal cord neurons (Patel et al. 2001; Dirig and Yaksh 1995; Naderi et al. 2005). Spinal application of baclofen reduced the activity of Aβ-, Aδ- and C-fibres in rats with carrageenan-inflamed hindpaws (Sokal and Chapman 2003) and considerably diminished c-Fos expression in dorsal horn neurons (Buritova et al. 1996) consistent with increased GABA$_B$ receptor-mediated neuronal inhibition. Finally, injection of baclofen in, for example, the ventrobasal complex of the thalamus (Potes et al. 2006a, b) shows the effectiveness of baclofen at the supraspinal level.

As chronic peripheral inflammation leads to enhanced excitability of dorsal horn neurons (Menetrey and Besson 1982; Takazawa et al. 2017), it is conceivable that the altered excitation/inhibition balance may lead to plastic changes in the expression of GABA$_B$ receptors. However, currently there is no clear picture on the regulation of GABA$_B$ receptor expression and its potential functional significance in inflammatory pain states. For example, GABA$_B$ receptor mRNA (McCarson and Enna 1999) and protein levels (Sands et al. 2003) appear to be upregulated in DRGs and the dorsal horn 24 h after formalin-induced hindpaw inflammation in rats. However, the increased GABA$_{B1}$ and GABA$_{B2}$ protein levels did not translate into increased GABA$_B$ receptor function in the dorsal horn as measured by [^{35}S] GTPγS binding (Sands et al. 2003). On the other hand, a decrease of GABA$_B$ receptors in the dorsal horn was observed 3–4 weeks after inducing inflammation (Castro-Lopes et al. 1995), and, finally, in a rat model of orofacial inflammation, GABA$_B$ receptors were downregulated in the trigeminal ganglion (Liu et al. 2019). Thus, there seems to be no general rule for the regulation of GABA$_B$ receptor expression in inflammatory pain states. Instead, a potential regulation of GABA$_B$ receptor expression might underlie specific temporal and spatial conditions.

Besides regulating neuronal excitability, GABA$_B$ receptors appear to be involved in other processes in inflammatory pain states. For instance, spinal GABA$_B$ receptors also appear to modulate the local inflammatory response as shown in a mouse model of arthritis (Bassi et al. 2016). Intrathecal injection of baclofen increased neutrophil recruitment to the inflamed knee joint in a p38 MAPK-dependent manner. Thus, in addition to the well-established analgesic activity, activation of spinal GABA$_B$ receptors also appears to exert a pro-inflammatory effect. This should be taken into account in the case of development analgesic drugs for the treatment of chronic inflammation targeting GABA$_B$ receptors.

Another example is the control of interleukin 1β release from satellite glial cells in trigeminal ganglia. GABA$_B$ receptors were recently found to be expressed in satellite glial cells in trigeminal ganglia where they control the activity of inwardly rectifying K^+ channels (Takeda et al. 2015). Satellite glial cells ensheath trigeminal neurons, are activated by neurotransmitters released from trigeminal neurons and appear to play an important role in the development of hyperalgesia in response to inflammatory stimuli via release of interleukin 1β (Takeda et al. 2009). Injection of baclofen into the trigeminal ganglion after inducing orofacial inflammation reduced mechanical allodynia, prevented downregulation of GABA$_B$ receptors in satellite glial cells and inhibited satellite glial cell activation as well as interleukin 1β release (Liu et al. 2019). This study provides intriguing indications that glial $GABA_B$ receptors adjacent to nociceptors may play a role in the regulation of orofacial inflammation. However, this mechanism might be restricted to the trigeminal ganglia since satellite glia cells in DRG neurons appear not to express $GABA_B$ receptors (Engle et al. 2012).

3.2 $GABA_B$ Receptors and Neuropathic Pain

Neuropathic pain originates from nerve injury and diseases that lead to peripheral or central neuronal damage such as diabetes (Schreiber et al. 2015), cancer (Davis 2018), multiple sclerosis (Khan and Smith 2014) and HIV (Aziz-Donnelly and Harrison 2017). Neuropathic pain is associated with a multitude of plastic alterations in the CNS leading to enhanced excitation and central sensitization (Gangadharan and Kuner 2013; Sandkuhler 2009). These include a prominent reduction of GABAergic inhibitory control (Zeilhofer et al. 2012a). Peripheral nerve damage significantly reduced GABA levels as well as GABA release in the dorsal horn (Eaton et al. 1998; Vaysse et al. 2011; Lever et al. 2003; Ibuki et al. 1997; Somers and Clemente 2002). Restoration of GABA levels by viral overexpression of GAD65 or injection of GABA-expressing cells into the dorsal horn significantly reduced neuropathic pain in rats with peripheral nerve injury (Vaysse et al. 2011; Jergova et al. 2012; Lee et al. 2007). Intrathecal injection of $GABA_A$ receptor and $GABA_B$ receptor antagonists showed that both receptors mediated the antinociceptive activity of GABA-expressing grafts (Jergova et al. 2012). Therefore, enhancing $GABA_B$ receptor activity is a promising strategy to relieve neuropathic pain.

In fact, systemic (Smith et al. 1994; Patel et al. 2001; Magnaghi et al. 2014) and intrathecal administration (Bai et al. 2014; Liu et al. 2018; Zemoura et al. 2016; Dias and Prado 2016; Hwang and Yaksh 1997; Malan et al. 2002; Lee et al. 2010; Gwak et al. 2006) of baclofen as well as injecting it into supraspinal structures (Migita et al. 2018) alleviated nociceptive responses in animal models of neuropathic pain. This implies, as in the case of chronic inflammatory pain, that $GABA_B$ receptors at all levels of the pain pathway contribute to the analgesic effect of baclofen. Interestingly, chronic intrathecal application of baclofen in a rat model of diabetic neuropathic pain normalized elevated expression levels of NR2B-containing NMDA receptors, most likely in a CREB-dependent manner (Liu et al. 2014; Bai et al. 2014). This illustrates that $GABA_B$ receptor activity not only counterbalances overexcitation but can trigger diverse pathways to affect neuronal excitability at different levels.

As neuropathic pain leads to diminished neuronal inhibition, a plastic downregulation of $GABA_B$ receptors might be a contributing factor. However, since baclofen still exhibits analgesic activity in animal models of neuropathic pain, an excessive downregulation of $GABA_B$ receptors is very unlikely.

In the rat model of streptozotocin-induced diabetes, hyperalgesia and allodynia develop 3 weeks after a single injection of streptozotocin. This is accompanied by increased glutamatergic excitatory activity in spinal cord neurons without changes in GABAergic and glycinergic inhibitory postsynaptic currents, indicating a preserved GABAergic tone under these conditions (Wang et al. 2007). However, GABA$_B$ receptor-mediated inhibition of primary afferent terminals was significantly reduced, which might be explained by downregulation of GABA$_B$ receptors at central nociceptor terminals. Indeed, a robust loss of GABA$_B$ receptors in the spinal cord was observed by Western blotting 5–7 weeks after induction of diabetes with streptozotocin (Wang et al. 2011). Because the onset of nocifensive behaviour (after 3 weeks) was not exactly paralleled by this robust GABA$_B$ receptor downregulation (starting after 5 weeks), it is unlikely that the loss of GABA$_B$ receptors is essential for the development of pain symptoms in this model of neuropathic pain.

Another example for a considerable downregulation of GABA$_B$ receptors is cancer-induced bone pain in rats (Zhou et al. 2017). In this model downregulation of GABA$_B$ receptors in the spinal cord exactly followed the time course of mechanical allodynia development. Chronic intrathecal application of baclofen partially restored GABA$_B$ receptor expression, which was associated with a reduction of elevated PKA and pCREB levels (Zhou et al. 2017). As activation of the PKA/CREB pathway contributes to the development of cancer-induced bone pain (Hang et al. 2013), there might be a link between downregulation of GABA$_B$ receptors and the expression of chronic pain in cancer-induced bone pain.

Whereas there is clear evidence for downregulation of GABA$_B$ receptors in the spinal cord in diabetes-induced and cancer-induced bone pain, inconsistent results were reported from models of neuropathic pain induced by chronic constriction of the sciatic nerve (CCI) and spinal nerve ligation (SNL). There is convincing evidence that GABA$_B$ receptors are downregulated in DRG neurons after peripheral nerve ligation (Engle et al. 2012), but no change was found in the dorsal horn (Engle et al. 2006; Smith et al. 1994; Zemoura et al. 2016). However, Wu et al. (2011) observed a strong reduction specifically in GABA$_{B1a}$ expression (expression GABA$_{B1b}$ was not affected) 7 and 14 days after spinal nerve ligation. This suggests that presynaptic GABA$_B$ receptors are mainly affected, most probably at central nociceptor terminals. Downregulation of GABA$_{B1a}$ as well as nocifensive behaviour was prevented or reduced, respectively, by intrathecal application of a p38-MAPK inhibitor (Wu et al. 2011). As p38-MAPK is specifically activated in dorsal horn microglia after spinal nerve ligation (Tsuda et al. 2004), this observation may link microglia activation to the downregulation of GABA$_{B1a}$ so far by an unknown mechanism.

The mechanism(s) involved in downregulating GABA$_B$ receptors under chronic pain conditions are currently unknown. However, it is well established that neuronal over-excitation caused by sustained activity of AMPA and NMDA receptors rapidly downregulates GABA$_B$ receptors and thereby strongly reduces GABA$_B$ receptor-mediated neuronal inhibition by enhancing lysosomal degradation of the receptors (Maier et al. 2010; Guetg et al. 2010; Kantamneni et al. 2014; Terunuma et al. 2010;

Zemoura et al. 2019). As chronic pain states are associated with considerably enhanced glutamate receptor activity, this mechanism might be one factor contributing to $GABA_B$ receptor downregulation.

Another very intriguing mechanism impairing $GABA_B$ receptor activity under conditions of neuropathic pain was proposed by Laffray et al. (2012). In the spinal nerve ligation model of neuropathic pain, the $GABA_B$ receptor interacting protein 14-3-3ζ was found to be selectively upregulated in the ipsilateral dorsal horn. Binding of 14-3-3ζ to $GABA_B$ receptors occurred at the plasma membrane and interfered with heterodimerization of the receptors. This rendered the receptors non-functional and diminished $GABA_B$ receptor signalling (Laffray et al. 2012). Preventing the interaction of $GABA_B$ receptors with 14-3-3ζ using an interfering synthetic peptide enhanced baclofen-mediated analgesia and, interestingly, partially reversed nocifensive behaviours in neuropathic rats in the absence of baclofen. This observation supports a role for diminished $GABA_B$ receptor signalling in the development of neuropathic pain.

These examples suggest a rather complex regulation of $GABA_B$ receptors during chronic pain states.

4 $GABA_B$ Receptors as Target for Treating Chronic Pain

The $GABA_B$ receptor agonist baclofen has a long history in the clinic for treating severe spasticity caused by, e.g. multiple sclerosis, spinal cord injury and stroke. It efficiently resolves spasm and relieves associated musculoskeletal pain (Slonimski et al. 2004). In addition, a variety of small clinical studies and case reports demonstrated the analgesic activity of baclofen in chronic pain states not associated with spasticity, including trigeminal neuralgia, intractable post herpetic neuralgia, complex regional pain syndrome, chronic pain after cerebral stroke, painful spinal cord lesions and neuropathic pain after peripheral nerve injury (Fromm and Terrence 1987; Fromm et al. 1984; Zuniga et al. 2000; Hosny et al. 2004; van der Plas et al. 2013; Kopsky et al. 2015; Lind et al. 2008; Herman et al. 1992; Taira and Hori 2007; Taira et al. 1995; Goto et al. 2013; Harmer and Larson 2002). Although baclofen is often successful in patients not responding to common analgesics (e.g. opioids), there are several drawbacks associated with baclofen that prevents its widespread application. Baclofen has a relatively short half-life (2–4 h) and it poorly crosses the blood-brain barrier. These characteristics require high doses for systemic application, which can cause tolerance and severe side effects (Brennan and Whittle 2008). For these reasons, baclofen is given in severe cases intrathecally with pumps. But even then side effects such as drowsiness, confusion, disturbance of speech, gastrointestinal problems, nausea, hypotension, sexual dysfunction and many others were observed (Slonimski et al. 2004). Therefore, baclofen will be reserved for severe chronic pain states resistant to common analgesics or as an adjuvant in combination with analgesics, e.g. opioids (Gatscher et al. 2002; Zuniga et al. 2000).

Because $GABA_B$ receptors are ubiquitously expressed and are involved in numerous physiological functions, their global activation will invariably be associated with unwanted effects. It is therefore rather unlikely that orthosteric $GABA_B$ receptor agonists will be successful as first-line analgesics. Novel strategies like the recent development of $GABA_B$ receptor positive allosteric modulators (Froestl 2010; Urwyler 2011) may have some potential for the generation of analgesics. Positive allosteric modulators increase the affinity and efficacy of the receptor for GABA, resulting in "use-dependent" potentiation of $GABA_B$ receptor activity. This kind of activity enhancement appears to be associated with fewer side effects (in particular sedation) in various animal models.

So far, only a very limited number of positive allosteric modulators were evaluated in animal models of acute and chronic pain (Table 1). Regarding neuropathic pain, only rac-BHFF was tested in the CCI model of neuropathic pain (Zemoura et al. 2016). Although rac-BHFF increased the paw withdrawal threshold after mechanical stimulation at a non-sedating oral dose in naïve mice, it was inactive in neuropathic mice. However, after activation of $GABA_B$ receptors with a subsaturating intrathecal dose of baclofen, rac-BHFF potentiated the analgesic effect of baclofen (Zemoura et al. 2016). This observation indicates that the reduced GABAergic tone (see above) in neuropathic mice was too low to permit efficient allosteric modulation of $GABA_B$ receptors and pain relief. Thus, in pain conditions associated with a considerably reduced GABAergic tone, allosteric modulators are not expected to display sufficient analgesic activity. They may, however, be useful to enhance the activity of low-doses of baclofen, which might result in fewer side effects.

Regarding chronic inflammatory pain, the positive allosteric modulators ADX71441 and ADX71943 reduced hyperalgesia in the formalin test and the monosodium iodoacetate model of chronic osteoarthritis (Kalinichev et al. 2014, 2017). ADX71943 very poorly crosses the blood-brain barrier and therefore exhibits, in contrast to ADX71441, no central effects (Kalinichev et al. 2014). In the model of chronic osteoarthritis ADX71943 exhibited strongest analgesic activity in the early, more inflammatory related, phase of the model and vanished in later stages presumably related to a more centrally driven neuropathic pain state (Kalinichev et al. 2017). Thus, peripherally acting allosteric modulators might be effective in acute and early inflammatory pain without a prominent central component. Such drugs are expected to lack the severe central side effects associated with baclofen.

$GABA_B$ receptors are also involved in the regulation of visceral pain (Page et al. 2006; Loeza-Alcocer et al. 2019). In acute visceral pain, the positive allosteric modulators CGP7930 (colon pain) (Brusberg et al. 2009) and ADX17441 (colon and bladder pain) (Kannampalli et al. 2017) decreased nociceptive responses. The analgesic effect of ADX17441 appeared to be predominantly mediated via peripheral and supraspinal sites since systemic and intra-cerebroventricular but not intrathecal administration was effective (Kannampalli et al. 2017). The observation that

Finally, a potential for developing novel highly specific therapeutic interventions might be the targeting of pain-specific protein-protein interactions. As discussed above (see Sect. 3.2), the group of Marc Laundry provided an excellent example of such an approach regarding $GABA_B$ receptors and neuropathic pain (Laffray et al. 2012). Ideally, the development of interfering peptides to target pain-related protein-protein interactions involving $GABA_B$ receptors requires the discovery of interactions specifically associated with the pain state of interest. Future research needs to show whether such an approach can overcome the limitations of $GABA_B$ receptor agonists.

References

Albrecht PJ, Rice FL (2010) Role of small-fiber afferents in pain mechanisms with implications on diagnosis and treatment. Curr Pain Headache Rep 14(3):179–188. https://doi.org/10.1007/s11916-010-0105-y

Almeida A, Storkson R, Lima D, Hole K, Tjolsen A (1999) The medullary dorsal reticular nucleus facilitates pain behaviour induced by formalin in the rat. Eur J Neurosci 11(1):110–122

Almeida A, Tjolsen A, Lima D, Coimbra A, Hole K (1996) The medullary dorsal reticular nucleus facilitates acute nociception in the rat. Brain Res Bull 39(1):7–15

Andrade R, Malenka RC, Nicoll RA (1986) A G protein couples serotonin and $GABA_B$ receptors to the same channels in hippocampus. Science 234(4781):1261–1265

Ataka T, Kumamoto E, Shimoji K, Yoshimura M (2000) Baclofen inhibits more effectively C-afferent than Aδ-afferent glutamatergic transmission in substantia gelatinosa neurons of adult rat spinal cord slices. Pain 86(3):273–282

Aziz-Donnelly A, Harrison TB (2017) Update of HIV-associated sensory neuropathies. Curr Treat Options Neurol 19(10):36. https://doi.org/10.1007/s11940-017-0472-3

Bai HP, Liu P, Wu YM, Guo WY, Guo YX, Wang XL (2014) Activation of spinal $GABA_B$ receptors normalizes N-methyl-D-aspartate receptor in diabetic neuropathy. J Neurol Sci 341 (1–2):68–72. https://doi.org/10.1016/j.jns.2014.04.002

Basbaum AI, Bautista DM, Scherrer G, Julius D (2009) Cellular and molecular mechanisms of pain. Cell 139(2):267–284. https://doi.org/10.1016/j.cell.2009.09.028

Bassi GS, do Malvar CD, Cunha TM, Cunha FQ, Kanashiro A (2016) Spinal $GABA_B$ receptor modulates neutrophil recruitment to the knee joint in zymosan-induced arthritis. Naunyn Schmiedebergs Arch Pharmacol 389(8):851–861. https://doi.org/10.1007/s00210-016-1248-0

Bean BP (1989) Neurotransmitter inhibition of neuronal calcium currents by changes in channel voltage dependence. Nature 340(6229):153–156. https://doi.org/10.1038/340153a0

Bell A (2018) The neurobiology of acute pain. Vet J 237:55–62. https://doi.org/10.1016/j.tvjl.2018.05.004

Benke D, Honer M, Michel C, Bettler B, Mohler H (1999) γ-Aminobutyric acid type B receptor splice variant proteins GBR1a and GBR1b are both associated with GBR2 *in situ* and display differential regional and subcellular distribution. J Biol Chem 274:27323–27330

Bennett DL, Woods CG (2014) Painful and painless channelopathies. Lancet Neurol 13 (6):587–599. https://doi.org/10.1016/s1474-4422(14)70024-9

Berecki G, McArthur JR, Cuny H, Clark RJ, Adams DJ (2014) Differential Cav2.1 and Cav2.3 channel inhibition by baclofen and alpha-conotoxin Vc1.1 via $GABA_B$ receptor activation. J Gen Physiol 143(4):465–479. https://doi.org/10.1085/jgp.201311104

Brennan PM, Whittle IR (2008) Intrathecal baclofen therapy for neurological disorders: a sound knowledge base but many challenges remain. Br J Neurosurg 22(4):508–519. https://doi.org/10.1080/02688690802233364

Brewer CL, Baccei ML (2018) Enhanced Postsynaptic $GABA_B$ Receptor Signaling in Adult Spinal Projection Neurons after Neonatal Injury. Neuroscience 384:329–339. https://doi.org/10.1016/j.neuroscience.2018.05.046

Brusberg M, Ravnefjord A, Martinsson R, Larsson H, Martinez V, Lindström E (2009) The $GABA_B$ receptor agonist, baclofen, and the positive allosteric modulator, CGP7930, inhibit visceral pain-related responses to colorectal distension in rats. Neuropharmacology 56 (2):362–367

Buritova J, Chapman V, Honore P, Besson JM (1996) The contribution of $GABA_B$ receptor-mediated events to inflammatory pain processing: carrageenan oedema and associated spinal c-Fos expression in the rat. Neuroscience 73(2):487–496

Bussieres N, El Manira A (1999) $GABA_B$ receptor activation inhibits N- and P/Q-type calcium channels in cultured lamprey sensory neurons. Brain Res 847(2):175–185

Callaghan B, Haythornthwaite A, Berecki G, Clark RJ, Craik DJ, Adams DJ (2008) Analgesic alpha-conotoxins Vc1.1 and Rg1A inhibit N-type calcium channels in rat sensory neurons via $GABA_B$ receptor activation. J Neurosci 28(43):10943–10951. https://doi.org/10.1523/jneurosci.3594-08.2008

Castro-Lopes JM, Malcangio M, Pan BH, Bowery NG (1995) Complex changes of $GABA_A$ and $GABA_B$ receptor binding in the spinal cord dorsal horn following peripheral inflammation or neurectomy. Brain Res 679(2):289–297

Castro AR, Morgado C, Lima D, Tavares I (2006) Differential expression of NK1 and $GABA_B$ receptors in spinal neurones projecting to antinociceptive or pronociceptive medullary centres. Brain Res Bull 69(3):266–275. https://doi.org/10.1016/j.brainresbull.2005.12.004

Castro J, Harrington AM, Garcia-Caraballo S, Maddern J, Grundy L, Zhang J, Page G, Miller PE, Craik DJ, Adams DJ, Brierley SM (2017) α-Conotoxin Vc1.1 inhibits human dorsal root ganglion neuroexcitability and mouse colonic nociception via $GABA_B$ receptors. Gut 66 (6):1083–1094. https://doi.org/10.1136/gutjnl-2015-310971

Charles KJ, Evans ML, Robbins MJ, Calver AR, Leslie RA, Pangalos MN (2001) Comparative immunohistochemical localisation of $GABA_{B1a}$, $GABA_{B1b}$ and $GABA_{B2}$ subunits in rat brain, spinal cord and dorsal root ganglion. Neuroscience 106(3):447–467

Chen G, van den Pol AN (1998) Presynaptic $GABA_B$ autoreceptor modulation of P/Q-type calcium channels and GABA release in rat suprachiasmatic nucleus neurons. J Neurosci 18 (5):1913–1922

Chen Q, Shao C, Zhou H, Ma R, Jiang P, Yang K (2017) Differential sensitivity of presynaptic and postsynaptic $GABA_B$ receptors in rat ventrolateral periaqueductal gray. Neuroreport 28 (18):1221–1224. https://doi.org/10.1097/wnr.0000000000000906

Cuny H, de Faoite A, Huynh TG, Yasuda T, Berecki G, Adams DJ (2012) γ-Aminobutyric acid type B (GABA⍰) receptor expression is needed for inhibition of N-type (Cav2.2) calcium channels by analgesic alpha-conotoxins. J Biol Chem 287(28):23948–23957. https://doi.org/10.1074/jbc.M112.342998

Davis MP (2018) Cancer-Related Neuropathic Pain: Review and Selective Topics. Hematol Oncol Clin North Am 32(3):417–431. https://doi.org/10.1016/j.hoc.2018.01.005

Delaney AJ, Crane JW (2016) Presynaptic $GABA_B$ receptors reduce transmission at parabrachial synapses in the lateral central amygdala by inhibiting N-type calcium channels. Sci Rep 6:19255. https://doi.org/10.1038/srep19255

Desarmenien M, Feltz P, Occhipinti G, Santangelo F, Schlichter R (1984) Coexistence of $GABA_A$ and $GABA_B$ receptors on A delta and C primary afferents. Br J Pharmacol 81(2):327–333. https://doi.org/10.1111/j.1476-5381.1984.tb10082.x

Dias QM, Prado WA (2016) The lesion of dorsolateral funiculus changes the antiallodynic effect of the intrathecal muscimol and baclofen in distinct phases of neuropathic pain induced by spinal nerve ligation in rats. Brain Res Bull 124:103–115. https://doi.org/10.1016/j.brainresbull.2016.04.001

Dirig DM, Yaksh TL (1995) Intrathecal baclofen and muscimol, but not midazolam, are antinociceptive using the rat-formalin model. J Pharmacol Exp Ther 275(1):219–227

Duan B, Cheng L, Bourane S, Britz O, Padilla C, Garcia-Campmany L, Krashes M, Knowlton W, Velasquez T, Ren X, Ross S, Lowell BB, Wang Y, Goulding M, Ma Q (2014) Identification of spinal circuits transmitting and gating mechanical pain. Cell 159(6):1417–1432. https://doi.org/10.1016/j.cell.2014.11.003

Eaton MJ, Plunkett JA, Karmally S, Martinez MA, Montanez K (1998) Changes in GAD- and GABA- immunoreactivity in the spinal dorsal horn after peripheral nerve injury and promotion of recovery by lumbar transplant of immortalized serotonergic precursors. J Chem Neuroanat 16 (1):57–72

Engle MP, Gassman M, Sykes KT, Bettler B, Hammond DL (2006) Spinal nerve ligation does not alter the expression or function of $GABA_B$ receptors in spinal cord and dorsal root ganglia of the rat. Neuroscience 138(4):1277–1287

Engle MP, Merrill MA, Marquez De Prado B, Hammond DL (2012) Spinal nerve ligation decreases γ-aminobutyric $acid_B$ receptors on specific populations of immunohistochemically identified neurons in L5 dorsal root ganglion of the rat. J Comp Neurol 520(8):1663–1677. https://doi.org/10.1002/cne.23005

Enna SJ, McCarson KE (2006) The role of GABA in the mediation and perception of pain. Adv Pharmacol 54:1–27

Foster E, Wildner H, Tudeau L, Haueter S, Ralvenius WT, Jegen M, Johannssen H, Hosli L, Haenraets K, Ghanem A, Conzelmann KK, Bosl M, Zeilhofer HU (2015) Targeted ablation, silencing, and activation establish glycinergic dorsal horn neurons as key components of a spinal gate for pain and itch. Neuron 85(6):1289–1304. https://doi.org/10.1016/j.neuron.2015.02.028

Fritschy JM, Meskenaite V, Weinmann O, Honer M, Benke D, Mohler H (1999) $GABA_B$ receptor splice variants GB1a and GB1b in rat brain: developmental regulation, cellular distribution and extrasynaptic localization. Eur J Neurosci 11(3):761–768

Fritschy JM, Sidler C, Parpan F, Gassmann M, Kaupmann K, Bettler B, Benke D (2004) Independent maturation of the $GABA_B$ receptor subunits $GABA_{B1}$ and $GABA_{B2}$ during postnatal development in rodent brain. J Comp Neurol 477(3):235–252

Froestl W (2010) Novel $GABA_B$ receptor positive modulators: a patent survey. Expert Opin Ther Pat 20(8):1007–1017

Fromm GH, Terrence CF (1987) Comparison of L-baclofen and racemic baclofen in trigeminal neuralgia. Neurology 37(11):1725–1728. https://doi.org/10.1212/wnl.37.11.1725

Fromm GH, Terrence CF, Chattha AS (1984) Baclofen in the treatment of trigeminal neuralgia: double-blind study and long-term follow-up. Ann Neurol 15(3):240–244. https://doi.org/10.1002/ana.410150306

Fuchs PN, Peng YB, Boyette-Davis JA, Uhelski ML (2014) The anterior cingulate cortex and pain processing. Front Integr Neurosci 8:35. https://doi.org/10.3389/fnint.2014.00035

Gaillard S, Lo Re L, Mantilleri A, Hepp R, Urien L, Malapert P, Alonso S, Deage M, Kambrun C, Landry M, Low SA, Alloui A, Lambolez B, Scherrer G, Le Feuvre Y, Bourinet E, Moqrich A (2014) GINIP, a G-interacting protein, functions as a key modulator of peripheral GABA receptor-mediated analgesia. Neuron 84:123–136. https://doi.org/10.1016/j.neuron.2014.08.056

Gangadharan V, Agarwal N, Brugger S, Tegeder I, Bettler B, Kuner R, Kurejova M (2009) Conditional gene deletion reveals functional redundancy of GABAB receptors in peripheral nociceptors in vivo. Mol Pain 5:68

Gangadharan V, Kuner R (2013) Pain hypersensitivity mechanisms at a glance. Dis Model Mech 6 (4):889–895. https://doi.org/10.1242/dmm.011502

Gassmann M, Bettler B (2012) Regulation of neuronal $GABA_B$ receptor functions by subunit composition. Nat Rev Neurosci 13(6):380–394. https://doi.org/10.1038/nrn3249

Gassmann M, Shaban H, Vigot R, Sansig G, Haller C, Barbieri S, Humeau Y, Schuler V, Müller M, Kinzel B, Klebs K, Schmutz M, Froestl W, Heid J, Kelly PH, Gentry C, Jaton AL, Van der Putten H, Mombereau C, Lecourtier L, Mosbacher J, Cryan JF, Fritschy JM, Lüthi A, Kaupmann K, Bettler B (2004) Redistribution of $GABA_{B1}$ protein and atypical $GABA_B$ responses in $GABA_{B2}$-deficient mice. J Neurosci 24(27):6086–6097

Gatscher S, Becker R, Uhle E, Bertalanffy H (2002) Combined intrathecal baclofen and morphine infusion for the treatment of spasticity related pain and central deafferentiation pain. Acta Neurochir Suppl 79:75–76
Goto S, Taira T, Horisawa S, Yokote A, Sasaki T, Okada Y (2013) Spinal cord stimulation and intrathecal baclofen therapy: combined neuromodulation for treatment of advanced complex regional pain syndrome. Stereotact Funct Neurosurg 91(6):386–391. https://doi.org/10.1159/000350022
Greif GJ, Sodickson DL, Bean BP, Neer EJ, Mende U (2000) Altered regulation of potassium and calcium channels by $GABA_B$ and adenosine receptors in hippocampal neurons from mice lacking $G\alpha_o$. J Neurophysiol 83(2):1010–1018
Guetg N, Aziz SA, Holbro N, Turecek R, Rose T, Seddik R, Gassmann M, Moes S, Jenoe P, Oertner TG, Casanova E, Bettler B (2010) NMDA receptor-dependent $GABA_B$ receptor internalization via CaMKII phosphorylation of serine 867 in $GABA_{B1}$. Proc Natl Acad Sci U S A 107(31):13924–13929
Guyon A, Kussrow A, Olmsted IR, Sandoz G, Bornhop DJ, Nahon JL (2013) Baclofen and other $GABA_B$ receptor agents are allosteric modulators of the CXCL12 chemokine receptor CXCR4. J Neurosci 33(28):11643–11654. https://doi.org/10.1523/jneurosci.6070-11.2013
Gwak YS, Hulsebosch CE (2011) GABA and central neuropathic pain following spinal cord injury. Neuropharmacology 60(5):799–808. https://doi.org/10.1016/j.neuropharm.2010.12.030
Gwak YS, Tan HY, Nam TS, Paik KS, Hulsebosch CE, Leem JW (2006) Activation of spinal GABA receptors attenuates chronic central neuropathic pain after spinal cord injury. J Neurotrauma 23(7):1111–1124. https://doi.org/10.1089/neu.2006.23.1111
Hammond DL (1997) Inhibitory neurotransmitters and nociception: role of GABA and glycine. In: Dickenson A, Besson JM (eds) The pharmacology of pain. Springer, Berlin, pp 361–383. https://doi.org/10.1007/978-3-642-60777-6_14
Hanack C, Moroni M, Lima WC, Wende H, Kirchner M, Adelfinger L, Schrenk-Siemens K, Tappe-Theodor A, Wetzel C, Kuich PH, Gassmann M, Roggenkamp D, Bettler B, Lewin GR, Selbach M, Siemens J (2015) GABA blocks pathological but not acute TRPV1 pain signals. Cell 160(4):759–770. https://doi.org/10.1016/j.cell.2015.01.022
Hang LH, Yang JP, Shao DH, Chen Z, Wang H (2013) Involvement of spinal PKA/CREB signaling pathway in the development of bone cancer pain. Pharmacol Rep 65(3):710–716
Harmer JP, Larson BS (2002) Pain relief from baclofen analgesia in a neuropathic pain patient who failed opioid and pharmacotherapy: case report. J Pain Palliat Care Pharmacother 16(3):61–64
Hawrot E, Xiao Y, Shi QL, Norman D, Kirkitadze M, Barlow PN (1998) Demonstration of a tandem pair of complement protein modules in $GABA_B$ receptor 1a. FEBS Lett 432(3):103–108
Herman RM, D'Luzansky SC, Ippolito R (1992) Intrathecal baclofen suppresses central pain in patients with spinal lesions. A pilot study. Clin J Pain 8(4):338–345
Hosny A, Simopoulos T, Collins B (2004) Response of intractable post herpetic neuralgia to intrathecal baclofen. Pain Physician 7(3):345–347
Hu C, Zhang G, Zhao YT (2014) Fucoidan attenuates the existing allodynia and hyperalgesia in a rat model of neuropathic pain. Neurosci Lett 571:66–71. https://doi.org/10.1016/j.neulet.2014.04.030
Hu C, Zhao YT, Zhang G, Xu MF (2017) Antinociceptive effects of fucoidan in rat models of vincristine-induced neuropathic pain. Mol Med Rep 15(2):975–980. https://doi.org/10.3892/mmr.2016.6071
Hwang JH, Yaksh TL (1997) The effect of spinal GABA receptor agonists on tactile allodynia in a surgically-induced neuropathic pain model in the rat. Pain 70(1):15–22
Ibuki T, Hama AT, Wang XT, Pappas GD, Sagen J (1997) Loss of GABA-immunoreactivity in the spinal dorsal horn of rats with peripheral nerve injury and promotion of recovery by adrenal medullary grafts. Neuroscience 76(3):845–858
Iyadomi M, Iyadomi I, Kumamoto E, Tomokuni K, Yoshimura M (2000) Presynaptic inhibition by baclofen of miniature EPSCs and IPSCs in substantia gelatinosa neurons of the adult rat spinal dorsal horn. Pain 85(3):385–393

Jergova S, Hentall ID, Gajavelli S, Varghese MS, Sagen J (2012) Intraspinal transplantation of GABAergic neural progenitors attenuates neuropathic pain in rats: a pharmacologic and neurophysiological evaluation. Exp Neurol 234(1):39–49. https://doi.org/10.1016/j.expneurol.2011.12.005

Jones KA, Borowsky B, Tamm JA, Craig DA, Durkin MM, Dai M, Yao W-J, Johnson M, Gunwaldsen C, Huang L-Y, Tang C, Shen Q, Salon JA, Morse K, Laz T, Smith KE, Nagarathnam D, Noble SA, Branchek TA, Gerald C (1998) $GABA_B$ receptor function as a heteromeric assembly of the subunits $GABA_{BR1}$ and $GABA_{BR2}$. Nature 396:674–679

Jones TL, Sweitzer SM, Peters MC, Wilson SP, Yeomans DC (2005) $GABA_B$ receptors on central terminals of C-afferents mediate intersegmental Aδ-afferent evoked hypoalgesia. Eur J Pain 9 (3):233–242

Kalinichev M, Donovan-Rodriguez T, Girard F, Haddouk H, Royer-Urios I, Schneider M, Bate ST, Marker C, Pomonis JD, Poli S (2017) ADX71943 and ADX71441, novel positive allosteric modulators of the $GABA_B$ receptor with distinct central/peripheral profiles, show efficacy in the monosodium iodoacetate model of chronic osteoarthritis pain in the rat. Eur J Pharmacol 795:43–49. https://doi.org/10.1016/j.ejphar.2016.11.056

Kalinichev M, Donovan-Rodriguez T, Girard F, Riguet E, Rouillier M, Bournique B, Haddouk H, Mutel V, Poli S (2014) Evaluation of peripheral versus central effects of $GABA_B$ receptor activation using a novel, positive allosteric modulator of the $GABA_B$ receptor ADX71943, a pharmacological tool compound with a fully peripheral activity profile. Br J Pharmacol 171 (21):4941–4954. https://doi.org/10.1111/bph.12812

Kannampalli P, Poli SM, Bolea C, Sengupta JN (2017) Analgesic effect of ADX71441, a positive allosteric modulator (PAM) of $GABA_B$ receptor in a rat model of bladder pain. Neuropharmacology 126:1–11. https://doi.org/10.1016/j.neuropharm.2017.08.023

Kantamneni S, Gonzalez-Gonzalez IM, Luo J, Cimarosti H, Jacobs SC, Jaafari N, Henley JM (2014) Differential regulation of $GABA_B$ receptor trafficking by different modes of N-methyl-D-aspartate (NMDA) receptor signaling. J Biol Chem 289(10):6681–6694. https://doi.org/10.1074/jbc.M113.487348

Kaupmann K, Huggel K, Heid J, Flor PJ, Bischoff S, Mickel SJ, McMaster G, Angst C, Bittiger H, Froestl W, Bettler B (1997) Expression cloning of $GABA_B$ receptors uncovers similarity to metabotropic glutamate receptors. Nature 386:239–246

Kaupmann K, Malitschek B, Schuler V, Heid J, Froestl W, Beck P, Mosbacher J, Bischoff S, Kulik A, Shigemoto R, Karshin A, Bettler B (1998) $GABA_B$ receptor subtypes assemble into functional heteromeric complexes. Nature 396:683–687

Khan N, Smith MT (2014) Multiple sclerosis-induced neuropathic pain: pharmacological management and pathophysiological insights from rodent EAE models. Inflammopharmacology 22 (1):1–22. https://doi.org/10.1007/s10787-013-0195-3

Kim SE, Chang L (2012) Overlap between functional GI disorders and other functional syndromes: what are the underlying mechanisms? Neurogastroenterol Motil 24(10):895–913. https://doi.org/10.1111/j.1365-2982.2012.01993.x

Kopsky DJ, Keppel Hesselink JM, Casale R (2015) Walking with neuropathic pain: paradoxical shift from burden to support? Case Rep Med 2015:764950. https://doi.org/10.1155/2015/764950

Kulik A, Nakadate K, Nyiri G, Notomi T, Malitschek B, Bettler B, Shigemoto R (2002) Distinct localization of GABA(B) receptors relative to synaptic sites in the rat cerebellum and ventrobasal thalamus. Eur J Neurosci 15(2):291–307

Kulik A, Vida I, Fukazawa Y, Guetg N, Kasugai Y, Marker CL, Rigato F, Bettler B, Wickman K, Frotscher M, Shigemoto R (2006) Compartment-dependent colocalization of Kir3.2-containing K^+ channels and $GABA_B$ receptors in hippocampal pyramidal cells. J Neurosci 26 (16):4289–4297

Kulik A, Vida I, Lujan R, Haas CA, Lopez-Bendito G, Shigemoto R, Frotscher M (2003) Subcellular localization of metabotropic $GABA_B$ receptor subunits $GABA_{B1a/b}$ and $GABA_{B2}$ in the rat hippocampus. J Neurosci 23(35):11026–11035

Laffray S, Bouali-Benazzouz R, Papon MA, Favereaux A, Jiang Y, Holm T, Spriet C, Desbarats P, Fossat P, Le Feuvre Y, Decossas M, Heliot L, Langel U, Nagy F, Landry M (2012) Impairment of $GABA_B$ receptor dimer by endogenous 14-3-3ζ in chronic pain conditions. EMBO J 31 (15):3239–3251. https://doi.org/10.1038/emboj.2012.161

Lambert NA, Wilson WA (1996) High-threshold Ca^{2+} currents in rat hippocampal interneurones and their selective inhibition by activation of $GABA_B$ receptors. J Physiol 492:115–127

Lau BK, Vaughan CW (2014) Descending modulation of pain: the GABA disinhibition hypothesis of analgesia. Curr Opin Neurobiol 29C:159–164. https://doi.org/10.1016/j.conb.2014.07.010

Lee B, Kim J, Kim SJ, Lee H, Chang JW (2007) Constitutive GABA expression via a recombinant adeno-associated virus consistently attenuates neuropathic pain. Biochem Biophys Res Commun 357(4):971–976. https://doi.org/10.1016/j.bbrc.2007.04.061

Lee J, Back SK, Lim EJ, Cho GC, Kim MA, Kim HJ, Lee MH, Na HS (2010) Are spinal GABAergic elements related to the manifestation of neuropathic pain in rat? Korean J Physiol Pharmacol 14(2):59–69. https://doi.org/10.4196/kjpp.2010.14.2.59

Legrain V, Iannetti GD, Plaghki L, Mouraux A (2011) The pain matrix reloaded: a salience detection system for the body. Prog Neurobiol 93(1):111–124. https://doi.org/10.1016/j.pneurobio.2010.10.005

Lever I, Cunningham J, Grist J, Yip PK, Malcangio M (2003) Release of BDNF and GABA in the dorsal horn of neuropathic rats. Eur J Neurosci 18(5):1169–1174

Levy RA, Proudfit HK (1979) Analgesia produced by microinjection of baclofen and morphine at brain stem sites. Eur J Pharmacol 57(1):43–55. https://doi.org/10.1016/0014-2999(79)90102-x

Li G, Shao C, Chen Q, Wang Q, Yang K (2017) Accumulated GABA activates presynaptic $GABA_B$ receptors and inhibits both excitatory and inhibitory synaptic transmission in rat midbrain periaqueductal gray. Neuroreport 28(6):313–318. https://doi.org/10.1097/wnr.0000000000000756

Lind G, Schechtmann G, Winter J, Meyerson BA, Linderoth B (2008) Baclofen-enhanced spinal cord stimulation and intrathecal baclofen alone for neuropathic pain: Long-term outcome of a pilot study. Eur J Pain 12(1):132–136

Liu F, Zhang YY, Song N, Lin J, Liu MK, Huang CL, Zhou C, Wang H, Wang M, Shen JF (2019) $GABA_B$ receptor activation attenuates inflammatory orofacial pain by modulating interleukin-1beta in satellite glial cells: role of NF-kappaB and MAPK signaling pathways. Brain Res Bull 149:240–250. https://doi.org/10.1016/j.brainresbull.2019.04.018

Liu P, Guo WY, Zhao XN, Bai HP, Wang Q, Wang XL, Zhang YZ (2014) Intrathecal baclofen, a $GABA_B$ receptor agonist, inhibits the expression of p-CREB and NR2B in the spinal dorsal horn in rats with diabetic neuropathic pain. Can J Physiol Pharmacol 92(8):655–660. https://doi.org/10.1139/cjpp-2013-0463

Liu P, Yuan HB, Zhao S, Liu FF, Jiang YQ, Guo YX, Wang XL (2018) Activation of $GABA_B$ receptor suppresses diabetic neuropathic pain through toll-like receptor 4 signaling pathway in the spinal dorsal horn. Mediators Inflamm 2018:6016272. https://doi.org/10.1155/2018/6016272

Loeza-Alcocer E, McPherson TP, Gold MS (2019) Peripheral GABA receptors regulate colonic afferent excitability and visceral nociception. J Physiol 597(13):3425–3439. https://doi.org/10.1113/jp278025

Luscher C, Jan LY, Stoffel M, Malenka RC, Nicoll RA (1997) G protein-coupled inwardly rectifying K^+ channels (GIRKs) mediate postsynaptic but not presynaptic transmitter actions in hippocampal neurons. Neuron 19(3):687–695

Magnaghi V, Castelnovo LF, Faroni A, Cavalli E, Caffino L, Colciago A, Procacci P, Pajardi G (2014) Nerve regenerative effects of $GABA_B$ ligands in a model of neuropathic pain. Biomed Res Int 2014:368678. https://doi.org/10.1155/2014/368678

Maier PJ, Marin I, Grampp T, Sommer A, Benke D (2010) Sustained glutamate receptor activation down-regulates $GABA_B$ receptors by shifting the balance from recycling to lysosomal degradation. J Biol Chem 285(46):35606–35614. https://doi.org/10.1074/jbc.M110.142406

Malan TP, Mata HP, Porreca F (2002) Spinal $GABA_A$ and $GABA_B$ receptor pharmacology in a rat model of neuropathic pain. Anesthesiology 96(5):1161–1167

Malykhina AP, Wyndaele JJ, Andersson KE, De Wachter S, Dmochowski RR (2012) Do the urinary bladder and large bowel interact, in sickness or in health? ICI-RS 2011. NeurourolUrodyn 31(3):352–358. https://doi.org/10.1002/nau.21228

Martins I, Carvalho P, de Vries MG, Teixeira-Pinto A, Wilson SP, Westerink BH, Tavares I (2015) GABA acting on GABAB receptors located in a medullary pain facilitatory area enhances nociceptive behaviors evoked by intraplantar formalin injection. Pain 156(8):1555–1565. https://doi.org/10.1097/j.pain.0000000000000203

Martins I, Tavares I (2017) Reticular formation and pain: the past and the future. Front Neuroanat 11:51. https://doi.org/10.3389/fnana.2017.00051

May A (2007) Neuroimaging: visualising the brain in pain. Neurol Sci 28(Suppl 2):S101–S107. https://doi.org/10.1007/s10072-007-0760-x

McCarson KE, Enna SJ (1999) Nociceptive regulation of $GABA_B$ receptor gene expression in rat spinal cord. Neuropharmacology 38(11):1767–1773

McIntosh JM, Absalom N, Chebib M, Elgoyhen AB, Vincler M (2009) Alpha9 nicotinic acetylcholine receptors and the treatment of pain. Biochem Pharmacol 78(7):693–702. https://doi.org/10.1016/j.bcp.2009.05.020

Melin C, Jacquot F, Dallel R, Artola A (2013) Segmental disinhibition suppresses C-fiber inputs to the rat superficial medullary dorsal horn via the activation of $GABA_B$ receptors. Eur J Neurosci 37(3):417–428. https://doi.org/10.1111/ejn.12048

Melzack R, Wall PD (1965) Pain mechanisms: a new theory. Science 150(3699):971–979

Menetrey D, Besson JM (1982) Electrophysiological characteristics of dorsal horn cells in rats with cutaneous inflammation resulting from chronic arthritis. Pain 13(4):343–364

Migita K, Matsuzaki Y, Koga K, Matsumoto T, Mishima K, Hara S, Honda K (2018) Involvement of $GABA_B$ receptor in the antihypersensitive effect in anterior cingulate cortex of partial sciatic nerve ligation model. J Pharmacol Sci 137(2):233–236. https://doi.org/10.1016/j.jphs.2018.05.009

Mintz IM, Bean BP (1993) $GABA_B$ receptor inhibition of P-type Ca^{2+} channels in central neurons. Neuron 10(5):889–898

Moore C, Gupta R, Jordt SE, Chen Y, Liedtke WB (2018) Regulation of pain and itch by TRP channels. Neurosci Bull 34(1):120–142. https://doi.org/10.1007/s12264-017-0200-8

Naderi N, Shafaghi B, Khodayar MJ, Zarindast MR (2005) Interaction between gamma-aminobutyric acid $GABA_B$ and cannabinoid CB1 receptors in spinal pain pathways in rat. Eur J Pharmacol 514(2-3):159–164. https://doi.org/10.1016/j.ejphar.2005.03.037

Nashawi H, Masocha W, Edafiogho IO, Kombian SB (2016) Paclitaxel causes electrophysiological changes in the anterior cingulate cortex via modulation of the γ-aminobutyric acid-ergic system. Med Princ Pract 25(5):423–428. https://doi.org/10.1159/000447775

Neugebauer V, Li W, Bird GC, Han JS (2004) The amygdala and persistent pain. Neuroscientist 10 (3):221–234. https://doi.org/10.1177/1073858403261077

Ng GYK, Clark J, Coulombe N, Ethier N, Hebert TE, Sullivan R, Kargman S, Chateauneuf A, Tsukamoto N, McDonald T, Whiting P, Mezey E, Johnson MP, Liu QY, Kolakowski LF, Evans JF, Bonner TI, O'Neill GP (1999) Identification of a $GABA_B$ receptor subunit, gb2, required for functional $GABA_B$ receptor activity. J Biol Chem 274(12):7607–7610

Otis TS, De Koninck Y, Mody I (1993) Characterization of synaptically elicited $GABA_B$ responses using patch-clamp recordings in rat hippocampal slices. J Physiol 463:391–407

Page AJ, O'Donnell TA, Blackshaw LA (2006) Inhibition of mechanosensitivity in visceral primary afferents by $GABA_B$ receptors involves calcium and potassium channels. Neuroscience 137 (2):627–636. https://doi.org/10.1016/j.neuroscience.2005.09.016

Patel S, Naeem S, Kesingland A, Froestl W, Capogna M, Urban L, Fox A (2001) The effects of $GABA_B$ agonists and gabapentin on mechanical hyperalgesia in models of neuropathic and inflammatory pain in the rat. Pain 90(3):217–226

Pérez-Garci E, Gassmann M, Bettler B, Larkum ME (2006) The $GABA_{B1b}$ isoform mediates long-lasting inhibition of dendritic Ca^{2+} spikes in layer 5 somatosensory pyramidal neurons. Neuron 50(4):603–616
Pinto M, Sousa M, Lima D, Tavares I (2008) Participation of mu-opioid, $GABA_B$, and NK1 receptors of major pain control medullary areas in pathways targeting the rat spinal cord: implications for descending modulation of nociceptive transmission. J Comp Neurol 510 (2):175–187
Potes CS, Neto FL, Castro-Lopes JM (2006a) Administration of baclofen, a γ-aminobutyric acid type B agonist in the thalamic ventrobasal complex, attenuates allodynia in monoarthritic rats subjected to the ankle-bend test. J Neurosci Res 83(3):515–523. https://doi.org/10.1002/jnr.20737
Potes CS, Neto FL, Castro-Lopes JM (2006b) Inhibition of pain behavior by $GABA_B$ receptors in the thalamic ventrobasal complex: effect on normal rats subjected to the formalin test of nociception. Brain Res 1115(1):37–47. https://doi.org/10.1016/j.brainres.2006.07.089
Price GW, Kelly JS, Bowery NG (1987) The location of $GABA_B$ receptor binding sites in mammalian spinal cord. Synapse 1(6):530–538. https://doi.org/10.1002/syn.890010605
Price GW, Wilkin GP, Turnbull MJ, Bowery NG (1984) Are baclofen-sensitive $GABA_B$ receptors present on primary afferent terminals of the spinal cord? Nature 307(5946):71–74
Reis GM, Duarte ID (2006) Baclofen, an agonist at peripheral $GABA_B$ receptors, induces antinociception via activation of TEA-sensitive potassium channels. Br J Pharmacol 149 (6):733–739
Ren K, Anseloni V, Zou SP, Wade EB, Novikova SI, Ennis M, Traub RJ, Gold MS, Dubner R, Lidow MS (2004) Characterization of basal and re-inflammation-associated long-term alteration in pain responsivity following short-lasting neonatal local inflammatory insult. Pain 110 (3):588–596. https://doi.org/10.1016/j.pain.2004.04.006
Sandkuhler J (2009) Models and mechanisms of hyperalgesia and allodynia. Physiol Rev 89 (2):707–758. https://doi.org/10.1152/physrev.00025.2008
Sands SA, McCarson KE, Enna SJ (2003) Differential regulation of $GABA_B$ receptor subunit expression and function. J Pharmacol Exp Ther 305(1):191–196
Schreiber AK, Nones CF, Reis RC, Chichorro JG, Cunha JM (2015) Diabetic neuropathic pain: Physiopathology and treatment. World J Diabetes 6(3):432–444. https://doi.org/10.4239/wjd.v6.i3.432
Schuler V, Luscher C, Blanchet C, Klix N, Sansig G, Klebs K, Schmutz M, Heid J, Gentry C, Urban L, Fox A, Spooren W, Jaton AL, Vigouret J, Pozza M, Kelly PH, Mosbacher J, Froestl W, Kaslin E, Korn R, Bischoff S, Kaupmann K, van der Putten H, Bettler B (2001) Epilepsy, hyperalgesia, impaired memory, and loss of pre- and postsynaptic $GABA_B$ responses in mice lacking $GABA_{B1}$. Neuron 31(1):47–58
Shaban H, Humeau Y, Herry C, Cassasus G, Shigemoto R, Ciocchi S, Barbieri S, van der Putten H, Kaupmann K, Bettler B, Lüthi A (2006) Generalization of amygdala LTP and conditioned fear in the absence of presynaptic inhibition. Nat Neurosci 9(8):1028–1035
Shafizadeh M, Semnanian S, Zarrindast MR, Hashemi B (1997) Involvement of $GABA_B$ receptors in the antinociception induced by baclofen in the formalin test. Gen Pharmacol 28(4):611–615
Slonimski M, Abram SE, Zuniga RE (2004) Intrathecal baclofen in pain management. Reg Anesth Pain Med 29(3):269–276
Smith GD, Harrison SM, Birch PJ, Elliott PJ, Malcangio M, Bowery NG (1994) Increased sensitivity to the antinociceptive activity of (+/−)-baclofen in an animal model of chronic neuropathic, but not chronic inflammatory hyperalgesia. Neuropharmacology 33(9):1103–1108
Soares Potes C, Lourenca Neto F, Manuel Castro-Lopes J (2006) Inhibition of pain behavior by $GABA_B$ receptors in the thalamic ventrobasal complex: Effect on normal rats subjected to the formalin test of nociception. Brain Res 1115(1):37–47
Sokal DM, Chapman V (2003) Inhibitory effects of spinal baclofen on spinal dorsal horn neurones in inflamed and neuropathic rats in vivo. Brain Res 987(1):67–75

Somers DL, Clemente FR (2002) Dorsal horn synaptosomal content of aspartate, glutamate, glycine and GABA are differentially altered following chronic constriction injury to the rat sciatic nerve. Neurosci Lett 323(3):171–174. https://doi.org/10.1016/s0304-3940(02)00157-x

Steiger JL, Bandyopadhyay S, Farb DH, Russek SJ (2004) cAMP response element-binding protein, activating transcription factor-4, and upstream stimulatory factor differentially control hippocampal $GABA_{BR1a}$ and $GABA_{BR1b}$ subunit gene expression through alternative promoters. J Neurosci 24:6115–6126

Taira T, Hori T (2007) Intrathecal baclofen in the treatment of post-stroke central pain, dystonia, and persistent vegetative state. Acta Neurochir Suppl 97(Pt 1):227–229

Taira T, Kawamura H, Tanikawa T, Iseki H, Kawabatake H, Takakura K (1995) A new approach to control central deafferentation pain: spinal intrathecal baclofen. Stereotact Funct Neurosurg 65 (1–4):101–105

Takazawa T, Choudhury P, Tong CK, Conway CM, Scherrer G, Flood PD, Mukai J, MacDermott AB (2017) Inhibition mediated by glycinergic and GABAergic receptors on excitatory neurons in mouse superficial dorsal horn is location-specific but modified by inflammation. J Neurosci 37(9):2336–2348. https://doi.org/10.1523/jneurosci.2354-16.2017

Takeda M, Nasu M, Kanazawa T, Shimazu Y (2015) Activation of $GABA_B$ receptors potentiates inward rectifying potassium currents in satellite glial cells from rat trigeminal ganglia: in vivo patch-clamp analysis. Neuroscience 288:51–58. https://doi.org/10.1016/j.neuroscience.2014.12.024

Takeda M, Takahashi M, Matsumoto S (2009) Contribution of the activation of satellite glia in sensory ganglia to pathological pain. Neurosci Biobehav Rev 33(6):784–792. https://doi.org/10.1016/j.neubiorev.2008.12.005

Terunuma M, Vargas KJ, Wilkins ME, Ramirez OA, Jaureguiberry-Bravo M, Pangalos MN, Smart TG, Moss SJ, Couve A (2010) Prolonged activation of NMDA receptors promotes dephosphorylation and alters postendocytic sorting of $GABA_B$ receptors. Proc Natl Acad Sci U S A 107(31):13918–13923

Todd AJ (2010) Neuronal circuitry for pain processing in the dorsal horn. Nat Rev Neurosci 11 (12):823–836. https://doi.org/10.1038/nrn2947

Towers S, Princivalle A, Billinton A, Edmunds M, Bettler B, Urban L, Castro-Lopes J, Bowery NG (2000) $GABA_B$ receptor protein and mRNA distribution in rat spinal cord and dorsal root ganglia. Eur J Neurosci 12(9):3201–3210

Tsuda M, Mizokoshi A, Shigemoto-Mogami Y, Koizumi S, Inoue K (2004) Activation of p38 mitogen-activated protein kinase in spinal hyperactive microglia contributes to pain hypersensitivity following peripheral nerve injury. Glia 45(1):89–95. https://doi.org/10.1002/glia.10308

Urwyler S (2011) Allosteric modulation of family C G-protein-coupled receptors: from molecular insights to therapeutic perspectives. Pharmacol Rev 63(1):59–126. https://doi.org/10.1124/pr.109.002501

van der Plas AA, van Rijn MA, Marinus J, Putter H, van Hilten JJ (2013) Efficacy of intrathecal baclofen on different pain qualities in complex regional pain syndrome. Anesth Analg 116 (1):211–215. https://doi.org/10.1213/ANE.0b013e31826f0a2e

Vaysse L, Sol JC, Lazorthes Y, Courtade-Saidi M, Eaton MJ, Jozan S (2011) GABAergic pathway in a rat model of chronic neuropathic pain: modulation after intrathecal transplantation of a human neuronal cell line. Neurosci Res 69(2):111–120. https://doi.org/10.1016/j.neures.2010.10.006

Vigot R, Barbieri S, Bräuner-Osborne H, Turecek R, Shigemoto R, Zhang YP, Lujan R, Jacobson LH, Biermann B, Fritschy JM, Vacher CM, Müller M, Sansig G, Guetg N, Cryan JF, Kaupmann K, Gassmann M, Oertner TG, Bettler B (2006) Differential compartmentalization and distinct functions of $GABA_B$ receptor variants. Neuron 50(4):589–601

Wang N, Zhang T, Su YL, Wang JY, Luo F (2016) Differential modulation of electrical stimulation of periaqueductal gray and thalamus on nociceptive behaviors of rats. Sheng Li Xue Bao 68 (2):115–125

Wang XL, Zhang HM, Chen SR, Pan HL (2007) Altered synaptic input and $GABA_B$ receptor function in spinal superficial dorsal horn neurons in rats with diabetic neuropathy. J Physiol 579 (Pt 3):849–861. https://doi.org/10.1113/jphysiol.2006.126102

Wang XL, Zhang Q, Zhang YZ, Liu YT, Dong R, Wang QJ, Guo YX (2011) Downregulation of $GABA_B$ receptors in the spinal cord dorsal horn in diabetic neuropathy. Neurosci Lett 490 (2):112–115. https://doi.org/10.1016/j.neulet.2010.12.038
Wang Y, Xing M, Cao Q, Ji A, Liang H, Song S (2019) Biological activities of fucoidan and the factors mediating its therapeutic effects: A review of recent studies. Mar Drugs 17(3):E183. https://doi.org/10.3390/md17030183
White JH, Wise A, Main MJ, Green A, Fraser NJ, Disney GH, Barnes AA, Emson P, Foord SM, Marshall FH (1998) Heterodimerization is required for the formation of functional $GABA_B$ receptors. Nature 396:679–682
White K, Targett M, Harris J (2018) Gainfully employing descending controls in acute and chronic pain management. Vet J 237:16–25. https://doi.org/10.1016/j.tvjl.2018.05.005
Whitehead RA, Puil E, Ries CR, Schwarz SK, Wall RA, Cooke JE, Putrenko I, Sallam NA, Macleod BA (2012) $GABA_B$ receptor-mediated selective peripheral analgesia by the non-proteinogenic amino acid, isovaline. Neuroscience 213:154–160. https://doi.org/10.1016/j.neuroscience.2012.04.026
Wu J, Xu Y, Pu S, Jiang W, Du D (2011) p38/MAPK inhibitor modulates the expression of dorsal horn $GABA_B$ receptors in the spinal nerve ligation model of neuropathic pain. Neuroimmunomodulation 18(3):150–155. https://doi.org/10.1159/000323141
Yam MF, Loh YC, Tan CS, Khadijah Adam S, Abdul Manan N, Basir R (2018) General pathways of pain sensation and the major neurotransmitters involved in pain regulation. Int J Mol Sci 19 (8):E2164. https://doi.org/10.3390/ijms19082164
Yang K, Furue H, Kumamoto E, Dong YX, Yoshimura M (2003) Pre- and postsynaptic inhibition mediated by $GABA_B$ receptors in rat ventrolateral periaqueductal gray neurons. Biochem Biophys Res Commun 302(2):233–237. https://doi.org/10.1016/s0006-291x(03)00156-6
Yang K, Ma H (2011) Blockade of $GABA_B$ receptors facilitates evoked neurotransmitter release at spinal dorsal horn synapse. Neuroscience 193:411–420. https://doi.org/10.1016/j.neuroscience.2011.07.033
Yang K, Ma R, Wang Q, Jiang P, Li YQ (2015) Optoactivation of parvalbumin neurons in the spinal dorsal horn evokes GABA release that is regulated by presynaptic $GABA_B$ receptors. Neurosci Lett 594:55–59. https://doi.org/10.1016/j.neulet.2015.03.050
Yang K, Wang D, Li YQ (2001) Distribution and depression of the $GABA_B$ receptor in the spinal dorsal horn of adult rat. Brain Res Bull 55(4):479–485
Zeilhofer HU, Benke D, Yevenes GE (2012a) Chronic pain states: pharmacological strategies to restore diminished inhibitory spinal pain control. Annu Rev Pharmacol Toxicol 52:111–133. https://doi.org/10.1146/annurev-pharmtox-010611-134636
Zeilhofer HU, Wildner H, Yevenes GE (2012b) Fast synaptic inhibition in spinal sensory processing and pain control. Physiol Rev 92(1):193–235
Zemoura K, Balakrishnan K, Grampp T, Benke D (2019) Ca^{2+}/Calmodulin-dependent protein kinase II (CaMKII) β-dependent phosphorylation of $GABA_{B1}$ triggers lysosomal degradation of $GABA_B$ Receptors via mind bomb-2 (MIB2)-mediated Lys-63-linked ubiquitination. Mol Neurobiol 56(2):1293–1309. https://doi.org/10.1007/s12035-018-1142-5
Zemoura K, Ralvenius WT, Malherbe P, Benke D (2016) The positive allosteric $GABA_B$ receptor modulator rac-BHFF enhances baclofen-mediated analgesia in neuropathic mice. Neuropharmacology 108:172–178. https://doi.org/10.1016/j.neuropharm.2016.04.028
Zheng J, Lu Y, Perl ER (2010) Inhibitory neurones of the spinal substantia gelatinosa mediate interaction of signals from primary afferents. J Physiol 588(Pt 12):2065–2075. https://doi.org/10.1113/jphysiol.2010.188052
Zhou YQ, Chen SP, Liu DQ, Manyande A, Zhang W, Yang SB, Xiong BR, Fu QC, Song ZP, Rittner H, Ye DW, Tian YK (2017) The role of spinal $GABA_B$ receptors in cancer-induced bone pain in rats. J Pain 18(8):933–946. https://doi.org/10.1016/j.jpain.2017.02.438
Zuniga RE, Schlicht CR, Abram SE (2000) Intrathecal baclofen is analgesic in patients with chronic pain. Anesthesiology 92(3):876–880

Keywords Anxiety · Depression · $GABA_B$ receptor · Hippocampal neurogenesis · Mood · Stress · Stress resilience

1 Introduction

The inhibitory action of GABA is mediated by the ionotropic $GABA_A$ and $GABA_C$ receptors, and the metabotropic $GABA_B$ receptor. The $GABA_A$ receptor is bicuculline-sensitive and the subsequent opening of its transmembrane channel which is permeable to chloride mediates rapid neuronal inhibition in the adult brain. In 1979, Norman Bowery and colleagues published the discovery of a novel type of GABA receptor that was described as being "atypical" and insensitive to the $GABA_A$ receptor antagonist bicuculline (Bowery et al. 1979). Baclofen was identified to be a potent and selective agonist of this novel receptor, and in 1980 it was demonstrated that baclofen acting on this novel receptor decreased neurotransmitter release in the central nervous system (Bowery et al. 1980). This atypical receptor described by Bowery and colleagues would later be referred to as the $GABA_B$ receptor (Hill and Bowery 1981). The $GABA_B$ receptor is a G-protein-coupled receptor that inhibits adenylate cyclase activity and mediates the slow and prolonged component of synaptic inhibition (Bowery et al. 2004). $GABA_B$ receptors are localized in most brain regions, and $GABA_{B(1)}$ receptor mRNA is detectable in almost all neuronal cell populations and is highly expressed in the limbic system (Bettler et al. 2004; McDonald et al. 2004). The receptor consists of two subunits, $GABA_{B(1)}$ and $GABA_{B(2)}$, which heterodimerise to form the functional $GABA_B$ receptor (Bettler et al. 2004). The $GABA_{B(1)}$ subunit contains the orthosteric ligand binding site, while the $GABA_{B(2)}$ subunit is responsible for G-protein activation and contains binding sites for positive allosteric modulators (Galvez et al. 2001; Bettler et al. 2004; Binet et al. 2004; Gassmann and Bettler 2012). Isoforms of the $GABA_{B(1)}$ receptor subunit have been identified (Lee et al. 2010) and the two main isoforms expressed in the brain are $GABA_{B(1a)}$ and $GABA_{B(1b)}$ which form $GABA_{B(1a,2)}$ and $GABA_{B(1b,2)}$ receptors, respectively (Lee et al. 2010). Structurally, $GABA_{B(1)}$ isoforms differ only by the presence of a sushi domain in the N-terminal ectodomain of the $GABA_{B(1a)}$ receptor subunit isoform (see Fig. 1).

Since its discovery, there has been a long-standing interest in the therapeutic potential of the $GABA_B$ receptor. In this review, we will summarize the studies assessing the role of the $GABA_B$ receptor in mood disorders, specifically in depression, and in anxiety disorders. Clinical and preclinical evidences supporting the role for $GABA_B$ receptors in the pathophysiology of depression and anxiety disorders will be summarized in addition to the preclinical evidence of the antidepressant and anxiolytic effects of pharmacological and genetic modulation of $GABA_B$ receptor activity. Unless otherwise stated, most of the preclinical studies discussed in this chapter have been conducted in male rodents. Since most of this evidence is from

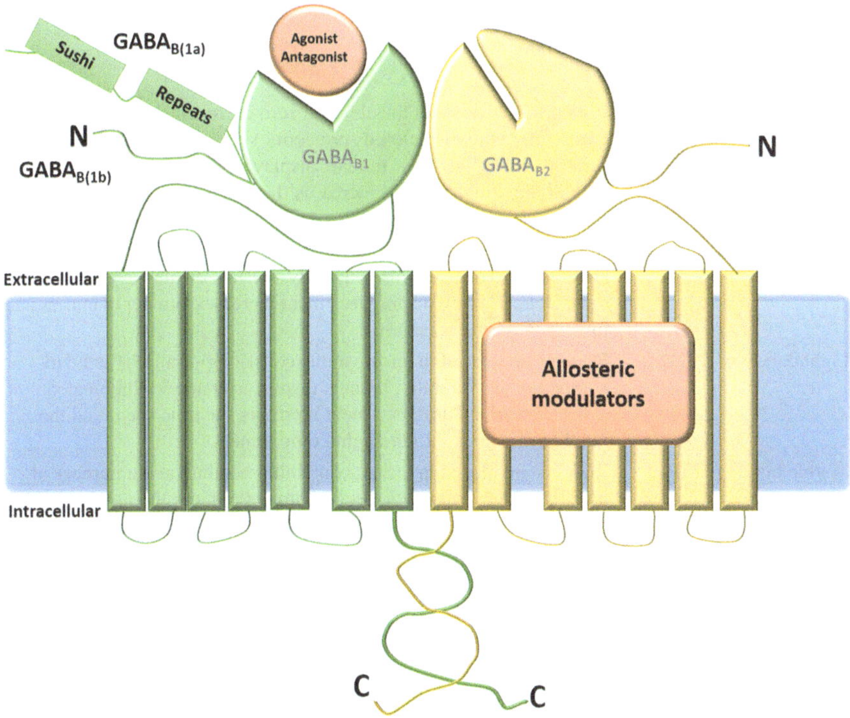

Fig. 1 Schematic representation of the GABA$_B$ receptor. GABA$_B$ receptors are composed of GABA$_{B(1)}$ and GABA$_{B(2)}$ receptor subunits that form an active heterodimer. The GABA$_{B(1)}$ receptor subunit is essential for the binding of GABA and GABA$_B$ receptor agonists and antagonists. GABA$_{B(1)}$ receptor subunit presents as two main isoforms, namely GABA$_{B(1a)}$ and GABA$_{B(1b)}$ that differ by the presence of a sushi domain in the N-terminal of the GABA$_{B(1a)}$ isoform. Adapted from Cryan and Kaupmann (2005)

preclinical studies, readers outside this research field are advised to first read Table 1, which summarizes the behavioural tests used to assess depression-, antidepressant- and anxiety-like behaviour in rodents (Cryan and Slattery 2007), prior to reading the review.

2 Role of the GABA$_B$ Receptor in the Modulation of Anxiety

2.1 Effects of GABA$_B$ Receptor Agonists and Positive Allosteric Modulators on Anxiety-Like Behaviour

Baclofen is the first described GABA$_B$ receptor agonist (Bowery et al. 1980), which was synthetized in 1962 by Heinrich Keberle in CIBA (Basel, Switzerland).

Table 1 Behavioural tests used in anxiety and depression research

Behavioural tests	Description
Anxiety	
Elevated zero maze (EZM)	Rodents are placed on the elevated zero maze consisting of two open (stressful) and two enclosed (protecting) elevated areas that form a zero or circle. Naturally, rodents display aversion to open areas. Anxiolytic drug treatment increases the number of entries and time spent in the open areas of the maze
Elevated plus maze (EPM)	Rodents are placed on the elevated plus maze consisting of two open arms and two closed arms. Naturally, rodents display aversion versus open areas. Anxiolytic drug treatment increases number of entries and time spent in the open arms
Light/dark box (LDB)	Rodents are placed in an apparatus consisting of a light and dark compartment. Naturally, rodents display aversion to illuminated areas. Anxiolytic drug treatment increases the time spent and the number of entries into the light compartment
Marble burying (MB)	Rodents are placed in a cage containing bedding and a number of novel marbles. Anxiolytic drug treatment reduces the number of marbles buried
Stress-induced hyperthermia (SIH)	Temperature is measured twice at an interval of 15 min. The difference between the first and second measurements is a physiological measure of anxiety. Anxiolytic drug treatment reduces the magnitude of the increase in temperature which is observed in the second measurement
Ultrasonic vocalizations	Rodent pups when separated from their dams produce alarm calls that are an index of anxiety. Anxiolytic drug treatment reduces the number of alarm calls
Vogel conflict test	Rodents are punished by electric shocks when trying to get either food or water, and thus the number of times the animal goes to get food or water decreases. Anxiolytic drugs increase the number of punished responses in the presence of shock as compared to unpunished responses
Staircase test	Rodents are placed in an enclosed staircase with five steps. The number of steps climbed and rearings made in a 3-min period are observed. Anxiolytic drugs reduce rearing at doses that do not reduce the number of steps climbed
Light-enhanced startle (LES)	Startle reactivity is increased by presentation of a bright light. Because LES is based on the innate aversion of rodents for bright light, it does not require training sessions. Anxiolytic drug treatment reduces startle potentiation
Fear-potentiated startle (FPS)	The test consists of two training sessions in which an aversive foot shock is paired with a neutral cue light. In the test session, presentation of this cue light is used to elicit startle potentiation. Anxiolytic drugs reduce startle potentiation
Fear conditioning (FC)	Rodents are placed in a FC box in which an aversive stimulus (electric shock) is paired with a neutral context (such as a location) or stimulus (such as a tone). This results in the expression of a fear response (freezing behaviour) in the presence of the neutral context or neutral stimulus alone. Anxiolytics reduce freezing behaviour

(continued)

Table 1 (continued)

Behavioural tests	Description
Four-plate test	Animals are exposed to a novel environment and exploratory behaviour is suppressed by the delivery of a mild electric foot shock contingent on quadrant crossing. Animals can only escape from this aversive situation by remaining motionless (passive avoidance). Anxiolytic drugs increase exploratory behaviour
Depression	
Forced swim test (FST)	Rodents placed in an inescapable container of water engage in escape oriented behaviours. After few minutes, the animal adopts an immobile posture just making sufficient movements to keep its head above water. Antidepressant drug treatment reduces time spent immobile
Tail suspension test (TST)	Mice, when suspended from the tail, will adopt an immobile posture. Antidepressant treatment decreases time spent immobile
Learned helplessness (LH)	Animals exposed to inescapable shocks subsequently fail to escape when an escape option is presented. Antidepressants increase the number of escapes
Olfactory bulbectomy (OB)	Removal of the olfactory bulbs causes several behavioural and neurochemical alterations, which are only reversed by chronic antidepressant treatment
Maternal deprivation	When rodents are separated from their dams during early postnatal life they can develop a number of depression-like behaviours. Chronic antidepressant treatment reverses those abnormal behaviours
Chronic mild stress (CMS)	Rodents are subjected to a variety of unpredictable stressors which leads to several behavioural alterations. Chronic treatment with antidepressants reverses those alterations
Social defeat stress (SDS)	Rodents are exposed to physical interaction with an unfamiliar aggressive animal, combined with sensory contact throughout the stress procedure. This protocol, repeated daily, induces depression-like behaviours that are reversed by antidepressant treatment

Adapted from Cryan and Slattery (2007)

Baclofen was formulated as an antiepileptic drug and marketed in 1972 as Lioresal. Currently, baclofen is indicated primarily to treat spasticity but it also has beneficial effects in treating pain, is used off-label in the treatment of alcohol use disorder and has been shown to inhibit the re-enforcing effects of many other addictive drugs (Bowery et al. 2002). However, there is also much preclinical evidence suggesting that $GABA_B$ receptor agonists such as baclofen may be potential therapeutic approaches to treat anxiety disorders (Cryan et al. 2005; Cryan and Slattery 2010; Felice et al. 2016) (summarized in Table 2).

Acute baclofen administration has been shown to reduce anxiety-like behaviour in several rat and mouse models (Ketelaars et al. 1988; File et al. 1991, 1992; Nastiti et al. 1991; Shephard et al. 1992; Andrews and File 1993; Amikishieva and Semendyaeva 2007; Lu et al. 2016), although some conflicting findings have also been reported. For instance, while one study reported that baclofen was effective in the Vogel conflict test (Ketelaars et al. 1988), another study reported no such effect

Table 2 Effects of $GABA_B$ receptor agonists and positive allosteric modulators (PAMs) on anxiety-like behaviours in rodents

Drug	Paradigm	Finding	References
Agonists			
Baclofen	Vogel conflict	Acute treatment ↓ anxiety in rats	File et al. (1991)
		Acute treatment ↓ anxiety in rats	Ketelaars et al. (1988)
		Acute treatment ↓ anxiety in rats	Shephard et al. (1992)
		Acute treatment ↔ in rats	Agmo et al. (1991)
		Acute treatment ↔ in rats	Li et al. (2013)
	EPM	Acute treatment ↓ anxiety in ethanol withdrawal model in rats	File et al. (1992)
		Acute treatment ↓ anxiety in mice	Amikishieva and Semendyaeva (2007)
		Injection in the AcbSh ↔ in 24 h food-deprived rats	Lopes et al. (2012)
		Acute treatment ↔ in rats	Li et al. (2013)
		Acute treatment ↔ mice	Dalvi and Rodgers (1996)
		Acute treatment ↔ in nicotine-induced anxiety in mice	Varani and Balerio (2012)
		PND 14–28 treatment ↑ anxiety in adult mice	Sweeney et al. (2014)
	EZM	Acute treatment ↔ in rats	Frankowska et al. (2007)
	LDB	Acute treatment ↔ in mice	Li et al. (2013)
	SI	Microinjection in the BLA ↔ in rats	Sanders and Shekhar (1995)
	SIH	Acute treatment ↔ in rats	Li et al. (2015)
	LES	Acute treatment ↔ in rats	Li et al. (2015)
CGP44532	Four plate	Acute treatment ↑ anxiety in mice	Partyka et al. (2007)
SKF97541	EZM	Acute treatment ↔ in rats	Frankowska et al. (2007)
PAMS			
GS39783	EZM	Acute treatment ↓ anxiety in rats, mice	Cryan et al. (2004)
		Acute, chronic treatment ↓ anxiety in mice	Mombereau et al. (2004)
	EPM	Acute, chronic treatment ↓ anxiety in rats	Cryan et al. (2004)
	SIH	Acute, chronic treatment ↓ anxiety in mice	Cryan et al. (2004)
	LDB	Acute, chronic treatment ↓ anxiety in mice	Mombereau et al. (2004)
CGP7930	EZM	Acute treatment ↓ anxiety in rats	Frankowska et al. (2007)

(continued)

Table 2 (continued)

Drug	Paradigm	Finding	References
		Acute treatment ↓ anxiety in mice	Jacobson and Cryan (2008)
	EPM	Acute treatment ↔ in rats	Jacobson and Cryan (2008)
	SIH	Acute treatment ↓ anxiety in mice	Jacobson and Cryan (2008)
	Staircase	Acute treatment ↓ anxiety in mice	Jacobson and Cryan (2008)
rac-BHFF	SIH	Acute treatment ↓ anxiety in mice	Malherbe et al. (2008)
BHF177	EPM	Acute treatment ↔ in mice	Li et al. (2013)
	LDB	Acute treatment ↔ in mice	Li et al. (2013)
	Vogel conflict	Acute treatment ↔ in mice	Li et al. (2013)
	LES	Acute treatment ↔ in rats (anxiolytic-like effects in high but not low LES-responding rats)	Li et al. (2015)
ADX71441	MB	Acute treatment ↓ anxiety in mice	Kalinichev et al. (2017)
	EPM	Acute treatment ↓ anxiety in mice, rats	Kalinichev et al. (2017)

AcbSh nucleus accumbens shell, *BLA* basolateral amygdala, *EPM* elevated plus maze, *EZM* elevated zero maze, *LDB* light/dark box, *LES* light-enhanced startle, *MB* marble burying, *PND* postnatal day, *SI* social interaction, *SIH* stress-induced hyperthermia
↓ = decreased; ↔ = no effects; ↑ = increased

(Agmo et al. 1991). However, the latter study also reported that higher doses of baclofen induced motor deficits in rats which may have reduced the number of licks, thus resulting in a potentially false negative finding in this test (Agmo et al. 1991). Similarly, Li and colleagues have reported that baclofen had sedative but not anxiolytic effects in rats in several behavioural tests (Li et al. 2015). Conflicting findings have also been reported in mice whereby baclofen was anxiolytic in some studies (Nastiti et al. 1991; Amikishieva and Semendyaeva 2007) but not in others (Dalvi and Rodgers 1996; Varani and Balerio 2012; Li et al. 2013). In one such study, baclofen increased punished drinking in the Vogel conflict test which would be indicative of an anxiolytic effect, but the authors suggest that this finding may also be due to analgesic effects of baclofen (Li et al. 2013). Motor impairing and hypothermic effects are characteristic side-effects of GABA$_B$ receptor agonists, and this likely confounds the interpretation of anxiety-related behavioural tests that are dependent on motor activity (e.g. elevated plus maze, Vogel conflict test, etc.) or body temperature (e.g. stress-induced hyperthermia) (Cryan et al. 2004). In addition, the effects of baclofen on anxiety may depend upon the developmental stage of the brain. For example, we have found that chronic treatment with R-baclofen during early postnatal life (Postnatal day (PND) 14- PND 28) in mice induced anxiety-like behaviour in adulthood in the elevated plus maze (EPM) but not in the stress-induced hyperthermia (SIH) and marble burying (MB) tests (Sweeney et al. 2014). This suggests that during early life $GABA_B$ receptor signalling might play a functional role in programming anxiety behaviour in adulthood (Sweeney et al. 2014), although this effect might also be test-specific.

Importantly, baclofen has several side-effects including sedation or somnolence, hypothermia, vertigo and muscle relaxation (Agabio et al. 2013). Moreover, repeated administration of $GABA_B$ agonists such as baclofen can induce receptor tolerance/desensitization resulting in a reduced therapeutic response following chronic administration (Lehmann et al. 2003). Thus, there has been great interest in developing drugs that target the $GABA_B$ receptor but with a reduced side-effect profile and that would not result in tolerance. As such, positive allosteric modulators (PAMs) offer several advantages over receptor agonists such as baclofen (Christopoulos 2002): (1) PAMs target more diverse sites that are distinct from the highly evolutionary conserved orthosteric site thus potentially contributing to greater selectivity; (2) PAM binding leads to potentiation of GABA-mediated effects on the receptor rather than direct activation of the receptor; (3) saturation of allosteric binding sites does not induce downregulation or overstimulation of the target receptor; (4) PAMs are active only in tissues where the endogenous agonist is present giving a more specific drug activity. Essentially, PAMs of $GABA_B$ receptors offer the advantage of reduced risk for receptor desensitization/tolerance when compared with classical $GABA_B$ receptor agonists such as baclofen (Gjoni and Urwyler 2008, 2009).

The first $GABA_B$ receptor PAMs that were identified and characterized were CGP7930 (Urwyler et al. 2001; Adams and Lawrence 2007) and GS39783 (Urwyler et al. 2003), shortly followed by rac-BHFF (Malherbe et al. 2008), BHF177 (compound # 27 (Guery et al. 2007)), CMPPE (Perdona et al. 2011), COR627 and COR628 (Castelli et al. 2012).

Several preclinical studies have interrogated the effects of some of these $GABA_B$ receptor PAMs on anxiety-like behaviour (summarized in Table 2). Chronic and acute administration of GS39783 has been shown to induce anxiolytic-like effects with no effects on locomotion, cognition, temperature or narcosis (Cryan et al. 2004; Mombereau et al. 2004). A recent study identified the brain structures that are modulated by GS39783 under either basal or mild stress (anxiogenic) conditions which were induced by exposing mice to the open arm of an EPM (Pizzo et al. 2018). Under basal conditions, GS39783 increased c-Fos expression in the amygdala nuclei, cortical areas and periaqueductal gray (PAG) subregions, while it inhibited c-Fos expression in the dorsal raphe nucleus (DRN) (Pizzo et al. 2018). Under stress conditions (open arm exposure), GS39783 reversed stress-induced c-Fos expression in the granular cell layer of the dentate gyrus of the hippocampus, no longer increased c-Fos expression in the amygdala nor did it reduce c-Fos expression in the DRN (Pizzo et al. 2018). Together, this suggests that GS39783 modulation of anxiety may involve neural circuits involving the dentate gyrus of the hippocampus, the amygdala and the DRN.

CGP7930 has only modest anxiolytic-like effects in mice but a superior side-effect profile than $GABA_B$ receptor agonists (Frankowska et al. 2007; Jacobson and Cryan 2008). Specifically, CGP7930 was effective in the elevated zero maze (EZM) in rats (Frankowska et al. 2007) and exhibited modest anxiolytic effects in the SIH, staircase test and EZM in mice (Jacobson and Cryan 2008). However, CGP7930 had no anxiolytic effects in the EPM in mice (Jacobson and Cryan 2008).

Both Rac-BHFF and BHF177 induce anxiolytic effects in some tests but not others. Specifically, Rac-BHFF and BHF177 induced anxiolytic-like effects in the SIH test in mice and rats, a test of the physiological anxiety response (Malherbe et al. 2008; Vinkers et al. 2010; Li et al. 2015). BHF177 induced anxiolytic-like effects on light-enhanced startle (LES; a test based on the innate aversion of rodents for bright light) in high-, but not low-LES responding rats in the staircase test (Li et al. 2015) but was inactive in the EPM and light dark box test in mice (Li et al. 2013). Importantly, BHF177, at doses over 40 mg/kg, caused hypothermia in contrast to other $GABA_B$ receptor PAMs including CGP7930 and Rac-BHFF (Vinkers et al. 2010) which may have confounded findings in SIH test. On the other hand, Rac-BHFF at the same dose that induced anxiolytic-like effects in the SIH (100 mg/kg) did not enhance baclofen- and γ-hydroxybutyric acid (GHB)-induced hypothermia (Koek et al. 2010), suggesting that its effects in the SIH test are not confounded by effects of $GABA_B$ receptor modulation of body temperature. A novel $GABA_B$ receptor PAM ADX71441 has also been shown to be effective in the MB test in mice and in the EPM in mice and rats (Kalinichev et al. 2017). Recently, Rondard and colleagues (Lecat-Guillet et al. 2017) developed time-resolved fluorescence resonance energy transfer (trFRET) sensors which represent an innovative tool to screen and identify new $GABA_B$ receptors PAMs with lower side-effect profiles. Interestingly, trFRET revealed that GS39783 exhibits low intrinsic agonist activity (as expected by a PAM), whereas CGP7930 and rac-BHFF display agonist-PAMs characteristics (Lecat-Guillet et al. 2017). This finding is in agreement with behavioural studies outlined above demonstrating that GS39783 induced anxiolytic-like behavioural effects without affecting locomotion, cognition, temperature or narcosis, and suggests that this drug may be a good target for clinical development. Effects of PAMs on conditioned anxiety have also been examined. BHF177 did not affect conditioned fear responses in the fear-potentiated startle (FPS) test in rats (Li et al. 2015) and was ineffective in the Vogel conflict test (Li et al. 2013). Similarly, treatment with GS39783 did not affect conditioned fear responses in mice (Sweeney et al. 2013).

Taken together, preclinical evidence suggests that activation of the $GABA_B$ receptor may induce anxiolytic-like effects particularly in tests of innate anxiety whereby PAMs decrease innate anxiety in some tests but not others, and thus perhaps do so in a test-specific manner. Importantly, these findings may also be confounded by motor impairing and hypothermic effects.

2.2 Effects of $GABA_B$ Receptor Loss of Function and $GABA_B$ Receptor Antagonists on Anxiety-Like Behaviour

Given the evidence that agonists and PAMs of the $GABA_B$ receptor can exert anxiolytic effects, several studies have also interrogated the impact of genetically induced $GABA_B$ receptor loss of function and $GABA_B$ receptor antagonists on anxiety-like behaviour (summarized in Table 3).

Mice lacking either the $GABA_{B(1)}$ or $GABA_{B(2)}$ receptor subunits exhibit an anxious phenotype. Specifically, $GABA_{B(1)}{}^{-/-}$ mice were more anxious in the light dark box (LDB) test and the staircase test (Mombereau et al. 2004). In addition, these mice exhibited anxiety/panic-like behaviour in the EZM actively jumping off the maze (Mombereau et al. 2004). Similarly, mice lacking the $GABA_{B(2)}$ receptor subunit also exhibit anxiety-like behaviour in the LDB (Mombereau et al. 2005). Anxiety behaviour has also been assessed in mice lacking specific isoforms of the $GABA_{B(1)}$ receptor subunit. $GABA_{B(1a)}{}^{-/-}$ and $GABA_{B(1b)}{}^{-/-}$ mice did not exhibit altered behaviour in innate tests of anxiety including in the EPM, SIH and MB tests (Jacobson et al. 2007; O'Leary et al. 2014). Similarly, $GABA_{B(1a)}{}^{-/-}$ and $GABA_{B(1b)}{}^{-/-}$ mice that underwent early life stress (via maternal separation) or chronic stress in adulthood (via social defeat stress) did not exhibit differences in innate anxiety behaviour when compared to wild type mice (O'Leary et al. 2014). On the other hand, $GABA_{B(1a)}{}^{-/-}$ mice were unable to acquire conditioned taste aversion (CTA), whereas $GABA_{B(1b)}{}^{-/-}$ mice were unable to extinguish aversive taste memories in this test (Jacobson et al. 2006). Taken together, this suggests that loss of function of either the $GABA_{B(1)}$ or $GABA_{B(2)}$ receptor subunit increases innate anxiety, while loss of function of just one $GABA_{B(1)}$ receptor subunit isoform is not sufficient to affect innate anxiety-like behaviour. However, changes in locomotor activity can be a confounding factor of the behavioural tests, for instance $GABA_{B(1)}{}^{-/-}$ and $GABA_{B(1b)}{}^{-/-}$ (but not $GABA_{B(1a)}{}^{-/-}$) mice display hyperlocomotor activity in a new environment (Mombereau et al. 2004; O'Leary et al. 2014).

In contrast to the findings in genetically altered mice, the effects of $GABA_B$ receptor antagonists on anxiety behaviour are less clear (Table 3). Overall, however, the findings suggest that $GABA_B$ receptor antagonists can induce anxiolytic-like effects in rats (Zarrindast et al. 2001; Frankowska et al. 2007; Partyka et al. 2007) but less so in mice (Dalvi and Rodgers 1996; Mombereau et al. 2004; Sweeney et al. 2014). When given systemically to rats, $GABA_B$ receptor antagonists were effective in the EPM, EZM, Vogel conflict test and four-plate test (Zarrindast et al. 2001, Frankowska et al. 2007, Partyka et al. 2007) but were ineffective when locally administered into the basolateral amygdala or the shell of the nucleus accumbens (Sanders and Shekhar 1995; Lopes et al. 2012). In mice, chronic treatment with the $GABA_B$ receptor antagonist CGP56433A had no effect in the LDB test (Mombereau et al. 2004). Similarly, acute treatment with the $GABA_B$ receptor antagonist CGP 52432 did not have anxiolytic effects in the EPM, MB and SIH tests (Dalvi and Rodgers 1996; Sweeney et al. 2014) or in cued auditory fear conditioning (Sweeney et al. 2013). However, the $GABA_B$ receptor antagonist CGP36742 induced

Table 3 Effects of $GABA_B$ receptor inhibition or loss of function on anxiety-like behaviours in rodents

Mice	Paradigm	Finding	References
Genetic			
$GABA_{B(1)}^{-/-}$	LDB	↑ anxiety	Mombereau et al. (2004)
	Staircase test	↑ anxiety	Mombereau et al. (2004)
$GABA_{B(2)}^{-/-}$	LDB	↑ anxiety	Mombereau et al. (2005)
$GABA_{B(1a)}^{-/-}$	EPM	↔ anxiety	Jacobson et al. (2007)
		↔ anxiety	O'Leary et al. (2014)
	SIH	↔ anxiety	Jacobson et al. (2007)
		↔ anxiety	O'Leary et al. (2014)
	MB	↔ anxiety	Jacobson et al. (2007)
	FC	Generalized fear to a neutral context 24 h after training	Lynch et al. (2017)
$GABA_{B(1b)}^{-/-}$	EPM	↔ anxiety	Jacobson et al. (2007)
		↔ anxiety	O'Leary et al. (2014)
	SIH	↔ anxiety	Jacobson et al. (2007)
		↔ anxiety	O'Leary et al. (2014)
	MB	↔ anxiety	Jacobson et al. (2007)
Antagonists			
CGP 35348	EPM	Acute treatment ↔ anxiety in mice	Dalvi and Rodgers (1996)
		Acute treatment ↓ anxiety in rats	Zarrindast et al. (2001)
CGP56433A	LDB	Chronic treatment ↔ anxiety in mice	Mombereau et al. (2004)
SCH 50911	EZM	Acute treatment ↓ anxiety in rats	Frankowska et al. (2007)
CGP 36742	EPM	Acute treatment ↓ anxiety in rats	Partyka et al. (2007)
	Vogel conflict	Acute treatment ↓ anxiety in rats	Partyka et al. (2007)
	Four plate	Acute treatment ↓ anxiety in rats	Partyka et al. (2007)
CGP51176	Four plate	Acute treatment anxiety ↔ in mice	Partyka et al. (2007)
Saclofen	EPM	Injection in the AcbSh ↔ anxiety in 24 h food-deprived rats	Lopes et al. (2012)
		Acute treatment prevented nicotine-induced anxiety in mice	Varani and Balerio (2012)

(continued)

Table 3 (continued)

Mice	Paradigm	Finding	References
CGP52432	FC	Acute treatment ↔ anxiety in mice	Sweeney et al. (2013)
	EPM	PND 14–28 treatment ↔ anxiety in adult mice	Sweeney et al. (2014)

EPM elevated plus maze, *EZM* elevated zero maze, *FC* fear conditioning, *LDB* light/dark box, *LES* light-enhanced startle, *MB* marble burying, *PND* postnatal day, *SI* social interaction, *SIH* stress-induced hyperthermia, *AcbSh* nucleus accumbens shell
↓ = decreased; ↔ = no effects; ↑ = increased

anxiolytic-like effects in the four-plate test in mice (Partyka et al. 2007), and the $GABA_B$ receptor antagonist 2OH-Saclofen reversed the effects of nicotine treatment on anxiety-like behaviours in mice (Varani and Balerio 2012). In addition, the $GABA_B$ receptor antagonist CGP 36216 when administered intracerebroventricularly (ICV) or in the dorsal hippocampus or ventral hippocampus induced fear generalization in mice treated after fear memory consolidation (Lynch et al. 2017). Importantly, the clinical use of $GABA_B$ receptor antagonists has been limited mainly by their potential side-effects including pain, gastroesophageal reflux disease, drug addiction and proconvulsive action (Vergnes et al. 1997; Ghose et al. 2011).

In summary, $GABA_B$ receptor agonists and PAMs exert anxiolytic-like effects, while loss of function of the $GABA_B$ receptor ($GABA_{B(1)}{}^{-/-}$ and $GABA_{B(2)}{}^{-/-}$ mice) induced anxiogenic-like effects. However, loss of function of either the $GABA_{B(1a)}$ or $GABA_{B(1b)}$ receptor subunit isoform alone did not affect anxiety-like behaviour, likely because these mice still express functional $GABA_B$ receptors ($GABA_{B(1b,2)}$ or $GABA_{B(1a,2)}$, respectively). The impact of $GABA_B$ receptor antagonists on anxiety is at present somewhat less clear but sometimes similar to agonists/PAMs appears to be anxiolytic. The precise mechanisms underlying the anxiolytic effects of both $GABA_B$ receptor antagonists, and agonists/PAMS which would be expected to have opposing pharmacological effects are not yet fully understood but may be a function of the fact that $GABA_B$ receptors are found both pre-synaptically and post-synaptically and that drugs might differ in their efficacy at these different receptor sites and at different subunits of the receptor (Cryan and Kaupmann 2005; Sun et al. 2016; Freyd et al. 2017). Nevertheless, the evidence overwhelmingly supports the $GABA_B$ receptor as a valid drug development target for the treatment of anxiety disorders.

3 Role of the $GABA_B$ Receptor in Depression and Antidepressant Action

One of the first indications that the $GABA_B$ receptor may play a role in depression came from preclinical studies reporting that chronic treatment with antidepressant drugs or repeated electroconvulsive shock upregulated $GABA_B$ receptor binding and function in the mouse and rat frontal cortex (Pilc and Lloyd 1984; Lloyd et al. 1985; Suzdak and Gianutsos 1986; Gray and Green 1987; Szekely et al. 1987; Pratt and Bowery 1993). More recently, it has been reported that chronic treatment with antidepressants (fluoxetine, phenelzine, desipramine and tranylcypromine) increased the expression of the $GABA_{B(1a)}$ receptor subunit isoform in the rat hippocampus (Sands et al. 2004). As outlined below, it has since been shown that pharmacological or genetic blockade of $GABA_B$ receptor activity exerts antidepressant-like effects. While these effects of $GABA_B$-receptor antagonist induction of antidepressant-like behaviour seem to be opposing to antidepressant-induced upregulation of the $GABA_B$ receptor, they might be due to drug selective effects on either or both presynaptic and postsynaptic $GABA_B$ receptors (Cryan and Kaupmann 2005, Sun et al. 2016, Freyd et al. 2017).

3.1 Effects of $GABA_B$ Receptor Agonists on Depression-Like Behaviour

The effects of $GABA_B$ receptor agonists on depression-related behaviours in rodents are summarized in Table 4. Several studies have reported that baclofen induced antidepressant-like behaviour in the forced swimming test (FST) in both mice and rats (Aley and Kulkarni 1989, 1990; Car and Wisniewska 2006; Frankowska et al. 2007; Khan et al. 2016). In agreement, it has also been reported that acute treatment with the $GABA_B$ receptor agonist SKF 97541, or the $GABA_B$ receptor PAM, CGP 7930, induced antidepressant-like effects in the rat FST (Frankowska et al. 2007). However, negative findings have also been reported. Indeed, the $GABA_B$ receptor agonists Phaclofen and CGP 44532, and the PAM, GS39783, did not exhibit antidepressant-like activity in the FST in mice or rats (Mombereau et al. 2004; Slattery et al. 2005; Nowak et al. 2006; Araki et al. 2016; Pesarico et al. 2016). Moreover, it was reported that chronic administration of baclofen exacerbated learned helplessness in rats (Nakagawa et al. 1996b) and that baclofen attenuated the effects of several antidepressants in the rat FST and in the learned helplessness model (Nakagawa et al. 1996a, 1996b). More recently, a study showed the baclofen inhibited the antidepressant-like effects of ketamine (which has rapid antidepressant effects) in the mouse tail suspension test (TST) (Rosa et al. 2016). Taken together, it is not yet entirely clear whether pharmacological activation of the $GABA_B$ receptor has antidepressant-like effects.

Table 4 Effects of $GABA_B$ receptor agonists and positive allosteric modulators (PAMs) on depression-like behaviours in rodents

Drug	Paradigm	Finding	References
Agonists			
Baclofen	FST	Acute treatment ↓ immobility in mice	Aley and Kulkarni (1989)
		Acute treatment ↓ immobility in mice	Aley and Kulkarni (1990)
		Acute treatment ↓ immobility in mice	Khan et al. (2016)
		Acute treatment ↓ immobility in rats	Car and Wisniewska (2006)
		Acute treatment ↓ immobility in rats	Frankowska et al. (2007)
		Acute treatment ↔ immobility in isolation reared mice	Araki et al. (2016)
	LH	Chronic treatment ↑ escape failures in rats	Nakagawa et al. (1996a)
		Chronic treatment ↑ escape failures in desipramine-treated rats	Nakagawa et al. (1996b)
	TST	Acute treatment ↑ immobility in mice treated with ketamine/ascorbic acid	Rosa et al. (2016)
PAMs			
SKF 97541	FST	Acute treatment ↓ immobility in rats	Frankowska et al. (2007)
CGP7930		Acute treatment ↓ immobility in rats	Frankowska et al. (2007)
GS39783		Acute treatment ↔ immobility in mice	Mombereau et al. (2004)
		Acute treatment ↔ immobility in mice	Slattery et al. (2005)
CGP 44532		Acute treatment ↔ immobility in mice	Nowak et al. (2006)

FST forced swim test, *LH* learned helplessness
↓ = decreased; ↔ = no effects; ↑ = increased

3.2 *Effects of $GABA_B$ Receptor Blockade or Loss of Function on Depression-Like Behaviour*

In contrast to the data on $GABA_B$ receptor agonists and PAMs, we have much stronger evidence that $GABA_B$ receptor blockade (either pharmacologically or genetically) induces antidepressant-like behaviour (see Table 5). Most studies report that chronic or acute treatment with $GABA_B$ receptor antagonists has antidepressant-like effects in both mice and rats. For instance, the $GABA_B$ receptor antagonist, CGP36742, exhibits antidepressant-like activity in mice in several behavioural tests including the FST, chronic mild stress paradigm, olfactory bulbectomy model and

Table 5 Effects of GABA$_B$ receptor inhibition or loss of function on depression-like behaviours in rodents

Mice	Paradigm	Finding	References
Genetic			
GABA$_{B(1)}$$^{-/-}$	FST	↓ immobility	Mombereau et al. (2004)
GABA$_{B(2)}$$^{-/-}$		↓ immobility	Mombereau et al. (2005)
GABA$_{B(1a)}$$^{-/-}$	FST	↓ immobility	O'Leary et al. (2014)
		↓immobility	Jacobson et al. (2017)
	MS	↑ susceptibility	O'Leary et al. (2014)
	SDS	↑ susceptibility	Jacobson et al. (2017)
GABA$_{B(1b)}$$^{-/-}$	FST	↓ immobility	O'Leary et al. (2014)
	MS	↑ resilience	O'Leary et al. (2014)
	SDS	↑ resilience	O'Leary et al. (2014)
Antagonists			
CGP36742	LH	Acute treatment ↓ depression-like behaviour in rats	Nakagawa et al. (1999)
	FST	Acute treatment ↓ immobility in mice	Nowak et al. (2006)
CGP56433A	FST	Acute treatment ↓ immobility in mice	Mombereau et al. (2004)
		Acute treatment ↓ immobility in mice	Slattery et al. (2005)
CGP51176	FST	Acute treatment ↓ immobility in mice	Nowak et al. (2006)
	Sucrose preference	Chronic treatment ↓ anhedonia in the CMS rat model	Nowak et al. (2006)
SCH 50911	FST	Acute treatment ↓ immobility in mice	Frankowska et al. (2007)
CGP52432	FST	Acute, subchronic and chronic treatment ↓ immobility in mice	Felice et al. (2012)

CMS chronic mild stress, *FST* forced swim test, *LH* learned helplessness, *MS* maternal separation, *SDS* social defeat stress

↓ = decreased; ↔ = no effects; ↑ = increased

the learned helplessness paradigm (Nakagawa et al. 1999; Nowak et al. 2006). Similarly, the GABA$_B$ receptor antagonists CGP51176, CGP51176A, CGP56433A, SCH50911 and CGP52432 also induced antidepressant-like effects in both the mouse and rat FST (Mombereau et al. 2004; Slattery et al. 2005; Frankowska et al. 2007; Felice et al. 2012). In addition, CGP51176A has also been shown to reduce stress-induced anhedonia as measured by increased sucrose consumption in the chronic mild stress rat model (Nowak et al. 2006).

Studies in genetically modified GABA$_B$ receptor mice have revealed findings similar to that observed with receptor antagonists. GABA$_{B(1)}$$^{-/-}$ and GABA$_{B(2)}$$^{-/-}$ mice exhibit an antidepressant-like phenotype in the FST (Mombereau et al. 2004,

2005). In the TST, male but not female $GABA_{B(1b)}^{-/-}$ mice displayed decreased immobility suggesting antidepressant-like phenotype whereas male and female $GABA_{B(1a)}^{-/-}$ mice exhibited increased immobility, suggesting a depression-like phenotype. In the FST, both $GABA_{B(1a)}^{-/-}$ and $GABA_{B(1b)}^{-/-}$ mice exhibited an antidepressant-like phenotype (O'Leary et al. 2014). However, male but not female $GABA_{B(1b)}^{-/-}$ mice are hyperactive in the open field test which may have contributed to the reduced immobility of males in the FST and TST (O'Leary et al. 2014). Interestingly, $GABA_{B(1a)}^{-/-}$ mice are more susceptible whereas $GABA_{B(1b)}^{-/-}$ mice are more resilient to early life stress (via maternal separation) and social defeat stress in adulthood (O'Leary et al. 2014). Specifically, $GABA_{B(1a)}^{-/-}$ mice are more susceptible to stress (maternal separation or social defeat stress) -induced anhedonia as measured in the saccharin preference and female urine sniffing tests, and were also more susceptible to social defeat stress-induced social avoidance (O'Leary et al. 2014). On the other hand, $GABA_{B(1b)}^{-/-}$ mice were resilient to stress-induced anhedonia and psychosocial stress-induced social withdrawal (O'Leary et al. 2014). In addition, $GABA_{B(1a)}^{-/-}$ but not $GABA_{B(1b)}^{-/-}$ mice exhibited a blunted 8-OH-DPAT-induced corticosterone and adrenocorticotropic hormone (ACTH) release, thus suggesting disrupted regulation of the hypothalamic-pituitary-adrenal (HPA) axis which is the neuroendocrine stress response system (Jacobson et al. 2017).

Taken together, preclinical pharmacological studies and studies using genetically altered $GABA_B$ receptor mice strongly suggest that inhibition of $GABA_B$ receptors has therapeutic potential in the treatment of depression (Alexander 2017; Jacobson et al. 2018). As described earlier, sometimes, the $GABA_B$ receptor agonist baclofen has also been shown to have antidepressant-like effects in the forced swim test (FST). The precise mechanisms underlying how opposing pharmacological manipulations (agonist vs. antagonist) could exert similar antidepressant-like effects is unknown. However, it may be a function of the fact that $GABA_B$ receptors are found both pre-synaptically and post-synaptically, and that drugs might differ in their selectivity for these differentially located $GABA_B$ receptors. The subunit composition of affected receptors might also influence behavioural responses to pharmacological agents. For example, it has been shown that mice lacking $GABA_{B(1b)}$ receptor subunit isoform exhibit a stress-resilient phenotype, while mice lacking the $GABA_{B(1a)}$ subunit are more stress-susceptible (O'Leary et al. 2014).

4 Clinical Evidence of the Role of $GABA_B$ Receptor in Mood Disorders

The preclinical evidence of the therapeutic potential of $GABA_B$ receptor modulation in the treatment of depression is also supported by clinical evidence. One of the first clinical indications of a role for the $GABA_B$ receptor in depression comes from a small study reporting that baclofen may worsen depressive like-symptoms (Post et al. 1991). In that study, patients with primary affective disorder were chronically treated with baclofen (10–55 mg/day). Out of five patients, three exhibited increased depression during baclofen treatment and these depressive symptoms improved during baclofen withdrawal (Post et al. 1991). This baclofen-induced worsening of depressive symptoms seems counterintuitive to its antidepressant-like effects in preclinical studies. The reasons underlying this discrepancy are unclear but may relate to the fact that preclinical assessments of baclofen were not done in animal models of depression per se, e.g. stress-induced anhedonia, but were conducted using "normal" animals in the FST which is a behavioural test of antidepressant-drug-like activity and not a model of depression. Nevertheless, several studies also reported that depressed patients displayed blunted baclofen-induced growth hormone release (Marchesi et al. 1991; O'Flynn and Dinan 1993), further suggesting a role for the $GABA_B$ receptor in depression. The effects of baclofen on depression and anxiety-related clinical measures are contradictory, however, as summarized in a recent review on its off-label use to treat alcohol use disorder (Agabio and Leggio 2018).

Postmortem studies have reported regional alterations in $GABA_B$ receptor subunit expression in brains from depressed suicide victims (Ghose et al. 2011) and depressed individuals (Klempan et al. 2009). Specifically, it was reported that depressed suicide victims exhibited upregulation of the $GABA_{B(2)}$ receptor subunit in cortical and subcortical brain regions compared with non-depressed suicide victims (Klempan et al. 2009). More recently, it was reported that $GABA_{B(1)}$ and $GABA_{B(2)}$ receptor subunit expression was reduced in the superior frontal cortex of subjects with bipolar disorder (Fatemi et al. 2017). In the hippocampus of depressed patients, $GABA_{B(2)}$ gene expression was reported to be increased by 50% (Ghose et al. 2011). In addition, in the dentate gyrus of the hippocampus of these depressed patients, there was a 30% decrease in the expression of the $GABA_{B(1a)}$ receptor subunit isoform when compared with controls (Ghose et al. 2011). Interestingly, the dentate gyrus is one of just a few brain areas where neurogenesis, the birth of new neurons, occurs throughout life (Altman 1962a, b; Spalding et al. 2013; Boldrini et al. 2018; Moreno-Jimenez et al. 2019). Hippocampal neurogenesis has been implicated in the mechanism of antidepressant action (Santarelli et al. 2003; David et al. 2009; O'Leary and Cryan 2014; Miller and Hen 2015) and recently we and others reported that $GABA_B$ receptor antagonists that have antidepressant-like behavioural effects increase hippocampal neurogenesis (Felice et al. 2012; Giachino et al. 2014). We have also found that the stress-resilient behavioural phenotype of $GABAB_{(1b)}^{-/-}$ mice is accompanied by resilience to stress-induced decreases in adult hippocampal neurogenesis (O'Leary et al. 2014).

There is also evidence from human transcranial magnetic stimulation (TMS) studies that there are alterations in $GABA_B$ receptor activity in depression. The first such study suggested that $GABA_B$ neurophysiological deficits are closely related to the pathophysiology of major depressive disorder (Levinson et al. 2010). In that study, patients with major depressive disorder (MDD) exhibited decreased cortical silence, a measure of intracortical inhibition thought to be a marker of $GABA_B$ receptor neurotransmission. Other more recent studies have confirmed that depressed patients exhibit a decreased cortical silent period (a TMS measure of $GABA_B$ receptor activity) (Veronezi et al. 2016). Accordingly, adolescents with depression and a lifetime history of suicidal behaviours exhibited impaired long-interval intracortical inhibition (LICI; which is a TMS measure of $GABA_B$ receptor-mediated inhibition) when compared to healthy adolescents and to depressed adolescents without a history of suicidal behaviour (Lewis et al. 2018). A follow-up small study by the same group reported an association between increases in $GABA_B$-mediated cortical inhibition and a reduction in suicidal ideation over time in adolescents treated for depression (Lewis et al. 2019). A paired-pulse TMS (ppTMS) study revealed that patients with treatment resistant depression (TRD) exhibit more reduced $GABA_A$ and $GABA_B$ receptor-mediated cortical inhibition compared to non-TRD patients and healthy subjects (Jeng et al. 2019), thus suggesting a potential role for $GABA_B$ receptor function in TRD. In addition, selective serotonin reuptake inhibitor (SSRI) antidepressants were shown to modulate $GABA_B$ receptor-mediated long-interval intracortical inhibition in non-TRD patients (Jeng et al. 2019), thus providing clinical evidence for a role of $GABA_B$ receptors in antidepressant action.

In contrast to depression, clinical studies interrogating a role for the $GABA_B$ receptor in anxiety disorders are sparse and the evidence is largely indirect. Nevertheless, there is strong evidence that GABAergic neurotransmission plays a role in the treatment and pathophysiology of anxiety disorders as benzodiazepines (which act on the $GABA_A$ receptor) are used to treat anxiety disorders (Nemeroff 2003). In terms of a potential role for $GABA_B$ receptors, baclofen has been shown to attenuate the anxiety that is associated with alcohol withdrawal, post-traumatic stress, panic disorder and traumatic spinal-cord lesions (Cryan et al. 2005).

In summary, both clinical and preclinical evidence strongly support a role for the $GABA_B$ receptor in depression and anxiety disorders. However, the involvement of the $GABA_B$ receptor in the pathophysiology of anxiety disorders is less explored in clinical studies when compared with depression. Indeed, the majority of clinical studies on the role of the GABAergic system in anxiety disorders are focused on the $GABA_A$ receptor. However, it is worth noting that the $GABA_B$ receptor can contribute to inhibition by also modulating $GABA_A$ receptor activity at presynaptic and postsynaptic sites (Cryan et al. 2005; Tao et al. 2013), thus suggesting a potential upstream modulating role for the $GABA_B$ receptor in anxiety disorders. Moreover, preclinical studies suggest that agonists and PAMs of the $GABA_B$ receptor have anxiolytic effects.

5 Conclusions and Perspectives

Although both preclinical and clinical studies suggest the $GABA_B$ receptor as a potential target for the development of new therapeutic approaches for mood and anxiety disorders, only one $GABA_B$ receptor-based compound, SGS272 (CGP36742, a $GABA_B$ receptor antagonist), progressed to Phase II clinical trials and was investigated as a potential treatment for cognitive deficits (Ghose et al. 2011). To date, however, no clinical trials assessing the effects of GABA$_B$ receptor antagonists in depressed patients have ever been conducted. The development of such antagonists of the GABA$_B$ receptor for the treatment of mood disorders is mainly hampered by its potential side-effects, particularly the potential risk of proconvulsive action. However, the abundance of preclinical evidence of the antidepressant-like effects of GABA$_B$ receptor antagonists cannot be ignored, and thus novel and more selective GABA$_B$ receptor antagonists with a better side-effect profile could lead to new therapeutic approaches in the clinic. In 2014, the first negative allosteric modulator (NAM) of the GABA$_B$ receptor was generated. This was a CGP7930 analogue, called CLH304a (also named Compound 14) (Chen et al. 2014). In 2016, two additional novel NAMs, CLH391 and CLH393, were synthetized based on the structure of CLH304a (Sun et al. 2016). It would be expected that NAMs would have a better side-effect profile than antagonists and as such, the discovery of these NAMs is very promising for the development of innovative drugs that negatively modulate GABA$_B$ receptor action, and thus might have antidepressant potential with a reduced side-effect profile.

The GABA$_B$ receptor plays a key role in anxiety disorders as demonstrated by a plethora of preclinical evidence. PAMs represent promising drugs to treat anxiety-like disorders with safer side-effect profiles than GABA$_B$ receptor agonists. ADX 71441 is the first GABA$_B$ receptor PAM approved for phase I clinical trial (Kalinichev et al. 2017) indicated for alcohol use disorder, Charcot-Marie-Tooth disease and nicotine dependence. However, future clinical trials are required to evaluate the effects of PAMs in anxiety disorders.

Overall, the GABA$_B$ receptor represents a promising target to develop new therapeutic treatments for depression and anxiety disorders. Since Bowery and colleagues' discovery of the GABA$_B$ receptor in 1979, thousands of studies investigating its role in mammals and non-mammals such as the drosophila model (Manev and Dzitoyeva 2010) have been published. The introduction of genetic tools has allowed the further study of the role of GABA$_B$ receptor subunits and their isoforms in mice. Despite the drive of scientists to study the GABA$_B$ receptor, there is still a lot unknown. In particular, side-effects associated with GABA$_B$ receptor modulation hamper its path to become a relevant drug target. However, the introduction of novel tools to study the GABA$_B$ receptor (e.g. FRET-Based Sensors) and the discovery of novel GABA$_B$ receptor PAMs and NAMs will pave the way towards GABA$_B$ receptor therapeutics in human disorders such as depression and anxiety disorders. However, NAMs have yet to be tested in vivo. Intracellular GABA$_B$ receptor-associated proteins may also be important targets to modulate GABA$_B$ receptor activity because protein–protein interaction may allow more precise and temporal GABA$_B$ receptor activity modulation. Particularly, the K^+ channel tetramerization domain (KCTD) that is associated with the GABA$_{B2}$ receptor C-terminus is envisaged to be a promising target (Sereikaite et al. 2019).

References

Adams CL, Lawrence AJ (2007) CGP7930: a positive allosteric modulator of the GABAB receptor. CNS Drug Rev 13:308–316

Agabio R, Leggio L (2018) Baclofen in the treatment of patients with alcohol use disorder and other mental health disorders. Front Psych 9:464

Agabio R, Preti A, Gessa GL (2013) Efficacy and tolerability of baclofen in substance use disorders: a systematic review. Eur Addict Res 19:325–345

Agmo A, Pruneda R, Guzman M, Gutierrez M (1991) GABAergic drugs and conflict behavior in the rat: lack of similarities with the actions of benzodiazepines. Naunyn Schmiedeberg's Arch Pharmacol 344:314–322

Alexander RC (2017) The potential efficacy of GABAB antagonists in depression. Curr Opin Pharmacol 35:101–104

Aley KO, Kulkarni SK (1989) GABA-mediated modification of despair behavior in mice. Naunyn Schmiedeberg's Arch Pharmacol 339:306–311

Aley KO, Kulkarni SK (1990) Effect of baclofen, a GABAB-agonist, on forced swimming-induced immobility in mice. Arch Int Pharmacodyn Ther 307:18–31

Altman J (1962a) Are new neurons formed in the brains of adult mammals? Science 135:1127–1128

Altman J (1962b) Autoradiographic study of degenerative and regenerative proliferation of neuroglia cells with tritiated thymidine. Exp Neurol 5:302–318

Amikishieva AV, Semendyaeva SN (2007) Effects of baclofen on anxiety, sexual motivation, and olfactory perception in male mice in different psychoemotional states. Neurosci Behav Physiol 37:929–937

Andrews N, File SE (1993) Increased 5-HT release mediates the anxiogenic response during benzodiazepine withdrawal: a review of supporting neurochemical and behavioural evidence. Psychopharmacology 112:21–25

Araki R, Hiraki Y, Nishida S, Kuramoto N, Matsumoto K, Yabe T (2016) Epigenetic regulation of dorsal raphe GABA(B1a) associated with isolation-induced abnormal responses to social stimulation in mice. Neuropharmacology 101:1–12

Bettler B, Kaupmann K, Mosbacher J, Gassmann M (2004) Molecular structure and physiological functions of GABA(B) receptors. Physiol Rev 84:835–867

Binet V, Brajon C, Le Corre L, Acher F, Pin JP, Prezeau L (2004) The heptahelical domain of GABA(B2) is activated directly by CGP7930, a positive allosteric modulator of the GABA (B) receptor. J Biol Chem 279:29085–29091

Boldrini M, Fulmore CA, Tartt AN, Simeon LR, Pavlova I, Poposka V, Rosoklija GB, Stankov A, Arango V, Dwork AJ, Hen R, Mann JJ (2018) Human hippocampal neurogenesis persists throughout aging. Cell Stem Cell 22:589–599.e585

Bowery NG, Doble A, Hill DR, Hudson AL, Shaw JS, Turnbull MJ (1979) Baclofen: a selective agonist for a novel type of GABA receptor proceedings. Br J Pharmacol 67:444P–445P

Bowery NG, Hill DR, Hudson AL, Doble A, Middlemiss DN, Shaw J, Turnbull M (1980) (−) Baclofen decreases neurotransmitter release in the mammalian CNS by an action at a novel GABA receptor. Nature 283:92–94

Bowery NG, Bettler B, Froestl W, Gallagher JP, Marshall F, Raiteri M, Bonner TI, Enna SJ (2002) International union of pharmacology. XXXIII. Mammalian gamma-aminobutyric acid (B) receptors: structure and function. Pharmacol Rev 54:247–264

Bowery N, Enna SJ, Olsen RW (2004) Six decades of GABA. Biochem Pharmacol 68:1477–1478

Car H, Wisniewska RJ (2006) Antidepressant-like effects of baclofen and LY367385 in the forced swim test in rats. Pharmacol Rep 58:758–764

Castelli MP, Casu A, Casti P, Lobina C, Carai MA, Colombo G, Solinas M, Giunta D, Mugnaini C, Pasquini S, Tafi A, Brogi S, Gessa GL, Corelli F (2012) Characterization of COR627 and COR628, two novel positive allosteric modulators of the GABA(B) receptor. J Pharmacol Exp Ther 340:529–538

Chen LH, Sun B, Zhang Y, Xu TJ, Xia ZX, Liu JF, Nan FJ (2014) Discovery of a negative allosteric modulator of GABAB receptors. ACS Med Chem Lett 5:742–747

Christopoulos A (2002) Allosteric binding sites on cell-surface receptors: novel targets for drug discovery. Nat Rev Drug Discov 1:198–210

Cryan JF, Kaupmann K (2005) Don't worry 'B' happy!: a role for GABA(B) receptors in anxiety and depression. Trends Pharmacol Sci 26:36–43

Cryan JF, Slattery DA (2007) Animal models of mood disorders: recent developments. Curr Opin Psychiatry 20:1–7

Cryan JF, Slattery DA (2010) GABAB receptors and depression. Curr Status Adv Pharmacol 58:427–451

Cryan JF, Kelly PH, Chaperon F, Gentsch C, Mombereau C, Lingenhoehl K, Froestl W, Bettler B, Kaupmann K, Spooren WP (2004) Behavioral characterization of the novel GABAB receptor-positive modulator GS39783 (N,N'-dicyclopentyl-2-methylsulfanyl-5-nitro-pyrimidine-4,6-diamine): anxiolytic-like activity without side effects associated with baclofen or benzodiazepines. J Pharmacol Exp Ther 310:952–963

Cryan JF, Page ME, Lucki I (2005) Differential behavioral effects of the antidepressants reboxetine, fluoxetine, and moclobemide in a modified forced swim test following chronic treatment. Psychopharmacology 182:335–344

Dalvi A, Rodgers RJ (1996) GABAergic influences on plus-maze behaviour in mice. Psychopharmacology 128:380–397

David DJ, Samuels BA, Rainer Q, Wang JW, Marsteller D, Mendez I, Drew M, Craig DA, Guiard BP, Guilloux JP, Artymyshyn RP, Gardier AM, Gerald C, Antonijevic IA, Leonardo ED, Hen R (2009) Neurogenesis-dependent and -independent effects of fluoxetine in an animal model of anxiety/depression. Neuron 62:479–493

Fatemi SH, Folsom TD, Thuras PD (2017) GABAA and GABAB receptor dysregulation in superior frontal cortex of subjects with schizophrenia and bipolar disorder. Synapse 71

Felice D, O'Leary OF, Pizzo RC, Cryan JF (2012) Blockade of the GABA(B) receptor increases neurogenesis in the ventral but not dorsal adult hippocampus: relevance to antidepressant action. Neuropharmacology 63:1380–1388

Felice D, O'Leary OF, Cryan JF (2016) Targeting the GABAB receptor for the treatment of depression and anxiety disorders. In: Colombo G (ed) GABAB receptor the receptors, vol 29. Humana Press, Cham

File SE, Zharkovsky A, Gulati K (1991) Effects of baclofen and nitrendipine on ethanol withdrawal responses in the rat. Neuropharmacology 30:183–190

File SE, Zharkovsky A, Hitchcott PK (1992) Effects of nitrendipine, chlordiazepoxide, flumazenil and baclofen on the increased anxiety resulting from alcohol withdrawal. Prog Neuro-Psychopharmacol Biol Psychiatry 16:87–93

Frankowska M, Filip M, Przegalinski E (2007) Effects of GABAB receptor ligands in animal tests of depression and anxiety. Pharmacol Rep 59:645–655

Freyd T, Warszycki D, Mordalski S, Bojarski AJ, Sylte I, Gabrielsen M (2017) Ligand-guided homology modelling of the GABAB2 subunit of the GABAB receptor. PLoS One 12:e0173889

Galvez T, Duthey B, Kniazeff J, Blahos J, Rovelli G, Bettler B, Prezeau L, Pin JP (2001) Allosteric interactions between GB1 and GB2 subunits are required for optimal GABA(B) receptor function. EMBO J 20:2152–2159

Gassmann M, Bettler B (2012) Regulation of neuronal GABA(B) receptor functions by subunit composition. Nat Rev Neurosci 13:380–394

Ghose S, Winter MK, McCarson KE, Tamminga CA, Enna SJ (2011) The GABAβ receptor as a target for antidepressant drug action. Br J Pharmacol 162:1–17

Giachino C, Barz M, Tchorz JS, Tome M, Gassmann M, Bischofberger J, Bettler B, Taylor V (2014) GABA suppresses neurogenesis in the adult hippocampus through GABAB receptors. Development 141:83–90

Gjoni T, Urwyler S (2008) Receptor activation involving positive allosteric modulation, unlike full agonism, does not result in GABAB receptor desensitization. Neuropharmacology 55:1293–1299

Gjoni T, Urwyler S (2009) Changes in the properties of allosteric and orthosteric GABAB receptor ligands after a continuous, desensitizing agonist pretreatment. Eur J Pharmacol 603:37–41

Gray JA, Green AR (1987) Increased GABAB receptor function in mouse frontal cortex after repeated administration of antidepressant drugs or electroconvulsive shocks. Br J Pharmacol 92:357–362

Guery S, Floersheim P, Kaupmann K, Froestl W (2007) Syntheses and optimization of new GS39783 analogues as positive allosteric modulators of GABA B receptors. Bioorg Med Chem Lett 17:6206–6211

Hill DR, Bowery NG (1981) 3H-baclofen and 3H-GABA bind to bicuculline-insensitive GABA B sites in rat brain. Nature 290:149–152

Jacobson LH, Cryan JF (2008) Evaluation of the anxiolytic-like profile of the GABAB receptor positive modulator CGP7930 in rodents. Neuropharmacology 54:854–862

Jacobson LH, Kelly PH, Bettler B, Kaupmann K, Cryan JF (2006) GABA(B(1)) receptor isoforms differentially mediate the acquisition and extinction of aversive taste memories. J Neurosci 26:8800–8803

Jacobson LH, Bettler B, Kaupmann K, Cryan JF (2007) Behavioral evaluation of mice deficient in GABA(B(1)) receptor isoforms in tests of unconditioned anxiety. Psychopharmacology 190:541–553

Jacobson LH, Hoyer D, Fehlmann D, Bettler B, Kaupmann K, Cryan JF (2017) Blunted 5-HT1A receptor-mediated responses and antidepressant-like behavior in mice lacking the GABAB1a but not GABAB1b subunit isoforms. Psychopharmacology 234:1511–1523

Jacobson LH, Vlachou S, Slattery DA, Li X, Cryan JF (2018) The gamma-aminobutyric acid B receptor in depression and reward. Biol Psychiatry 83:963–976

Jeng JS, Li CT, Lin HC, Tsai SJ, Bai YM, Su TP, Chang YW, Cheng CM (2019) Antidepressant-resistant depression is characterized by reduced short- and long-interval cortical inhibition. Psychol Med:1–7

Kalinichev M, Girard F, Haddouk H, Rouillier M, Riguet E, Royer-Urios I, Mutel V, Lutjens R, Poli S (2017) The drug candidate, ADX71441, is a novel, potent and selective positive allosteric modulator of the GABAB receptor with a potential for treatment of anxiety, pain and spasticity. Neuropharmacology 114:34–47

Ketelaars CE, Bollen EL, Rigter H, Bruinvels J (1988) GABA-B receptor activation and conflict behaviour. Life Sci 42:933–942

Khan MI, Ostadhadi S, Zolfaghari S, Ejtemaei Mehr S, Hassanzadeh G, Dehpour AR (2016) The involvement of NMDA receptor/NO/cGMP pathway in the antidepressant like effects of baclofen in mouse force swimming test. Neurosci Lett 612:52–61

Klempan TA, Sequeira A, Canetti L, Lalovic A, Ernst C, Ffrench-Mullen J, Turecki G (2009) Altered expression of genes involved in ATP biosynthesis and GABAergic neurotransmission in the ventral prefrontal cortex of suicides with and without major depression. Mol Psychiatry 14:175–189

Koek W, France CP, Cheng K, Rice KC (2010) GABAB receptor-positive modulators: enhancement of GABAB receptor agonist effects in vivo. J Pharmacol Exp Ther 335:163–171

Lecat-Guillet N, Monnier C, Rovira X, Kniazeff J, Lamarque L, Zwier JM, Trinquet E, Pin JP, Rondard P (2017) FRET-based sensors unravel activation and allosteric modulation of the GABAB receptor. Cell Chem Biol 24:360–370

Lee C, Mayfield RD, Harris RA (2010) Intron 4 containing novel GABAB1 isoforms impair GABAB receptor function. PLoS One 5:e14044

Lehmann A, Mattsson JP, Edlund A, Johansson T, Ekstrand AJ (2003) Effects of repeated administration of baclofen to rats on GABAB receptor binding sites and subunit expression in the brain. Neurochem Res 28:387–393

Levinson AJ, Fitzgerald PB, Favalli G, Blumberger DM, Daigle M, Daskalakis ZJ (2010) Evidence of cortical inhibitory deficits in major depressive disorder. Biol Psychiatry 67:458–464
Lewis CP, Nakonezny PA, Blacker CJ, Vande Voort JL, Port JD, Worrell GA, Jo HJ, Daskalakis ZJ, Croarkin PE (2018) Cortical inhibitory markers of lifetime suicidal behavior in depressed adolescents. Neuropsychopharmacology 43:1822–1831
Lewis CP, Camsari DD, Sonmez AI, Nandakumar AL, Gresbrink MA, Daskalakis ZJ, Croarkin PE (2019) Preliminary evidence of an association between increased cortical inhibition and reduced suicidal ideation in adolescents treated for major depression. J Affect Disord 244:21–24
Li X, Risbrough VB, Cates-Gatto C, Kaczanowska K, Finn MG, Roberts AJ, Markou A (2013) Comparison of the effects of the GABAB receptor positive modulator BHF177 and the GABAB receptor agonist baclofen on anxiety-like behavior, learning, and memory in mice. Neuropharmacology 70:156–167
Li X, Kaczanowska K, Finn MG, Markou A, Risbrough VB (2015) The GABA(B) receptor positive modulator BHF177 attenuated anxiety, but not conditioned fear, in rats. Neuropharmacology 97:357–364
Lloyd KG, Thuret F, Pilc A (1985) Upregulation of gamma-aminobutyric acid (GABA) B binding sites in rat frontal cortex: a common action of repeated administration of different classes of antidepressants and electroshock. J Pharmacol Exp Ther 235:191–199
Lopes AP, Ganzer L, Borges AC, Kochenborger L, Januario AC, Faria MS, Marino-Neto J, Paschoalini MA (2012) Effects of GABA ligands injected into the nucleus accumbens shell on fear/anxiety-like and feeding behaviours in food-deprived rats. Pharmacol Biochem Behav 101:41–48
Lu Y, Li CJ, Chen C, Luo P, Zhou M, Li C, Xu XL, Lu Q, He Z, Guo LJ (2016) Activation of GABAB2 subunits alleviates chronic cerebral hypoperfusion-induced anxiety-like behaviours: a role for BDNF signalling and Kir3 channels. Neuropharmacology 110:308–321
Lynch JF, Winiecki P, Gilman TL, Adkins JM, Jasnow AM (2017) Hippocampal GABAB (1a) receptors constrain generalized contextual fear. Neuropsychopharmacology 42:914–924
Malherbe P, Masciadri R, Norcross RD, Knoflach F, Kratzeisen C, Zenner MT, Kolb Y, Marcuz A, Huwyler J, Nakagawa T, Porter RH, Thomas AW, Wettstein JG, Sleight AJ, Spooren W, Prinssen EP (2008) Characterization of (R,S)-5,7-di-tert-butyl-3-hydroxy-3-trifluoromethyl-3H-benzofuran-2-one as a positive allosteric modulator of GABAB receptors. Br J Pharmacol 154:797–811
Manev H, Dzitoyeva S (2010) GABA-B receptors in Drosophila. Adv Pharmacol 58:453–464
Marchesi C, Chiodera P, De Ferri A, De Risio C, Dasso L, Menozzi P, Volpi R, Coiro V (1991) Reduction of GH response to the GABA-B agonist baclofen in patients with major depression. Psychoneuroendocrinology 16:475–479
McDonald AJ, Mascagni F, Muller JF (2004) Immunocytochemical localization of GABABR1 receptor subunits in the basolateral amygdala. Brain Res 1018:147–158
Miller BR, Hen R (2015) The current state of the neurogenic theory of depression and anxiety. Curr Opin Neurobiol 30:51–58
Mombereau C, Kaupmann K, Froestl W, Sansig G, van der Putten H, Cryan JF (2004) Genetic and pharmacological evidence of a role for GABA(B) receptors in the modulation of anxiety- and antidepressant-like behavior. Neuropsychopharmacology 29:1050–1062
Mombereau C, Kaupmann K, Gassmann M, Bettler B, van der Putten H, Cryan JF (2005) Altered anxiety and depression-related behaviour in mice lacking GABAB(2) receptor subunits. Neuroreport 16:307–310
Moreno-Jimenez EP, Flor-Garcia M, Terreros-Roncal J, Rabano A, Cafini F, Pallas-Bazarra N, Avila J, Llorens-Martin M (2019) Adult hippocampal neurogenesis is abundant in neurologically healthy subjects and drops sharply in patients with Alzheimer's disease. Nat Med 25:554–560
Nakagawa Y, Ishima T, Ishibashi Y, Tsuji M, Takashima T (1996a) Involvement of GABAB receptor systems in action of antidepressants. II: baclofen attenuates the effect of desipramine

whereas muscimol has no effect in learned helplessness paradigm in rats. Brain Res 728:225–230
Nakagawa Y, Ishima T, Ishibashi Y, Tsuji M, Takashima T (1996b) Involvement of GABAB receptor systems in experimental depression: baclofen but not bicuculline exacerbates helplessness in rats. Brain Res 741:240–245
Nakagawa Y, Sasaki A, Takashima T (1999) The GABA(B) receptor antagonist CGP36742 improves learned helplessness in rats. Eur J Pharmacol 381:1–7
Nastiti K, Benton D, Brain PF (1991) The effects of compounds acting at the benzodiazepine receptor complex on the ultrasonic calling of mouse pups. Behav Pharmacol 2:121–128
Nemeroff CB (2003) The role of GABA in the pathophysiology and treatment of anxiety disorders. Psychopharmacol Bull 37:133–146
Nowak G, Partyka A, Palucha A, Szewczyk B, Wieronska JM, Dybala M, Metz M, Librowski T, Froestl W, Papp M, Pilc A (2006) Antidepressant-like activity of CGP 36742 and CGP 51176, selective GABAB receptor antagonists, in rodents. Br J Pharmacol 149:581–590
O'Flynn K, Dinan TG (1993) Baclofen-induced growth hormone release in major depression: relationship to dexamethasone suppression test result. Am J Psychiatry 150:1728–1730
O'Leary OF, Cryan JF (2014) A ventral view on antidepressant action: roles for adult hippocampal neurogenesis along the dorsoventral axis. Trends Pharmacol Sci 35:675–687
O'Leary OF, Felice D, Galimberti S, Savignac HM, Bravo JA, Crowley T, El Yacoubi M, Vaugeois JM, Gassmann M, Bettler B, Dinan TG, Cryan JF (2014) GABAB(1) receptor subunit isoforms differentially regulate stress resilience. Proc Natl Acad Sci U S A 111:15232–15237
Partyka A, Klodzinska A, Szewczyk B, Wieronska JM, Chojnacka-Wojcik E, Librowski T, Filipek B, Nowak G, Pilc A (2007) Effects of GABAB receptor ligands in rodent tests of anxiety-like behavior. Pharmacol Rep 59:757–762
Perdona E, Costantini VJ, Tessari M, Martinelli P, Carignani C, Valerio E, Mok MH, Zonzini L, Visentini F, Gianotti M, Gordon L, Rocheville M, Corsi M, Capelli AM (2011) In vitro and in vivo characterization of the novel GABAB receptor positive allosteric modulator, 2-{1-[2-(4-chlorophenyl)-5-methylpyrazolo[1,5-a]pyrimidin-7-yl]-2-piperidinyl}ethanol (CMPPE). Neuropharmacology 61:957–966
Pesarico AP, Stangherlin EC, Rosa SG, Mantovani AC, Zeni G, Nogueira CW (2016) Contribution of NMDA, GABAA and GABAB receptors and l-arginine-NO-cGMP, MEK1/2 and CaMK-II pathways in the antidepressant-like effect of 7-fluoro-1,3-diphenylisoquinoline-1-amine in mice. Eur J Pharmacol 782:6–13
Pilc A, Lloyd KG (1984) Chronic antidepressants and GABA "B" receptors: a GABA hypothesis of antidepressant drug action. Life Sci 35:2149–2154
Pizzo R, O'Leary OF, Cryan JF (2018) Elucidation of the neural circuits activated by a GABAB receptor positive modulator: relevance to anxiety. Neuropharmacology 136:129–145
Post RM, Ketter TA, Joffe RT, Kramlinger KL (1991) Lack of beneficial effects of l-baclofen in affective disorder. Int Clin Psychopharmacol 6:197–207
Pratt GD, Bowery NG (1993) Repeated administration of desipramine and a GABAB receptor antagonist, CGP 36742, discretely up-regulates GABAB receptor binding sites in rat frontal cortex. Br J Pharmacol 110:724–735
Rosa PB, Neis VB, Ribeiro CM, Moretti M, Rodrigues AL (2016) Antidepressant-like effects of ascorbic acid and ketamine involve modulation of GABAA and GABAB receptors. Pharmacol Rep 68:996–1001
Sanders SK, Shekhar A (1995) Regulation of anxiety by GABAA receptors in the rat amygdala. Pharmacol Biochem Behav 52:701–706
Sands SA, Reisman SA, Enna SJ (2004) Effect of antidepressants on GABA(B) receptor function and subunit expression in rat hippocampus. Biochem Pharmacol 68:1489–1495
Santarelli L, Saxe M, Gross C, Surget A, Battaglia F, Dulawa S, Weisstaub N, Lee J, Duman R, Arancio O, Belzung C, Hen R (2003) Requirement of hippocampal neurogenesis for the behavioral effects of antidepressants. Science 301:805–809

Sereikaite V, Fritzius T, Kasaragod VB, Bader N, Maric HM, Schindelin H, Bettler B, Stromgaard K (2019) Targeting the gamma-aminobutyric acid type B (GABAB) receptor complex: development of inhibitors targeting the K(+) channel tetramerization domain (KCTD) containing proteins/GABAB receptor protein-protein interaction. J Med Chem 62:8819–8830

Shephard RA, Wedlock P, Wilson NE (1992) Direct evidence for mediation of an anticonflict effect of baclofen by GABAB receptors. Pharmacol Biochem Behav 41:651–653

Slattery DA, Desrayaud S, Cryan JF (2005) GABAB receptor antagonist-mediated antidepressant-like behavior is serotonin-dependent. J Pharmacol Exp Ther 312:290–296

Spalding KL, Bergmann O, Alkass K, Bernard S, Salehpour M, Huttner HB, Bostrom E, Westerlund I, Vial C, Buchholz BA, Possnert G, Mash DC, Druid H, Frisen J (2013) Dynamics of hippocampal neurogenesis in adult humans. Cell 153:1219–1227

Sun B, Chen L, Liu L, Xia Z, Pin JP, Nan F, Liu J (2016) A negative allosteric modulator modulates GABAB-receptor signalling through GB2 subunits. Biochem J 473:779–787

Suzdak PD, Gianutsos G (1986) Effect of chronic imipramine or baclofen on GABA-B binding and cyclic AMP production in cerebral cortex. Eur J Pharmacol 131:129–133

Sweeney FF, O'Leary OF, Cryan JF (2013) GABAB receptor ligands do not modify conditioned fear responses in BALB/c mice. Behav Brain Res 256:151–156

Sweeney FF, O'Leary OF, Cryan JF (2014) Activation but not blockade of GABAB receptors during early-life alters anxiety in adulthood in BALB/c mice. Neuropharmacology 81:303–310

Szekely AM, Barbaccia ML, Costa E (1987) Effect of a protracted antidepressant treatment on signal transduction and [3H](−)-baclofen binding at GABAB receptors. J Pharmacol Exp Ther 243:155–159

Tao W, Higgs MH, Spain WJ, Ransom CB (2013) Postsynaptic GABAB receptors enhance extrasynaptic GABAA receptor function in dentate gyrus granule cells. J Neurosci 33:3738–3743

Urwyler S, Mosbacher J, Lingenhoehl K, Heid J, Hofstetter K, Froestl W, Bettler B, Kaupmann K (2001) Positive allosteric modulation of native and recombinant gamma-aminobutyric acid (B) receptors by 2,6-Di-tert-butyl-4-(3-hydroxy-2,2-dimethyl-propyl)-phenol (CGP7930) and its aldehyde analog CGP13501. Mol Pharmacol 60:963–971

Urwyler S, Pozza MF, Lingenhoehl K, Mosbacher J, Lampert C, Froestl W, Koller M, Kaupmann K (2003) N,N'-Dicyclopentyl-2-methylsulfanyl-5-nitro-pyrimidine-4,6-diamine (GS39783) and structurally related compounds: novel allosteric enhancers of gamma-aminobutyric acid B receptor function. J Pharmacol Exp Ther 307:322–330

Varani AP, Balerio GN (2012) GABA(B) receptors involvement in the effects induced by nicotine on anxiety-related behaviour in mice. Pharmacol Res 65:507–513

Vergnes M, Boehrer A, Simler S, Bernasconi R, Marescaux C (1997) Opposite effects of GABAB receptor antagonists on absences and convulsive seizures. Eur J Pharmacol 332:245–255

Veronezi BP, Moffa AH, Carvalho AF, Galhardoni R, Simis M, Bensenor IM, Lotufo PA, Machado-Vieira R, Daskalakis ZJ, Brunoni AR (2016) Evidence for increased motor cortical facilitation and decreased inhibition in atypical depression. Acta Psychiatr Scand 134:172–182

Vinkers CH, Cryan JF, Berend Olivier B, Groenink L (2010) Elucidating GABAA and GABAB receptor functions in anxiety using the stress-induced hyperthermia paradigm: a review. Open Pharmacol J:1–14

Zarrindast M, Rostami P, Sadeghi-Hariri M (2001) GABA(A) but not GABA(B) receptor stimulation induces antianxiety profile in rats. Pharmacol Biochem Behav 69:9–15

$GABA_B$ Receptors in Neurodegeneration

Alessandra P. Princivalle

Contents

Abstract GABA is the main inhibitory neurotransmitter in the mammalian central nervous system (CNS) and acts via metabotropic $GABA_B$ receptors. Neurodegenerative diseases are a major burden and affect an ever increasing number of humans. The actual therapeutic drugs available are partially effective to slow down the progression of the diseases, but there is a clear need to improve pharmacological treatment thus find alternative drug targets and develop newer pharmaco-treatments. This chapter is dedicated to reviewing the latest evidence about $GABA_B$ receptors and their inhibitory mechanisms and pathways involved in the neurodegenerative pathologies.

Keywords Alzheimer's disease · $GABA_B$ receptors · Hippocampal sclerosis · Neurodegenerative diseases · Parkinson's disease · Temporal lobe epilepsy

A. P. Princivalle (✉)
Department of Bioscience and Chemistry, Biomolecular Research Centre, College of Health, Wellbeing and Life Sciences at Sheffield Hallam University, Sheffield, UK
e-mail: a.p.princivalle@shu.ac.uk

Curr Topics Behav Neurosci (2022) 52: 267–290
https://doi.org/10.1007/7854_2021_222
Published Online: 17 March 2021

1 Introduction

The $GABA_B$ receptor is the main inhibitory receptor in the mammalian brain (Curtis 1974; Krnjevic 1974). It was first described by Bowery et al. (1980) who used pharmacology techniques to identify it, but for almost 20 years afterwards no other study succeeded in confirming the presence of this receptor. In 1997 Kaupmann et al. characterised the sequence of the receptor gene, transcript, protein, and the molecular structure of the receptor. They also demonstrated the presence of alternative splice variants of the $GABA_B$ receptor; these were different in the N-terminus domain and named $GABA_{B1a}$ and $GABA_{B1b}$. One year later the same group (Kaupmann et al. 1998) and two others (Jones et al. 1998; White et al. 1998) demonstrated that this receptor was not fully functional in cells enriched with it, and they described a second $GABA_B$ receptor gene, transcript, protein, and the molecular structure of the mature fully functional receptor. These two proteins have therefore been defined as subunits of the fully functional receptor and given the names of $GABA_{B1}$ and $GABA_{B2}$. The same studies also demonstrated for the first time that, in order for a G-protein coupled receptor to be fully functional, it has to work as a dimer composed of these two subunits. In a short period of time other subunits were described (Isomoto et al. 1998) but with very minor roles and very low expression.

After the molecular characterisation of the two main subunits, DNA or RNA probes and antibodies became available or could be produced in order to study the distribution and the level of expression of both the proteins and the transcripts. Thus, many groups began to investigate distribution and expression levels of the $GABA_B$ receptors. Many of these studies were focused on specific areas of the brain and spinal cord in rodents, primates, and in humans. Animal models of various neurodegenerative diseases were used in order to shed light on the structure, expression, and physiological and pathological roles of $GABA_B$ receptors in these conditions.

Electrophysiological and pharmacological evidence demonstrated abnormalities of the $GABA_B$ receptor in many pathological conditions such as spasticity, epilepsy, anxiety, depression, and cognitive deficits (Marescaux et al. 1992; Mott and Lewis 1994; Olpe et al. 1993; Meeren et al. 2004; Stewart et al. 2009; Gassmann and Bettler 2012; Castelli and Gessa 2016). Further, more recently, involvement of the $GABA_B$ receptor has been demonstrated in neurodegenerative diseases such as Alzheimer's (Dal Prà et al. 2019; Tang 2019), amyotrophic lateral sclerosis (Schumacher et al. 2019), Huntington's (Rosas-Arellano et al. 2018; Rekik et al. 2011), Parkinson's (Hillman et al. 2012), essential tremors (Paris-Robidas et al. 2012), and autoimmune encephalitis (Moser et al. 2018; Maureille et al. 2019). In this chapter the attention is focused on the role that $GABA_B$ receptors play in epilepsy, and, more specifically, temporal lobe epilepsy associated with hippocampal sclerosis TLE-HS. Attention is also given to two major neurodegenerative diseases, Alzheimer's (AD) and Parkinson's disease (PD).

2 GABA$_B$ Receptor and Its Effects

It is well known that GABA$_B$ receptors belong to the G-protein coupled (guanine nucleotide binding protein) receptor family (Wojcik and Neff 1984; Hill et al. 1984; Hill 1985; Karbon and Enna 1985; Andrade et al. 1986) and thus are associated with slow synaptic neurotransmission. GABA$_B$ receptors were initially identified by their insensitivity to the GABA$_A$ antagonist bicuculline and their selective activation by (−)baclofen (Hill and Bowery 1981). Later, a number of compounds specific for GABA$_B$ receptors were identified, e.g., the antagonist 2-hydroxy-saclofen and the class of antagonists named CGP. Activation of GABA$_B$ receptors produces three major effects: (a) increases in postsynaptic neuronal K^+ conductance to generate long-lasting inhibitory postsynaptic potentials (Dutar and Nicoll 1988); (b) inhibition of adenylate cyclase activity, leading to a reduction in cAMP levels (Wojcik and Neff 1984; Hill et al. 1984; Hill 1985; Karbon and Enna 1985; Andrade et al. 1986; Rascol et al. 1989); (c) decrease of membrane Ca^{2+} flux GABA$_B$ receptor activation mediated by G-proteins that are members of the pertussis toxin-sensitive family $G_{i\alpha}/G_{o\alpha}$ (Odagaki et al. 2000; Odagaki and Koyama 2001). These actions are discussed separately below. More details about the history and structure of the GABA$_B$ receptors can be found in chapters "Historical Perspective on the GABAB Receptor" and "GABAB Receptor Structure", respectively.

2.1 *K^+ Channels*

When activated by an agonist, GABA$_B$ receptors increase K^+ conductance, producing hyperpolarisation of the cell membrane, which has been reported in various brain regions including the cortex (pyramidal cells; Connors et al. 1982; Karlsson and Olpe 1989; Luhmann and Prince 1991), hippocampus (granule cells and interneurons; Fujita 1979; Misgeld et al. 1984; Dutar and Nicoll 1988; Williams and Lacaille 1992), cerebellum (Schreurs et al. 1992; Vigot and Batini 1997), amygdala (Rainnie et al. 1991), and thalamus (Hirsch and Burnod 1987; Crunelli and Leresche 1991; Curró Dossi et al. 1992).

It has been reported that in K^+ subunit deletion, G protein-activated inwardly rectifying potassium (GIRK) channel 2 (GIRK2) mutant mice, in hippocampal neurons, postsynaptic K^+ currents induced by the GABA$_B$ receptor agonist baclofen are reduced or absent, and it was demonstrated that deletion of GIRK2 did not involve presynaptic inhibition. Therefore, GIRK-containing channels were shown not to be responsible for presynaptic effects (Lüscher et al. 1997). In contrast, a K^+ current is shown to be coupled to GABA$_B$ receptors on presynaptic terminals in hippocampal cultures (Thompson and Gähwiler 1992), so changes in membrane K^+ flux appear to be due to postsynaptic GABA$_B$ receptor activation (Saint et al. 1990).

2.2 Ca^{2+} Channels

Baclofen and GABA depress somatic Ca^{2+} currents not only in peripheral neurons (Dolphin and Scott 1986, 1987, 1990) but also in cultured mammalian hippocampal and cerebellar neurons (Huston et al. 1990; Wojcik et al. 1990; Pfrieger et al. 1994). $GABA_B$ receptor-mediated blockage of Ca^{2+} channels and coupling mechanisms are involved (Scott et al. 1991). A reduction of Ca^{2+} currents can be considered responsible for the depression of synaptic transmission by presynaptic $GABA_B$ receptors (Huston et al. 1990).

More recent evidence showed that presynaptic neurotransmitter release is indeed modulated by $GABA_B$ receptors through Ca^{2+} channels (White et al. 1998). Patch clamp measurements from a presynaptic terminal indicate that baclofen reduced Ca^{2+} currents, but had no effect on presynaptic K^+ currents and this was G-protein dependent (Takahashi et al. 1998).

2.3 Inhibition of Adenylate Cyclase

$GABA_B$ receptor agonists inhibit basal and forskolin-stimulated neuronal adenylate cyclase in brain slices (Knight and Bowery 1996), through a G-protein dependent mechanism that results in a reduced level of intracellular cAMP. When $GABA_B$ receptors are activated, one α subunit is released from the G-protein and interacts with AC to inhibit cAMP formation. The G-protein involved has been demonstrated to be $G_{i\alpha}/G_{o\alpha}$, because ADP-ribosylation of the G-protein by pertussis toxin blocked any receptor interaction (Asano and Ogasawara 1986; Xu and Wojcik 1986). The βγ subunit of the G-protein interacts with K^+ and Ca^{2+} channels and can potentiate β-adrenoreceptor-mediated cAMP production (Knight and Bowery 1996), via cross talk mechanisms (Lefkowitz 1992). More details about the physiology of $GABA_B$ receptors can be found in chapter “GABAB Receptor Signal Transduction”.

3 $GABA_A$ Receptors

GABA acts also through $GABA_A$ receptors. $GABA_A$ receptors are ligand-gated Cl^- ion channels generating fast synaptic inhibition (Schofield et al. 1987; Smith and Olsen 1995). $GABA_A$ receptors can be pharmacologically distinguished by the competitive antagonist bicuculline, they are modulated by many therapeutic agents, such as benzodiazepines (BZD), and are a potential drug target for a number of neurological disorders. $GABA_A$ receptors are widely distributed in the CNS (Fritschy and Mohler 1995). The existence of multiple $GABA_A$ receptor subunits has been demonstrated by regional differences in affinity and distribution of binding sites for BZD receptor ligands (Niddam et al. 1987; Sieghart et al. 1987; Bureau and Olsen 1990, 1993; Ruano et al. 1992).

4 $GABA_B$ in Neurodegenerative Diseases

Neurodegeneration is defined as the progressive atrophy and loss of function of neurons, which is present in neurodegenerative diseases. Neurodegenerative diseases are also characterised by deposition of proteins, due to a fault in post-translational processing, specifically defective proteolysis, leading to overproduction of misfolded proteins. There are several such proteins that undergo incorrect post-translational processing; however, for the purpose of this chapter only the major proteins involved are considered: tau, amyloid-β (Aβ), and α-synuclein.

4.1 $GABA_B$ in Alzheimer's Disease

Alzheimer's disease (AD), as mentioned above, is a neurodegenerative disease described for the first time in 1907 by the Bavarian physician and pathologist Alois Alzheimer (1864–1915). Unfortunately, since then, and over a century later, the prevalence of AD has tremendously increased. It is now the fifth most common cause of death globally. About 44 million people worldwide are living with dementia, 70% due to AD (Dumurgier and Sabia 2020). The main symptoms of AD are: loss of recent memory, disorientation to time and place, sometimes antisocial behaviour –"loss of inhibitions", lack of outward physical signs.

The symptoms observed are due to the specific regions of the brain affected by neurodegeneration, which are the hippocampus and the cortex, where the loss of neurons becomes increasingly evident with the progression of the disease (Fig. 1). The hippocampus is the centre for processing and storing memories, and cortex is the centre for high cognitive function built on memories.

The neuropathological features of AD are extracellular senile plaques made up by amyloid-β protein, and intracellular neurofibrillary tangles (NFT) made up of paired filaments and hyperphosphorylated tau protein (Fig. 2).

The majority of studies on AD are focused on the pathological processing of the amyloid precursor peptide (APP) leading to the formation of amyloid-β, the resulting build-up of amyloid plaques (Fig. 3), and also on the development of the tangles due to hyperphosphorylation of the tau protein (Fig. 4). In addition to this, recent evidence has shown that $GABA_B$ receptors also play a role in the pathology of AD.

The earliest indications for the role of $GABA_B$ receptors in AD emerged from a quantitative autoradiography binding study. In this study, a significant decrease of Bmax for $GABA_B$ receptors was reported in the cortex and hippocampus, especially in the *stratum moleculare* of the dentate gyrus (DG), the *stratum lacunosum-moleculare,* and the *stratum pyramidale* of CA1 (Chu et al. 1987a, b). Following

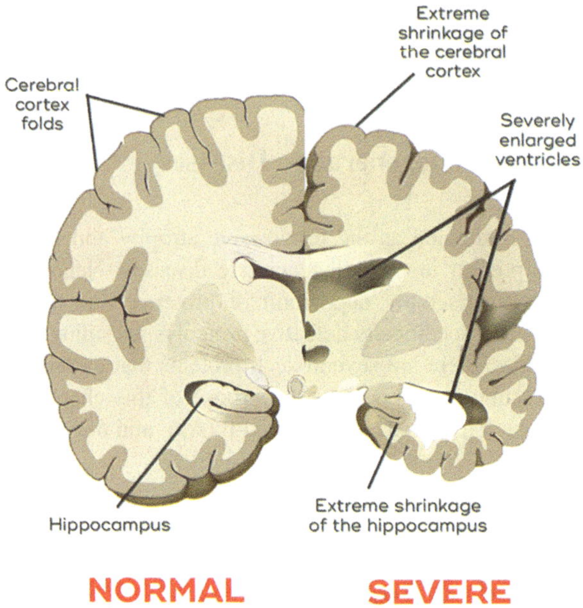

Fig. 1 Brain imaging showing a control brain on the left and a brain affected by Alzheimer's disease on the right. (Credit: Queensland Brain Institute, The University of Queensland. qbi.uq.edu.au/dementia)

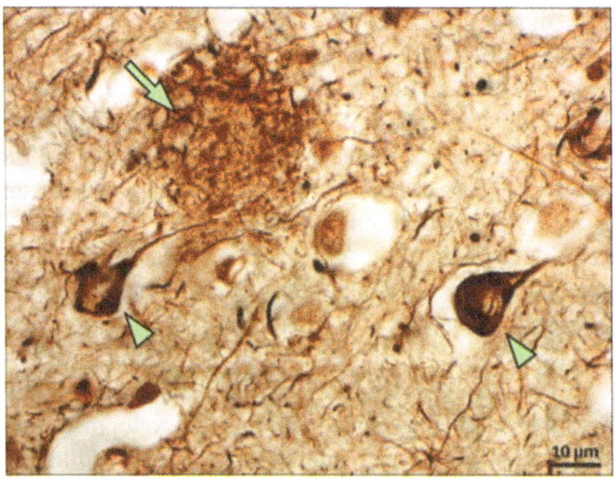

Fig. 2 Neuropathological hallmarks of Alzheimer's disease. Postmortem Bielschowsky silver staining of frontal cortex from a patient with Alzheimer's disease, showing the presence of a neuritic amyloid plaque (arrow), consisting of aggregated extracellular amyloid β fibrils, and intraneuronal neurofibrillary tangles (arrowheads), consisting of hyperphosphorylated tau protein. (Taken from Winblad et al. 2016 with permission from Elsevier journals License Number 4947750098122)

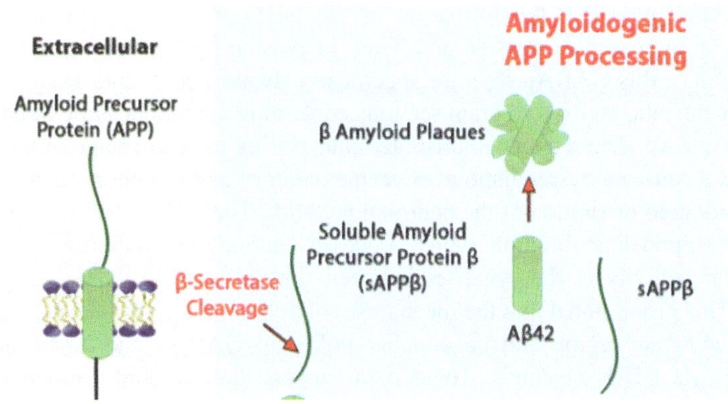

Fig. 3 Schematic representation of the amyloid-plaque formation. Adapted from https://www.biolegend.com/amyloid_precursor_protein

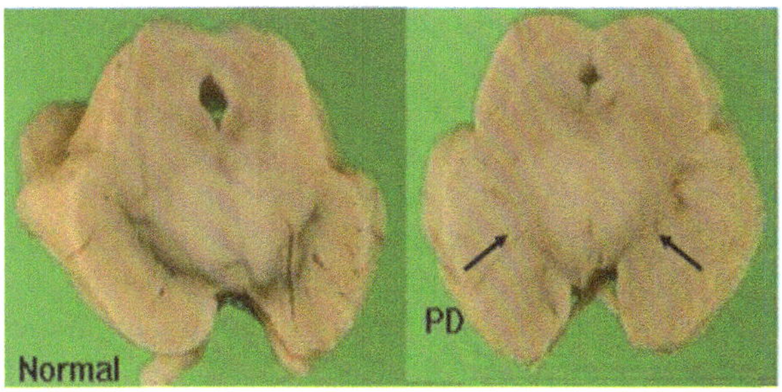

Fig. 4 Brain images showing a normal postmortem sample on the left and the loss of pigmented neurons in the pars compacta of the substantia nigra (SNpc) of a postmortem PD patient on the right (black arrows)

up from these early findings, it was reported that, in a mouse model of colchicine-infused hippocampus, the sensory memory of the mice was impaired and also the amount of GABA in the cortex was decreased. Conversely, in mice simultaneously treated with colchicine and the GABA$_B$ antagonist CGP36742, memory loss was not recorded. The researchers of that study therefore concluded that the GABA$_B$ receptor antagonist CGP36742 could be a treatment for AD (Yu et al. 1997). More experimental evidence obtained by immunohistochemistry emerged, corroborating the differential expression of GABA$_B$ receptors in varying stages of AD according to the Braak Staging. These data suggested that the expression of GABA$_{B1}$ is stable in CA1 through all the stages of the disease. In contrast, in the initial stages (Braak III/IV) of the pathology, the expression of GABA$_{B1}$ expression is higher in CA2-4, which could be interpreted as a compensatory (or self-defending) mechanism where the expression decreases with the progression of the disease (Braak V/VI), leading to neuronal death and impairment between excitation and inhibition. From these data it can be concluded that the formation of the NFTs in the hippocampus initially induces an increased expression and, later, increasing NFT accumulation stops the expression of this $GABA_B$ receptor specifically (Iwakiri et al. 2005).

In recent data, rat ex vivo brain sections containing the hippocampus were treated with excess Aβ. These data demonstrated that, during the early stage of the disease, amyloid-β causes a dysregulation between excitatory and inhibitory neurotransmission, leading to disruption of the neuronal network. These changes are significant in the septo-hippocampal region, which processes learning and memory, according to oscillatory activity at the synapses between fimbria and CA3 (Nava-Mesa et al. 2013). This group noted that the mechanism of action of amyloid-β was localised at the postsynaptic region and presumably linked to $GABA_B$ and its K^+ and Ca^{2+} channels via GIRK channels. These data suggest that amyloid-β modifies GIRK channels in CA3 pyramidal neurons in a way that is linked to the functioning of $GABA_B$ in the modulation of the hippocampal circuit. Another study on the effect of amyloid-beta (Aβ) on gene expression demonstrated that the level of expression of GIRK2, 3, and 4 subunits was decreased, but $GABA_B$ receptor expression was unaffected. These data corroborate the previous observations showing a relationship between the effect of Aβ and K^+ channels linked to $GABA_B$ receptors (Mayordomo-Cava et al. 2015). Another study showed that in a rat streptozotocin-induced diabetic (STZ) model of sporadic AD, baclofen enhanced memory, again showing a role for $GABA_B$ receptors in AD (Pilipenko et al. 2018). One of the latest pieces of evidence that $GABA_B$ receptors play a role in AD is the link between $GABA_B$/APP and the formation of Aβ, emerging from a study on sequence-related epitopes in APP with nanomolar affinity for the sushi-domain on the N-terminal site of presynaptic $GABA_{B1a}$ receptors. This study demonstrates, by using a proteomics approach, a multiprotein complex containing APP, c-Jun N-terminal kinase-interacting protein (JIP) and calsyntenin, together with $GABA_{B1a}$. This multiprotein complex facilitates Aβ formation and blocks the axonal trafficking of presynaptic of the $GABA_B$ receptor, decreasing its expression (Dinamarca et al. 2019). In a genetic mouse model of AD expressing a chimeric mouse/human (Mo/Hu) APP-695 with mutations linked to familial AD (Oh et al. 2009), the use of various immunohistochemical techniques demonstrated a decreased expression of $GABA_{B1}$ in the cell membrane surface of the *stratum lacunosum-moleculare* of CA1 pyramidal cells at 6 months of age. This reduced expression became more pronounced at 12 months of age and was coupled with an increase of the subunit in the intracellular compartment. Further, a reduction of $GABA_B$ receptors was observed in the axon terminal synapsing pyramidal CA1 cells (Martín-Belmonte et al. 2020a). The same group demonstrated a significant decrease of $GABA_B$ receptors in the *stratum moleculare* of the DG, and also in axon terminals synapsing dendritic spines of granule cells, more evident in the outer than in the inner molecular layer (Martín-Belmonte et al. 2020b).

All these data taken together, starting from the earliest indication (Chu et al. 1987a, b) up to the most recent data (Martín-Belmonte et al. 2020a, b), indicate that $GABA_B$ receptors, and particularly $GABA_{B1a}$, have a decreased expression in the hippocampus. The reported reductions in $GABA_B$ expression are specific to the hippocampal subregions; however, it seems a general trend extended to CA1, CA3, and DG (which make up the trisynaptic circuit). Functionally, due to the decrease of GABA$_B$ receptors, there is an augmented production of Aβ, because the lack of GABA$_B$ receptors promotes the proteolysis of APP. This further supports the conclusions that GABA$_B$-mediated synaptic transmission is a major contributor in AD and GABA$_B$ receptors may be a suitable target for more effective drugs to treat AD. All these studies have focused their attention on GABA$_{B1}$, firstly because its expression was higher in the neuronal bodies and proximal dendrites where NFT accumulates in AD (Iwakiri et al. 2005). Secondly, GABA$_{B1a}$ has been demonstrated to form a complex with APP, whereas GABA$_{B1b}$ does not. Furthermore, the GABA$_{B1a}$ knock-out mice model showed a lack "of GABA$_B$ axonal transport and deficit in GBR-mediated inhibition of glutamate release". This model also showed that secreted APP functions as a GABA$_{B1a}$ ligand to modulate synaptic neurotransmission (Dinamarca et al. 2019; Martín-Belmonte et al. 2020a, b).

4.2 *GABA$_B$ in Parkinson's Disease*

Parkinson's disease (PD) is another neurodegenerative disease, described for the first time by James Parkinson (1817) in "An Essay on the Shaking Palsy". Toodayan (2018) PD is also known as *paralysis agitans*; it was first called Parkinson's disease by Jean-Martin Charcot in 1884. PD affects about 0.1–0.2% of the whole population. The incidence of the disease increases with age affecting 1% of people over 60 years of age. The main symptoms of PD are tremor at rest, muscle rigidity, and bradykinesia. The symptoms observed are due to the specific region of the brain affected by the loss of dopaminergic neurons: the substantia nigra (SN) (Fig. 4).

The main neuropathological features are Lewy bodies, which contain α-synuclein (Fig. 5) in the SN and this is exhibited through impairment of voluntary movement (Braak et al. 2003).

When the disease progresses these features spread to the cortex and neocortex (Tysnes and Storstein 2017). Figure 6 below illustrates the whole circuit and the inhibitory and excitatory connection.

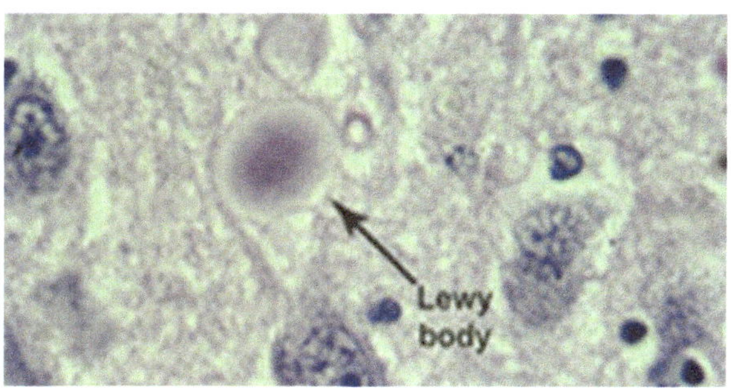

Fig. 5 Photomicrographs showing the presence of Lewy bodies containing α-synuclein. Taken from https://www.alz.org/alzheimers-dementia/what-is-dementia/types-of-dementia/lewy-body-dementia

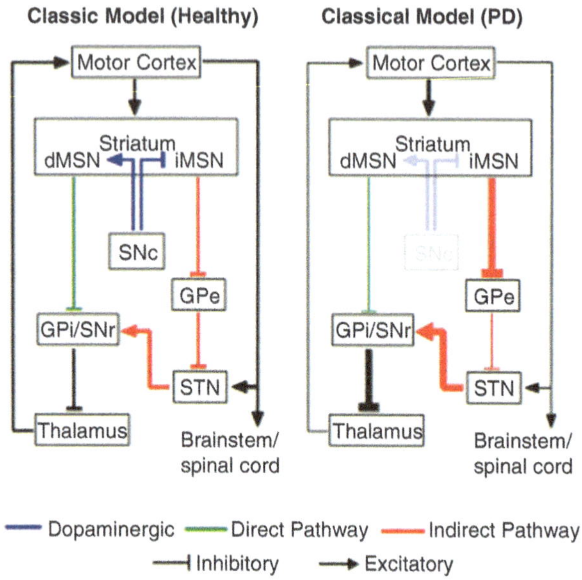

Fig. 6 Schematic representation of the nigrostriatal circuit in the physiology and pathology of Parkinson's disease. Adapted from (McGregor and Nelson 2019; with permission)

When the dopaminergic neurons in the substantia nigra pars compacta (SNpc) start to degenerate, the dopaminergic signal to the caudate and putamen is reduced. The response of the caudate and putamen therefore becomes modified, which results in an overall increase in the output of the interior globus pallidus (GPi). This increases results in the inhibition of the thalamus. The thalamic excitatory signal to the motor cortex is diminished, thus causing reduced motor control. Also, the subthalamic nucleus (STN) plays a critical role in the regulation of movement, and abnormal activity of its neurons is associated with basal ganglia motor symptoms (McGregor and Nelson 2019).

The first evidence of involvement of $GABA_B$ receptors in PD emerged from electrophysiological recordings in neurons isolated from the globus pallidus (GP) in the presence of baclofen. The data showed that the $GABA_B$-mediated effect was present only in one of the subtypes of GP neurons with a small soma, and the activation of $GABA_B$ modulated high-voltage-activated (HVA) calcium currents which may have an impact on the basal ganglia circuit (Stefani et al. 1999). A 40% decrease in the expression of $GABA_B$ receptors in the SNpc and in the GPi was reported in a binding study utilising a 1-methyl-4-phenyl-1,2,3,6-tetrahydropyridine (MPTP)-induced PD monkey model (Calon et al. 2000). The same group also demonstrated significant decreases in the mRNAs for $GABA_{B1}$ (−69%) and $GABA_{B2}$ (−66%) in the SNpc, and that the decreased expression of $GABA_{B1}$ mRNA was related to dopamine (DA) concentration (Calon et al. 2001). The same group also analysed, via binding experiments, the expression of $GABA_B$ receptors in human postmortem specimens and found a reduced binding in the putamen and external globus pallidus (GPe) in PD patients compared with controls (Calon et al. 2003). Also, in a rat model of PD with induced lesions of the nigrostriatal pathway, a reduction in $GABA_B$ mRNA was reported in the SNpc, whereas expression of the $GABA_{B1a}$ subunit was significantly increased in the substantia nigra pars reticulate (SNpr), entopeduncular nucleus, and the STN. Since these brain parts received reduced GABAergic innervation due to the lesion, this could indicate that the increased $GABA_{B1a}$ expression represents a compensatory mechanism (Johnston and Duty 2003). The fine-tuned localization of $GABA_{B1}$ receptors was investigated by Smith et al. (2000), who found an immunopositive signal in the striatopallidal complex in neuronal bodies and dendrites, striatal dendritic spine, axons, and axon terminals. Analysis of $GABA_{B1}$ receptor distribution via immunogold electron microscopy showed extrasynaptic sites on dendrites, spines, and somata in the striatopallidal complex and perisynaptically at the synapses in the GP.

Whole-cell patch-clamp recordings were used to investigate tonic activation of $GABA_B$ receptors at pre- and post-synaptic levels, and the data indicated a major tonic activation of presynaptic $GABA_B$ receptors on the STN terminals compared to postsynaptic $GABA_B$ receptors on STN neurons. Therefore the presynaptic $GABA_B$ receptors could be considered as a new therapeutic target for treating some of the PD symptoms (Chen and Yung 2005).

It was later demonstrated through the frequency-dependent activation of post-synaptic $GABA_B$ receptors that the GP regulates the activity of the STN. These results clarify a novel way in which burst activity can be generated in the STN and suggest that the effect of $GABA_B$ on STN neurons could generate abnormal burst activity in PD (Hallworth and Bevan 2005).

Further proof of the role that the $GABA_B$ receptor plays has come from studies in a rat model whereby the nigrostriatal pathway was depleted by treatment with 6-hydroxydopamine, and the rats were treated with the $GABA_B$ receptor antagonist CGP 56999A. The results showed that the antagonist treatment attenuated the lack of DA in the rat striatum (Enna et al. 2006).

A recently conducted investigation in an MPTP rat model demonstrated that baclofen reversed the effect of PD-like induced symptoms (Tyagi et al. 2015). Another recent study in a mouse model of PD proved that the loss of GABAergic inhibition in the striatonigral connection led to motor impairment (Borgkvist et al. 2015), corroborating once more the role of $GABA_B$ receptors in PD and moreover how it can be used as a potential drug target to treat certain parkinsonian symptoms.

It has long been established that DA plays a pivotal role in action selection and learning in the nigrostriatal pathway. However, any link between DA and $GABA_B$ receptors was not clearly defined until recently. The DA released into the striatum is influenced by local neurons, the majority of which are GABAergic, though it was not clear if it was a direct or indirect modulation via cholinergic innervation. Lopes et al. (2019) established that in the striatum GABA is capable of inhibiting release of DA via both ionotropic and metabotropic GABA receptors and that these actions are not mediated by acetylcholine. These results also demonstrated a tonic inhibition of DA release by striatal GABA, which occurs mainly via $GABA_B$ receptors. However, there is still a lack of evidence of whether GABA receptors are expressed on DA axons (Lopes et al. 2019).

Previously, the main neuropathological features of PD were mentioned: Lewy bodies containing α-synuclein and their accumulation in the intracellular space are major factors in the disease. Emmanouilidou et al. (2016) have examined the molecular pathway of α-synuclein secretion in mouse nucleus striatum and have found a new synaptic network that regulates α-synuclein release. They showed that α-synuclein secretion is a calcium-regulated mechanism depending on the activation of the sulfonylurea receptor 1 (SUR1), which is an inwardly rectifying potassium ion channel Kir6 subunit that senses intracellular levels of the nucleotide ATP. They also demonstrated that modulation of GABA release through SUR1 located on GABAergic neurons controls α-synuclein release through activation of the presynaptic $GABA_B$ receptors. This study suggests that GABA transmission via SUR1 in mouse striatum modulates the α-synuclein secretory pathway, providing new insights for potential therapeutics to treat PD (Emmanouilidou et al. 2016). Also, in a transgenic drosophila model carrying α-synuclein, it was shown that the transgenic flies lacked the capability of climbing, and this action was reversed by providing the drosophila with levodopa (L-DOPA_ or a $GABA_B$ (but not $GABA_A$) agonist in their food (Hillman et al. 2012).

There have been useful studies directed towards clarification of various mechanisms underlying the pathophysiology of PD, of which those involving the role of $GABA_B$ receptors have been summarised above. Taken together, all the evidence available to date not only shows a fundamental role of $GABA_B$ receptors in PD, but also that via more recently described $GABA_B$ receptor innervation and modulation pathways there could be further potentials for better targeted therapies which may treat PD symptoms in a more effective manner.

4.3 $GABA_B$ in Temporal Lobe Epilepsy

Different types of epilepsy are classified according to structural aetiology referring to abnormalities visible on structural neuroimaging such as magnetic resonance imaging (MRI). The structural malformations may be acquired or genetic. The majority of focal seizures originate in the temporal lobes (Zentner et al. 1995).

A well-known form of epilepsy linked with structural malformation is temporal lobe epilepsy associated with hippocampal sclerosis (TLE-HS). Temporal lobe epilepsy (TLE) is the most common form of focal epilepsy. About 6 out of 10 people with focal epilepsy have TLE. Seizures in TLE start in one point (*focus*) and then may involve both temporal lobes in the brain. TLE is subdivided in two types: mesial temporal lobe epilepsy (MTLE) and neocortical temporal lobe epilepsy (NTLE). MTLE encompasses the *medial* or internal structures of the temporal lobe. Seizures often begin in the hippocampus or surrounding area and account for almost 80% of all temporal lobe seizures. NTLE encompasses the outer part of the temporal lobe. About 30–40% of patients affected by MTLE-HS are pharmaco-resistant (Engel 2001). When seizures are prolonged and repeated they produce severe neuronal loss in the temporal lobe, mostly observed in the hippocampus, entorhinal cortex, amygdala, and other brain areas (Van Paesschen et al. 1997; Sutula and Hermann 1999).

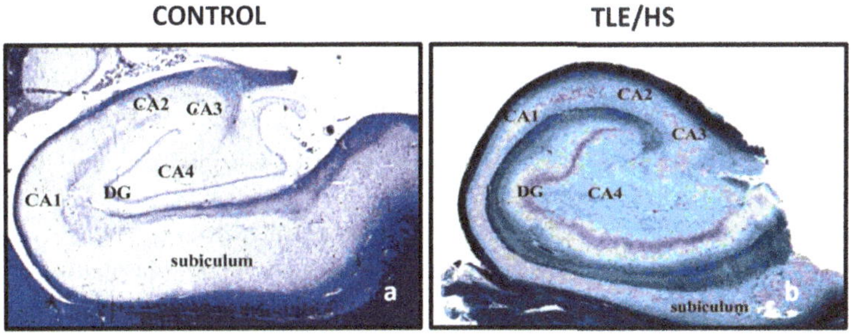

Fig. 7 Cresyl violet/Luxol fast blue stained sections of human hippocampus: (**a**) control specimen; (**b**) sclerotic specimen (not in scale). Taken from Principalle PhD thesis (2003)

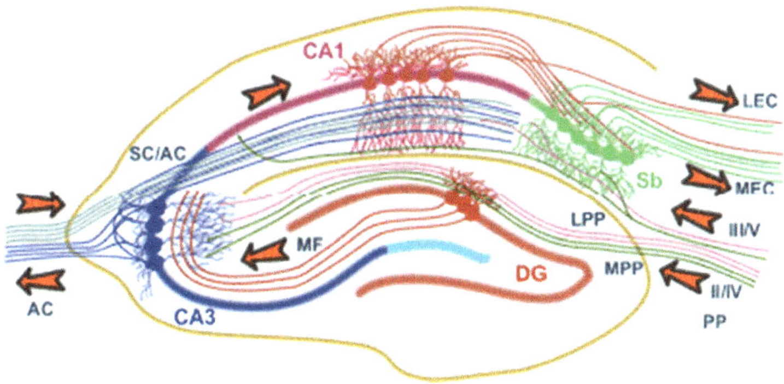

Fig. 8 The hippocampal Network: The hippocampus forms a principally uni-directional network, with input from the entorhinal cortex (EC) that forms connections with the dentate gyrus (DG) and cornu ammonis CA3 pyramidal neurons via the perforant path (PP – split into lateral and medial). CA3 neurons also receive input from the DG via the mossy fibres (MF). They send axons to CA1 pyramidal cells via the Schaffer collateral pathway (SC), as well as to CA1 cells in the contralateral hippocampus via the associational commisural (AC) pathway. CA1 neurons also receive inputs directly from the PP and send axons to the Subiculum (Sb). These neurons in turn send the main hippocampal output back to the EC, forming a loop. Taken from http://www.bristol.ac.uk/synaptic/pathways/

TLE-HS is not always considered or classified among the classical neurodegenerative diseases such as AD or PD. However, neurodegeneration in cornu ammonis (CA) subregions (Fig. 7), aberrant mossy fibre (MF) sprouting (Sutula et al. 1989), granule cell dispersion (Houser 1990), and astrogliosis (Seifert et al. 2010) have been reported in the hippocampus in individuals with TLE-HS. Since the temporal lobe is a major cortical structure involved in learning and memory (Halgren et al. 1991), recurrent spontaneous seizures (which are the primary triggering cause of TLE-HS) result in damage to this structure and therefore memory is impaired (Helmstaedter 2002).

Neurodegeneration associated with TLE-HS has been observed in the human hippocampi (Fig. 7) and subsequently reproduced in rodent models. It is characterised by affecting the so-called trisynaptic circuit (Fig. 8), specifically

CA1 and CA3 subregions of the hippocampus, but not the CA2, DG, or subiculum; the neurodegeneration reported is also associated with MF sprouting (Gloor 1991; Sloviter 1994).

This neurodegeneration is accompanied by loss in GABAergic cells and altered expression of inhibitory receptor subunits in the DG and other parts of the hippocampal formation (Sperk et al. 2004).

Inhibition mediated by GABA has been demonstrated to be reduced in neurons surviving hippocampal sclerosis (HS) associated with TLE (Mangan et al. 1995; Mangan and Lothman 1996). Evoked inhibitory post-synaptic potentials (IPSPs) in neurons from TLE-HS samples have been shown to be reduced compared with samples from patients with different structural lesions (Isokawa et al. 1991; Knowles et al. 1992).

In order to account for this finding, a number of research groups proposed different hypotheses for the decreased synthesis of GABA: (1) impairment in the GABA transporters (GAT) or the glutamate decarboxylase (GAD) enzyme, (2) reduced binding of GABA to the two receptor subtypes, $GABA_A$ and $GABA_B$, (3) reduced production of the receptors at transcriptional or translational level, or (4) post-translational modifications in the hippocampal area. More recent evidence emerged showing impairment in GABA (Thomas et al. 2003, 2005) and GABA transporters (Mathern et al. 1999; Schijns et al. 2015). Also $GABA_A$ receptor subunits were demonstrated to be differentially expressed in both in human TLE-HS hippocampal specimens (Loup et al. 2006) and in animal models (Pirker et al. 2003; Mazzuferi et al. 2010), and in the amygdala and the entorhinal cortex of human patients (Stefanits et al. 2019).

Most interesting for the purpose of this chapter is that anomalies in the expression of $GABA_B$ receptors have been reported both in human TLE-HS and animal models of it.

$GABA_B$ presynaptic receptor function has been demonstrated to be reduced in the DG granule cells of both kindled and kainate rat models of epilepsy (Buhl et al. 1996; Haas et al. 1996). Similar reduction was reported also in CA1 of partially (hippocampus)- or fully (amYgdala)-kindled rats (Asprodini et al. 1992; Wu and Leung 1997); none of these studies reported malfunctions in the $GABA_B$ receptor-mediated post-synaptic potentials. However, Mangan and Lothman (1996) observed a reduction in both pre- and post-synaptic $GABA_B$ receptor function in CA1 neurons in a rat hippocampal-kindling model. The $GABA_{B1}$ and $GABA_{B2}$ transcript expression patterns have been reported in great detail in the rat (Kaupmann et al. 1997, 1998; Muñoz et al. 1998; Bischoff et al. 1999; Benke et al. 1999; Liang et al. 2000; Jones et al. 1998; Kuner et al. 1999; Ng et al. 1999; Clark et al. 2000) and in human hippocampus (Berthele et al. 2001; Princivalle et al. 2003). The binding parameters of agonists and antagonists for $GABA_B$ receptors have also been reported extensively (Billinton et al. 2001; Princivalle et al. 2002; Furtinger et al. 2003). $GABA_B$ receptors have been demonstrated to be differentially expressed in the hippocampus of TLE-HS in rat and mouse models, as well as in human specimens (Princivalle et al. 2001, 2003; Nishimura et al. 2005;Teichgräber et al. 2009; Rocha et al. 2015; Sheilabi et al. 2018).

Electrophysiological evidence indicated that $GABA_B$ receptor expression may be an important factor for the onset of ictogenesis in the rat limbic system and, perhaps, in MTLE patients (Avoli et al. 2004). Lang et al. (2014) demonstrated that $GABA_B$ receptors regulate hippocampal hyperexcitability by inhibiting CA3 glutamatergic synapses. They postulate that positive allosteric modulation of $GABA_B$ receptors may be effective in reducing seizure-related hyperexcitability. All these data demonstrate the loss of $GABA_B$ receptor function in TLE in rodents and humans.

The main common feature emerging from all these pieces of research evidence is that the physiological role of GABA and $GABA_B$ receptors is to induce hyperpolarization. Later on, however, it came up to light that GABA not only has an inhibitory action but could also have a depolarizing action, suggesting that GABA transmission is also involved in promoting epilepsy (Köhling et al. 1998; Cohen et al. 2002). In fact, Kantrowitz and colleagues (2005) demonstrated, by using electrophysiological techniques, that $GABA_B$ receptors regulate the synaptic depolarization to GABA response, and also that blocking of $GABA_B$ receptors with the specific antagonist CGP 55845A caused the depolarizing GABA response to become excitatory and pro-convulsive. Additionally, in very recent years it has been demonstrated in a mouse model of TLE that inhibition of presynaptic $GABA_B$ receptors has a depolarising action on cholecystokinin-positive basket cells [CCK(+) BCs], in the hippocampus, specifically in CA3 (Dugladze et al. 2013).

All this body of evidence highlights the pivotal role that the $GABA_B$ receptor plays in TLE-HS, and the latest data particularly corroborate the importance that the reduced expression of $GABA_B$ receptors has in the pathophysiology of TLE. In the future, studies are needed to design, develop, and test innovative drugs which can target $GABA_B$ receptors, specifically in the trisynaptic circuit.

5 Conclusions

It has long been recognised that GABA is the main inhibitory neurotransmitter in the mammalian brain and that it acts via the $GABA_A$ and $GABA_B$ receptors. This chapter has focused on the review of the role and mechanisms of action of $GABA_B$ receptors in three neurological diseases, which appear similar in some aspects and dissimilar in others. They are similar because they all show neurodegeneration; they are dissimilar because the cerebral circuits involved in their pathophysiology are different in PD versus AD and TLE/HS, and because the main neuropathological features are different.

Altogether, the $GABA_B$ receptor plays a pivotal role in the inhibitory pathway in order to control the balance between excitatory/inhibitory signals in the trisynaptic circuit of the hippocampus, which has been described and demonstrated to have neuronal loss, in both AD and TLE/HS. On the other hand, in PD, $GABA_B$ has been shown to modulate excitatory/inhibitory signals via more newly described pathways different from the trisynaptic circuit.

References

Andrade R, Malenka RC, Nicoll RA (1986) A G protein couples serotonin and GABAB receptors to the same channels in hippocampus. Science 234(4781):1261–1265. https://doi.org/10.1126/science.2430334

Asano T, Ogasawara N (1986) Uncoupling of gamma-aminobutyric acid B receptors from GTP-binding proteins by N-ethylmaleimide: effect of N-ethylmaleimide on purified GTP-binding proteins. Mol Pharmacol 29(3):244–249

Asprodini EK, Rainnie DG, Shinnick-Gallagher P (1992) Epileptogenesis reduces the sensitivity of presynaptic gamma-aminobutyric acidB receptors on glutamatergic afferents in the amygdala. J Pharmacol Exp Ther 262(3):1011–1021

Avoli M, Benini R, de Guzman P, Omar A (2004) GABA(B) receptor activation and limbic network ictogenesis. Neuropharm 46(1):43–51. https://doi.org/10.1016/s0028-3908(03)00307-1

Benke D, Honer M, Michel C, Bettler B, Mohler H (1999) γ-Aminobutiric acid type B receptor splice variant proteins GBR1a and GBR1b are both associated with GBR2 in situ and display differential regional and subcellular distribution. J Biol Chem 274:27323–27330. https://doi.org/10.1074/jbc.274.38.27323

Berthele A, Platzer S, Weis S, Conrad B, Tölle TR (2001) Expression of GABA(B1) and GABA (B2) mRNA in the human brain. Neuroreport 12(15):3269–3275. https://doi.org/10.1097/00001756-200110290-00025

Billinton A, Baird VH, Thom M, Duncan JS, Upton N, Bowery NG (2001) GABA(B) receptor autoradiography in hippocampal sclerosis associated with human temporal lobe epilepsy. Br J Pharmacol 132(2):475–480. https://doi.org/10.1038/sj.bjp.0703854

Bischoff S, Leonhard S, Reymann N, Schuler V, Shigemoto R, Kaupmann K, Bettler B (1999) Spatial distribution of GABA(B)R1 receptor mRNA and binding sites in the rat brain. J Comp Neurol 412(1):1–16

Borgkvist A, Avegno EM, Wong MY, Kheirbek MA, Sonders MS, Hen R, Sulzer D (2015) Loss of striatonigral GABAergic presynaptic inhibition enables motor sensitization in parkinsonian mice. Neuron 87(5):976–988. https://doi.org/10.1016/j.neuron.2015.08.022

Bowery NG, Hill DR, Hudson AL, Doble A, Middlemiss DN, Shaw J, Turnbull M (1980) (−) Baclofen decreases neurotransmitter release in the mammalian CNS by an action at a novel GABA receptor. Nature 283(5742):92–94. https://doi.org/10.1038/283092a0

Braak H, Del Tredici K, Rüb U, de Vos RA, Jansen Steur EN, Braak E (2003) Staging of brain pathology related to sporadic Parkinson's disease. Neurobiol Aging 24(2):197–211. https://doi.org/10.1016/s0197-4580(02)00065-9

Buhl EH, Otis TS, Mody I (1996) Zinc-induced collapse of augmented inhibition by GABA in a temporal lobe epilepsy model. Science 271(5247):369–373. https://doi.org/10.1126/science.271.5247.369

Bureau M, Olsen RW (1990) Multiple distinct subunits of the gamma-aminobutyric acid-A receptor protein show different ligand-binding affinities. Mol Pharmacol 37(4):497–502

Bureau MH, Olsen RW (1993) GABAA receptor subtypes: ligand binding heterogeneity demonstrated by photoaffinity labeling and autoradiography. J Neurochem 61(4):1479–1491. https://doi.org/10.1111/j.1471-4159.1993.tb13643.x

Calon F, Morissette M, Goulet M, Grondin R, Blanchet PJ, Bédard PJ, Di Paolo T (2000) 125I-CGP 64213 binding to GABA(B) receptors in the brain of monkeys: effect of MPTP and dopaminomimetic treatments. Exp Neurol 163(1):191–199. https://doi.org/10.1006/exnr.2000.7366

Calon F, Lavertu N, Lemieux AM, Morissette M, Goulet M, Grondin R, Blanchet PJ, Bédard PJ, Di Paolo T (2001) Effect of MPTP-induced denervation on basal ganglia GABA(B) receptors: correlation with dopamine concentrations and dopamine transporter. Synapse 40(3):225–234. https://doi.org/10.1002/syn.1045

Calon F, Morissette M, Rajput AH, Hornykiewicz O, Bédard PJ, Di Paolo T (2003) Changes of GABA receptors and dopamine turnover in the postmortem brains of parkinsonians with

levodopa-induced motor complications. Mov Disord 18(3):241–253. https://doi.org/10.1002/mds.10343
Castelli MP, Gessa GL (2016) Distribution and localization of the GABA B receptor. In: GABAB receptor, 1st edn. Springer, Cham. ISBN: 978-3-319-46044-4
Chen L, Yung WH (2005) Tonic activation of presynaptic GABA(B) receptors on rat pallidosubthalamic terminals. Acta Pharmacol Sin 26(1):10–16. https://doi.org/10.1111/j.1745-7254.2005.00012.x
Chu DC, Penney JB Jr, Young AB (1987a) Cortical GABAB and GABAA receptors in Alzheimer's disease: a quantitative autoradiographic study. Neurology 37(9):1454–1459. https://doi.org/10.1212/wnl.37.9.1454
Chu DC, Penney JB Jr, Young AB (1987b) Quantitative autoradiography of hippocampal GABAB and GABAA receptor changes in Alzheimer's disease. Neurosci Lett 82(3):246–252. https://doi.org/10.1016/0304-3940(87)90264-3
Clark JA, Mezey E, Lam AS, Bonner TI (2000) Distribution of the GABAB receptor subunit gb2 in rat CNS. Brain Res 860:41–52. https://doi.org/10.1016/s0006-8993(00)01958-2
Cohen I, Navarro V, Clemenceau S, Baulac M, Miles R (2002) On the origin of interictal activity in human temporal lobe epilepsy in vitro. Science 298(5597):1418–1421. https://doi.org/10.1126/science.1076510
Connors BW, Gutnick MJ, Prince DA (1982) Electrophysiological properties of neocortical neurons in vitro. J Neurophysiol 48(6):1302–1320. https://doi.org/10.1152/jn.1982.48.6.1302
Crunelli V, Leresche N (1991) A role for GABAB receptors in excitation and inhibition of thalamocortical cells. Trends Neurosci 14(1):16–21. https://doi.org/10.1016/0166-2236(91)90178-w
Curró Dossi R, Paré D, Steriade M (1992) Various types of inhibitory postsynaptic potentials in anterior thalamic cells are differentially altered by stimulation of laterodorsal tegmental cholinergic nucleus. Neuroscience 47(2):279–289. https://doi.org/10.1016/0306-4522(92)90244-v
Curtis DR (1974) Amino acid neurotransmitters and the brain. Beattie Smith lecture, March 28, 1974. Med J Aust 2(20):723–731
Dal Prà I, Armato U, Chiarini A (2019) Family C G-protein-coupled receptors in Alzheimer's disease and therapeutic implications. Front Pharmacol 10:1282. https://doi.org/10.3389/fphar.2019.01282
Dinamarca MC, Raveh A, Schneider A, Fritzius T, Früh S, Rem PD, Stawarski M, Lalanne T, Turecek R, Choo M, Besseyrias V, Bildl W, Bentrop D, Staufenbiel M, Gassmann M, Fakler B, Schwenk J, Bettler B (2019) Complex formation of APP with GABAB receptors links axonal trafficking to amyloidogenic processing. Nat Commun 10(1):1331. https://doi.org/10.1038/s41467-019-09164-3
Dolphin AC, Scott RH (1986) Inhibition of calcium currents in cultured rat dorsal root ganglion neurones by (−)-baclofen. Br J Pharmacol 88(1):213–220. https://doi.org/10.1111/j.1476-5381.1986.tb09489.x
Dolphin AC, Scott RH (1987) Calcium channel currents and their inhibition by (−)-baclofen in rat sensory neurones: modulation by guanine nucleotides. J Physiol 386:1–17. https://doi.org/10.1113/jphysiol.1987.sp016518
Dolphin AC, Scott RH (1990) Activation of calcium channel currents in rat sensory neurons by large depolarizations: effect of guanine nucleotides and (−)-baclofen. Eur J Neurosci 2(1):104–108. https://doi.org/10.1111/j.1460-9568.1990.tb00386.x
Dugladze T, Maziashvili N, Börgers C, Gurgenidze S, Häussler U, Winkelmann A, Haas CA, Meier JC, Vida I, Kopell NJ, Gloveli T (2013) GABA(B) autoreceptor-mediated cell type-specific reduction of inhibition in epileptic mice. Proc Natl Acad Sci U S A 110(37):15073–15078. https://doi.org/10.1073/pnas.1313505110
Dumurgier J, Sabia S (2020) Nouvelles tendances épidémiologiques de la maladie d'Alzheimer [Epidemiology of Alzheimer's disease: latest trends]. Rev Prat 70(2):149–151. French
Dutar P, Nicoll RA (1988) A physiological role for GABAB receptors in the central nervous system. Nature 332(6160):156–158. https://doi.org/10.1038/332156a0
Emmanouilidou E, Minakaki G, Keramioti MV, Xylaki M, Balafas E, Chrysanthou-Piterou M, Kloukina I, Vekrellis K (2016) GABA transmission via ATP-dependent K+ channels regulates

α-synuclein secretion in mouse striatum. Brain J Neurol 139(Pt 3):871–890. https://doi.org/10.1093/brain/awv403
Engel J Jr (2001) Finally, a randomized, controlled trial of epilepsy surgery. N Engl J Med 345 (5):365–367. https://doi.org/10.1056/NEJM200108023450510
Enna SJ, Reisman SA, Stanford JA (2006) CGP 56999A, a GABA(B) receptor antagonist, enhances expression of brain-derived neurotrophic factor and attenuates dopamine depletion in the rat corpus striatum following a 6-hydroxydopamine lesion of the nigrostriatal pathway. Neurosci Lett 406(1–2):102–106. https://doi.org/10.1016/j.neulet.2006.07.004
Fritschy JM, Mohler H (1995) GABAA-receptor heterogeneity in the adult rat brain: differential regional and cellular distribution of seven major subunits. J Comp Neurol 359(1):154–194. https://doi.org/10.1002/cne.903590111
Fujita Y (1979) Evidence for the existence of inhibitory postsynaptic potentials in dendrites and their functional significance in hippocampal pyramidal cells of adult rabbits. Brain Res 175 (1):59–69. https://doi.org/10.1016/0006-8993(79)90514
Furtinger S, Pirker S, Czech T, Baumgartner C, Sperk G (2003) Increased expression of gamma-aminobutyric acid type B receptors in the hippocampus of patients with temporal lobe epilepsy. Neurosci Lett 352(2):141–145. https://doi.org/10.1016/j.neulet.2003.08.046
Gassmann M, Bettler B (2012) Regulation of neuronal GABA(B) receptor functions by subunit composition. Nature reviews. Neuroscience 13(6):380–394. https://doi.org/10.1038/nrn3249
Gloor P (1991) Mesial temporal sclerosis: historical background and an overview from a modern perspective. In: Lüders H (ed) Epilepsy surgery. Raven Press, New York, pp 689–703
Haas CA, Deller T, Naumann T, Frotscher M (1996) Selective expression of the immediate early gene c-jun in axotomized rat medial septal neurons is not related to neuronal degeneration. J Neurosci 16(5):1894–1903. https://doi.org/10.1523/JNEUROSCI.16-05-01894.1996
Halgren E, Stapleton J, Domalski P, Swartz BE, Delgado-Escueta AV, Walsh GO, Mandelkern M, Blahd W, Ropchan J (1991) Memory dysfunction in epilepsy patients as a derangement of normal physiology. Adv Neurol 55:385–410
Hallworth NE, Bevan MD (2005) Globus pallidus neurons dynamically regulate the activity pattern of subthalamic nucleus neurons through the frequency-dependent activation of postsynaptic GABAA and GABAB receptors. J Neurosci 25(27):6304–6315. https://doi.org/10.1523/JNEUROSCI.0450-05.2005
Helmstaedter C (2002) Effects of chronic epilepsy on declarative memory systems. Prog Brain Res 135:439–453. https://doi.org/10.1016/S0079-6123(02)35041-6
Hill DR (1985) GABAB receptor modulation of adenylate cyclase activity in rat brain slices. Br J Pharmacol 84(1):249–257
Hill DR, Bowery NG (1981) 3H-baclofen and 3H-GABA bind to bicuculline-insensitive GABA B sites in rat brain. Nature 290(5802):149–152. https://doi.org/10.1038/290149a0
Hill DR, Bowery NG, Hudson AL (1984) Inhibition of GABAB receptor binding by guanyl nucleotides. J Neurochem 42(3):652–657. https://doi.org/10.1111/j.1471-4159.1984.tb02732.x
Hillman R, Sinani J, Pendleton R (2012) The role of the GABA(B) receptor and calcium channels in a drosophila model of Parkinson's disease. Neurosci Lett 516(2):167–170. https://doi.org/10.1016/j.neulet.2012.03.034
Hirsch JC, Burnod Y (1987) A synaptically evoked late hyperpolarization in the rat dorsolateral geniculate neurons in vitro. Neuroscience 23(2):457–468. https://doi.org/10.1016/0306-4522(87)90069-8
Houser CR (1990) Granule cell dispersion in the dentate gyrus of humans with temporal lobe epilepsy. Brain Res 535(2):195–204. https://doi.org/10.1016/0006-8993(90)91601-c
Huston E, Scott RH, Dolphin AC (1990) A comparison of the effect of calcium channel ligands and GABAB agonists and antagonists on transmitter release and somatic calcium channel currents in cultured neurons. Neuroscience 38(3):721–729. https://doi.org/10.1016/0306-4522(90)90065-c
Isokawa M, Avanzini G, Finch DM, Babb TL, Levesque MF (1991) Physiologic properties of human dentate granule cells in slices prepared from epileptic patients. Epilepsy Res 9 (3):242–250. https://doi.org/10.1016/0920-1211(91)90058-n

Isomoto S, Kaibara M, Sakurai-Yamashita Y, Nagayama Y, Uezono Y, Yano K, Taniyama K (1998) Cloning and tissue distribution of novel splice variants of the rat GABAB receptor. Biochem Biophys Res Commun 253(1):10–15. https://doi.org/10.1006/bbrc.1998.9706

Iwakiri M, Mizukami K, Ikonomovic MD, Ishikawa M, Hidaka S, Abrahamson EE, DeKosky ST, Asada T (2005) Changes in hippocampal GABABR1 subunit expression in Alzheimer's patients: association with Braak staging. Acta Neuropathol 109(5):467–474. https://doi.org/10.1007/s00401-005-0985-9

Johnston T, Duty S (2003) Changes in GABA(B) receptor mRNA expression in the rodent basal ganglia and thalamus following lesion of the nigrostriatal pathway. Neuroscience 120 (4):1027–1035. https://doi.org/10.1016/s0306-4522(03)00418-4

Jones KA, Borowsky B, Tamm JA, Craig DA, Durkin MM, Dau M, Yao WY, Johnson M, Gunwaldsen C, Huang LY, Tang C, Shen Q, Salon JA, Morse K, Laz T, Smith KE, Nagarathnam D, Noble SA, Branchek TA, Gerald C (1998) GABAB receptors functional a heterodimeric assembly of the subunits GABABR1 and GABABR2. Nature 396:674–678. https://doi.org/10.1038/25348

Kantrowitz JT, Francis NN, Salah A, Perkins KL (2005) Synaptic depolarizing GABA response in adults is excitatory and proconvulsive when GABAB receptors are blocked. J Neurophysiol 93 (5):2656–2667. https://doi.org/10.1152/jn.01026.2004

Karbon EW, Enna SJ (1985) Characterization of the relationship between gamma-aminobutyric acid B agonists and transmitter-coupled cyclic nucleotide-generating systems in rat brain. Mol Pharmacol 27(1):53–59

Karlsson G, Olpe HR (1989) Late inhibitory postsynaptic potentials in rat prefrontal cortex may be mediated by GABAB receptors. Experientia 45(2):157–158. https://doi.org/10.1007/BF01954857

Kaupmann K, Huggel K, Heid J, Flor PJ, Bischoff S, Mickel SJ, McMaster G, Angst C, Bittiger H, Froestl W, Bettler B (1997) Expression cloning of GABA(B) receptors uncovers similarity to metabotropic glutamate receptors. Nature 386(6622):239–246. https://doi.org/10.1038/386239a0

Kaupmann K, Malitschek B, Schuler V, Heid J, Froestl W, Beck P, Mosbacher J, Bischoff S, Kulik A, Shigemoto R, Karschin A, Bettler B (1998) GABA(B)-receptor subtypes assemble into functional heteromeric complexes. Nature 396(6712):683–687. https://doi.org/10.1038/25360

Knight AR, Bowery NG (1996) The pharmacology of adenylyl cyclase modulation by GABAB receptors in rat brain slices. Neuropharmacology 35(6):703–712. https://doi.org/10.1016/0028-3908(96)84642-9

Knowles WD, Awad IA, Nayel MH (1992) Differences of in vitro electrophysiology of hippocampal neurons from epileptic patients with mesiotemporal sclerosis versus structural lesions. Epilepsia 33(4):601–609. https://doi.org/10.1111/j.1528-1157.1992.tb02335.x

Köhling R, Lücke A, Straub H, Speckmann EJ, Tuxhorn I, Wolf P, Pannek H, Oppel F (1998) Spontaneous sharp waves in human neocortical slices excised from epileptic patients. Brain J Neurol 121(Pt 6):1073–1087. https://doi.org/10.1093/brain/121.6.1073

Krnjević K (1974) Some neuroactive compounds in the substantia nigra. Adv Neurol 5:145–152

Kuner R, Köhr G, Grünewald S, Eisenhardt G, Bach A, Kornau HC (1999) Role of heteromer formation in GABAB receptor function. Science 283(5398):74–77. https://doi.org/10.1126/science.283.5398.74

Lang M, Moradi-Chameh H, Zahid T, Gane J, Wu C, Valiante T, Zhang L (2014) Regulating hippocampal hyperexcitability through GABAB receptors. Physiol Rep 2(4):e00278. https://doi.org/10.14814/phy2.278. PMID: 24771688; PMCID: PMC4001873

Lefkowitz RJ (1992) G proteins. The subunit story thickens. Nature 358(6385):372. https://doi.org/10.1038/358372a0

Liang F, Hatanaka Y, Saito H, Yamamori T, Hashikawa T (2000) Differential expression of γ-aminobutyric acid type B receptor-1a and -1b variants in GABA and non-GABAergic neurons of the rat brain. J Comp Neurol 416:475–495

Lopes EF, Roberts BM, Siddorn RE, Clements MA, Cragg SJ (2019) Inhibition of nigrostriatal dopamine release by striatal GABAA and GABAB receptors. J Neurosci 39(6):1058–1065. https://doi.org/10.1523/JNEUROSCI.2028-18.2018

Loup F, Picard F, André VM, Kehrli P, Yonekawa Y, Wieser HG, Fritschy JM (2006) Altered expression of alpha3-containing GABAA receptors in the neocortex of patients with focal epilepsy. Brain 129(Pt 12):3277–3289. https://doi.org/10.1093/brain/awl287

Luhmann HJ, Prince DA (1991) Postnatal maturation of the GABAergic system in rat neocortex. J Neurophysiol 65(2):247–263. https://doi.org/10.1152/jn.1991.65.2.247

Lüscher C, Jan LY, Stoffel M, Malenka RC, Nicoll RA (1997) G protein-coupled inwardly rectifying K+ channels (GIRKs) mediate postsynaptic but not presynaptic transmitter actions in hippocampal neurons. Neuron 19(3):687–695. https://doi.org/10.1016/s0896-6273(00)80381-5

Mangan PS, Lothman EW (1996) Profound disturbances of pre- and postsynaptic GABAB-receptor-mediated processes in region CA1 in a chronic model of temporal lobe epilepsy. J Neurophysiol 76(2):1282–1296. https://doi.org/10.1152/jn.1996.76.2.1282

Mangan PS, Rempe DA, Lothman EW (1995) Changes in inhibitory neurotransmission in the CA1 region and dentate gyrus in a chronic model of temporal lobe epilepsy. J Neurophysiol 74(2):829–840. https://doi.org/10.1152/jn.1995.74.2.829

Marescaux C, Vergnes M, Liu Z, Depaulis A, Bernasconi R (1992) GABAB receptor involvement in the control of genetic absence seizures in rats. Epilepsy Res Suppl 9:131–139

Martín-Belmonte A, Aguado C, Alfaro-Ruíz R, Moreno-Martínez AE, de la Ossa L, Martínez-Hernández J, Buisson A, Früh S, Bettler B, Shigemoto R, Fukazawa Y, Luján R (2020a) Reduction in the neuronal surface of post and presynaptic GABAB receptors in the hippocampus in a mouse model of Alzheimer's disease. Brain Pathol 30(3):554–575. https://doi.org/10.1111/bpa.12802

Martín-Belmonte A, Aguado C, Alfaro-Ruíz R, Moreno-Martínez AE, de la Ossa L, Martínez-Hernández J, Buisson A, Shigemoto R, Fukazawa Y, Luján R (2020b) Density of GABAB receptors is reduced in granule cells of the Hippocampus in a mouse model of Alzheimer's disease. Int J Mol Sci 21(7):2459. https://doi.org/10.3390/ijms21072459

Mathern GW, Mendoza D, Lozada A, Pretorius JK, Dehnes Y, Danbolt NC, Nelson N, Leite JP, Chimelli L, Born DE, Sakamoto AC, Assirati JA, Fried I, Peacock WJ, Ojemann GA, Adelson PD (1999) Hippocampal GABA and glutamate transporter immunoreactivity in patients with temporal lobe epilepsy. Neurology 52(3):453–472. https://doi.org/10.1212/wnl.52.3.453

Maureille A, Fenouil T, Joubert B et al (2019) Isolated seizures are a common early feature of paraneoplastic anti-$GABA_B$ receptor encephalitis. J Neurol 266(1):195–206. https://doi.org/10.1007/s00415-018-9132-0

Mayordomo-Cava J, Yajeya J, Navarro-López JD, Jiménez-Díaz L (2015) Amyloid-β(25-35) modulates the expression of GIRK and KCNQ Channel genes in the Hippocampus. PLoS One 10(7):e0134385. https://doi.org/10.1371/journal.pone.0134385

Mazzuferi M, Palma E, Martinello K, Maiolino F, Roseti C, Fucile S, Fabene PF, Schio F, Pellitteri M, Sperk G, Miledi R, Eusebi F, Simonato M (2010) Enhancement of GABA(A)-current run-down in the hippocampus occurs at the first spontaneous seizure in a model of temporal lobe epilepsy. Proc Natl Acad Sci U S A 107(7):3180–3185. https://doi.org/10.1073/pnas.0914710107

McGregor MM, Nelson AB (2019) Circuit mechanisms of Parkinson's disease. Neuron 101(6):1042–1056. https://doi.org/10.1016/j.neuron.2019.03.004

Meeren HK, van Luijtelaar EL, Lopes da Silva FH, Berdiev RK, Chepurnova NE, Chepurnov SA, Coenen AM (2004) Kortikotalamicheskaia teoriia proiskhozhdeniia generalizovannykh pik-volnovykh razriadov [the cortico-thalamic theory for generalised spike-wave discharges]. Usp Fiziol Nauk 35(1):3–19

Misgeld U, Klee MR, Zeise ML (1984) Differences in baclofen-sensitivity between CA3 neurons and granule cells of the guinea pig hippocampus in vitro. Neurosci Lett 47(3):307–311. https://doi.org/10.1016/0304-3940(84)90531-7

Moser A, Hanssen H, Wandinger KP (2018) Excessively increased CSF glutamate levels in $GABA_B$-receptor antibody associated encephalitis: a case report. J Neurol Sci 388:10–11. https://doi.org/10.1016/j.jns.2018.02.041
Mott DD, Lewis DV (1994) The pharmacology and function of central GABAB receptors. Int Rev Neurobiol 36:97–223. https://doi.org/10.1016/s0074-7742(08)60304-9
Muñoz A, Huntsman MM, Jones EG (1998) GABA(B) receptor gene expression in monkey thalamus. J Comp Neurol 394(1):118–126. https://doi.org/10.1002/(sici)1096-9861(19980427)394:1<118::aid-cne9>3.0.co;2-3
Nava-Mesa MO, Jiménez-Díaz L, Yajeya J, Navarro-Lopez JD (2013) Amyloid-β induces synaptic dysfunction through G protein-gated inwardly rectifying potassium channels in the fimbria-CA3 hippocampal synapse. Front Cell Neurosci 7:117. https://doi.org/10.3389/fncel.2013.00117
Ng GYK, Clark J, Coulombe N, Ethier N, Herbert TE, Sullivan R, Kargman S, Chateauneuf A, Tsukamoto N, Mezey E, Johnson MP, Liu Q, Kolakowski LF Jr, Evans JF, Bonner TI, O'Neill GP (1999) Identification of a GABAB receptor subunit, gB2, required for functional GABAB receptor activity. J Biol Chem 274:7607–7610. https://doi.org/10.1074/jbc.274.12.7607
Niddam R, Dubois A, Scatton B, Arbilla S, Langer SZ (1987) Autoradiographic localization of [3H] zolpidem binding sites in the rat CNS: comparison with the distribution of [3H]flunitrazepam binding sites. J Neurochem 49(3):890–899. https://doi.org/10.1111/j.1471-4159.1987.tb00977
Nishimura T, Schwarzer C, Gasser E, Kato N, Vezzani A, Sperk G (2005) Altered expression of GABA(A) and GABA(B) receptor subunit mRNAs in the hippocampus after kindling and electrically induced status epilepticus. Neuroscience 134(2):691–704. https://doi.org/10.1016/j.neuroscience.2005.04.013
Odagaki Y, Koyama T (2001) Identification of galpha subtype(s) involved in gamma-aminobutyric acid(B) receptor-mediated high-affinity guanosine triphosphatase activity in rat cerebral cortical membranes. Neurosci Lett 297(2):137–141. https://doi.org/10.1016/s0304-3940(00)01692-x
Odagaki Y, Nishi N, Koyama T (2000) Functional coupling of GABA(B) receptors with G proteins that are sensitive to N-ethylmaleimide treatment, suramin, and benzalkonium chloride in rat cerebral cortical membranes. J Neural Transm 107(10):1101–1116. https://doi.org/10.1007/s007020070024
Oh ES, Savonenko AV, King JF, Fangmark Tucker SM, Rudow GL, Xu G, Borchelt DR, Troncoso JC (2009) Amyloid precursor protein increases cortical neuron size in transgenic mice. Neurobiol Aging 30(8):1238–1244. https://doi.org/10.1016/j.neurobiolaging.2007.12.024
Olpe HR, Steinmann MW, Ferrat T, Pozza MF, Greiner K, Brugger F, Froestl W, Mickel SJ, Bittiger H (1993) The actions of orally active GABAB receptor antagonists on GABAergic transmission in vivo and in vitro. Eur J Pharmacol 233(2–3):179–186. https://doi.org/10.1016/0014-2999(93)90048-m
Paris-Robidas S, Brochu E, Sintes M et al (2012) Defective dentate nucleus GABA receptors in essential tremor. Brain 135(Pt 1):105–116. https://doi.org/10.1093/brain/awr301
Pfrieger FW, Gottmann K, Lux HD (1994) Kinetics of GABAB receptor-mediated inhibition of calcium currents and excitatory synaptic transmission in hippocampal neurons in vitro. Neuron 12(1):97–107. https://doi.org/10.1016/0896-6273(94)90155-4
Pilipenko V, Narbute K, Beitnere U, Rumaks J, Pupure J, Jansone B, Klusa V (2018) Very low doses of muscimol and baclofen ameliorate cognitive deficits and regulate protein expression in the brain of a rat model of streptozocin-induced Alzheimer's disease. Eur J Pharmacol 818:381–399. https://doi.org/10.1016/j.ejphar.2017.11.012
Pirker S, Schwarzer C, Czech T, Baumgartner C, Pockberger H, Maier H, Hauer B, Sieghart W, Furtinger S, Sperk G (2003) Increased expression of GABA(A) receptor beta-subunits in the hippocampus of patients with temporal lobe epilepsy. J Neuropathol Exp Neurol 62(8):820–834. https://doi.org/10.1093/jnen/62.8.820
Princivalle AP, Pangalos MN, Bowery NG, Spreafico R (2001) Distribution of GABA(B(1a)), GABA(B(1b)) and GABA(B2) receptor protein in cerebral cortex and thalamus of adult rats. Neuroreport 12(3):591–595. https://doi.org/10.1097/00001756-200103050-00032

Princivalle AP, Duncan JS, Thom M, Bowery NG (2002) Studies of GABA(B) receptors labelled with [(3)H]-CGP62349 in hippocampus resected from patients with temporal lobe epilepsy. Br J Pharmacol 136(8):1099–1106. https://doi.org/10.1038/sj.bjp.0704812

Princivalle AP, Duncan JS, Thom M, Bowery NG (2003) GABA(B1a), GABA(B1b) AND GABA (B2) mRNA variants expression in hippocampus resected from patients with temporal lobe epilepsy. Neuroscience 122(4):975–984. https://doi.org/10.1016/j.neuroscience.2003.08.044

Rainnie DG, Asprodini EK, Shinnick-Gallagher P (1991) Inhibitory transmission in the basolateral amygdala. J Neurophysiol 66(3):999–1009. https://doi.org/10.1152/jn.1991.66.3.999

Rascol O, Dutar P, Lamour Y (1989) Involvement of a pertussis toxin-sensitive G-protein in the pharmacological properties of septo-hippocampal neurones. Br J Pharmacol 96(4):956–960. https://doi.org/10.1111/j.1476-5381.1989.tb11907.x

Rekik L, Daguin-Nerrière V, Petit JY, Brachet P (2011) γ-Aminobutyric acid type B receptor changes in the rat striatum and substantia nigra following intrastriatal quinolinic acid lesions. J Neurosci Res 89(4):524–535. https://doi.org/10.1002/jnr.22574

Rocha L, Alonso-Vanegas M, Martínez-Juárez IE, Orozco-Suárez S, Escalante-Santiago D, Feria-Romero IA, Zavala-Tecuapetla C, Cisneros-Franco JM, Buentello-García RM, Cienfuegos J (2015) GABAergic alterations in neocortex of patients with pharmacoresistant temporal lobe epilepsy can explain the comorbidity of anxiety and depression: the potential impact of clinical factors. Front Cell Neurosci 8:442. https://doi.org/10.3389/fncel.2014.00442

Rosas-Arellano A, Estrada-Mondragón A, Mantellero CA, Tejeda-Guzmán C, Castro MA (2018) The adjustment of γ-aminobutyric $acid_A$ tonic subunits in Huntington's disease: from transcription to translation to synaptic levels into the neostriatum. Neural Regen Res 13(4):584–590. https://doi.org/10.4103/1673-5374.230270

Ruano D, Vizuete M, Cano J, Machado A, Vitorica J (1992) Heterogeneity in the allosteric interaction between the gamma-aminobutyric acid (GABA) binding site and three different benzodiazepine binding sites of the GABAA/benzodiazepine receptor complex in the rat nervous system. J Neurochem 58(2):485–493. https://doi.org/10.1111/j.1471-4159.1992.tb09747.x

Saint DA, Thomas T, Gage PW (1990) GABAB agonists modulate a transient potassium current in cultured mammalian hippocampal neurons. Neurosci Lett 118(1):9–13. https://doi.org/10.1016/0304-3940(90)90236-3

Schijns O, Karaca Ü, Andrade P, de Nijs L, Küsters B, Peeters A, Dings J, Pannek H, Ebner A, Rijkers K, Hoogland G (2015) Hippocampal GABA transporter distribution in patients with temporal lobe epilepsy and hippocampal sclerosis. J Chem Neuroanat 68:39–44. https://doi.org/10.1016/j.jchemneu.2015.07.004

Schofield PR, Darlison MG, Fujita N, Burt DR, Stephenson FA, Rodriguez H, Rhee LM, Ramachandran J, Reale V, Glencorse TA (1987) Sequence and functional expression of the GABA A receptor shows a ligand-gated receptor super-family. Nature 328(6127):221–227. https://doi.org/10.1038/328221a0

Schreurs BG, Sanchez-Andres JV, Alkon DL (1992) GABA-induced responses in Purkinje cell dendrites of the rabbit cerebellar slice. Brain Res 597(1):99–107. https://doi.org/10.1016/0006-8993(92)91510-I

Schumacher H, Meyer T, Prüss H (2019) $GABA_B$ receptor encephalitis in a patient diagnosed with amyotrophic lateral sclerosis. BMC Neurol 19(1):41. https://doi.org/10.1186/s12883-019-1269-7

Scott RH, Pearson HA, Dolphin AC (1991) Aspects of vertebrate neuronal voltage-activated calcium currents and their regulation. Prog Neurobiol 36(6):485–520. https://doi.org/10.1016/0301-0082(91)90014-r

Seifert G, Carmignoto G, Steinhäuser C (2010) Astrocyte dysfunction in epilepsy. Brain Res Rev 63(1–2):212–221. https://doi.org/10.1016/j.brainresrev.2009.10.004

Sheilabi MA, Battacharyya D, Caetano L, Thom M, Reuber M, Duncan JS, Princivalle AP (2018) Quantitative expression and localization of GABAB receptor protein subunits in hippocampi from patients with refractory temporal lobe epilepsy. Neuropharmacology 136(Pt A):117–128. https://doi.org/10.1016/j.neuropharm.2017.08.001

Sieghart W, Eichinger A, Richards JG, Möhler H (1987) Photoaffinity labeling of benzodiazepine receptor proteins with the partial inverse agonist [3H]Ro 15-4513: a biochemical and autoradiographic study. J Neurochem 48(1):46–52. https://doi.org/10.1111/j.1471-4159.1987.tb13125.x

Sloviter RS (1994) The functional organization of the hippocampal dentate gyrus and its relevance to the pathogenesis of temporal lobe epilepsy. Ann Neurol 35(6):640–654. https://doi.org/10.1002/ana.410350604

Smith GB, Olsen RW (1995) Functional domains of GABAA receptors. Trends Pharmacol Sci 16 (5):162–168. https://doi.org/10.1016/s0165-6147(00)89009-4

Smith Y, Charara A, Hanson JE, Paquet M, Levey AI (2000) GABA(B) and group I metabotropic glutamate receptors in the striatopallidal complex in primates. J Anat 196(Pt 4):555–576. https://doi.org/10.1046/j.1469-7580.2000.19640555.x

Sperk G, Furtinger S, Schwarzer C, Pirker S (2004) GABA and its receptors in epilepsy. Adv Exp Med Biol 548:92–103. https://doi.org/10.1007/978-1-4757-6376-8_7

Stefani A, Spadoni F, Giacomini P, Lavaroni F, Bernardi G (1999) The modulation of calcium current by GABA metabotropic receptors in a sub-population of pallidal neurons. Eur J Neurosci 11(11):3995–4005. https://doi.org/10.1046/j.1460-9568.1999.00836.x

Stefanits H, Milenkovic I, Mahr N, Pataraia E, Baumgartner C, Hainfellner JA, Kovacs GG, Kasprian G, Sieghart W, Yilmazer-Hanke D, Czech T (2019) Alterations in GABAA receptor subunit expression in the amygdala and entorhinal cortex in human temporal lobe epilepsy. J Neuropathol Exp Neurol 78(11):1022–1048. https://doi.org/10.1093/jnen/nlz085

Stewart LS, Wu Y, Eubanks JH, Han H, Leschenko Y, Perez Velazquez JL, Cortez MA, Snead OC 3rd (2009) Severity of atypical absence phenotype in GABAB transgenic mice is subunit specific. Epilepsy Behav 14(4):577–581. https://doi.org/10.1016/j.yebeh.2009.01.019

Sutula TP, Hermann B (1999) Progression in mesial temporal lobe epilepsy. Ann Neurol 45 (5):553–556

Sutula T, Cascino G, Cavazos J, Parada I, Ramirez L (1989) Mossy fiber synaptic reorganization in the epileptic human temporal lobe. Ann Neurol 26(3):321–330. https://doi.org/10.1002/ana.410260303

Takahashi T, Kajikawa Y, Tsujimoto T (1998) G-protein-coupled modulation of presynaptic calcium currents and transmitter release by a GABAB receptor. J Neurosci 18(9):3138–3146. https://doi.org/10.1523/JNEUROSCI.18-09-03138.1998

Tang BL (2019) Amyloid precursor protein (APP) and GABAergic neurotransmission. Cells 8 (6):550. https://doi.org/10.3390/cells8060550

Teichgräber LA, Lehmann TN, Meencke HJ, Weiss T, Nitsch R, Deisz RA (2009) Impaired function of GABA(B) receptors in tissues from pharmacoresistant epilepsy patients. Epilepsia 50(7):1697–1716. https://doi.org/10.1111/j.1528-1167.2009.02094

Thomas PM, Phillips JP, Delanty N, O'Connor WT (2003) Elevated extracellular levels of glutamate, aspartate and gamma-aminobutyric acid within the intraoperative, spontaneously epileptiform human hippocampus. Epilepsy Res 54(1):73–79. https://doi.org/10.1016/s0920-1211(03)00035-4

Thomas PM, Phillips JP, O'Connor WT (2005) Microdialysis of the lateral and medial temporal lobe during temporal lobe epilepsy surgery. Surg Neurol 63(1):70–79. https://doi.org/10.1016/j.surneu.2004.02.031

Thompson SM, Gähwiler BH (1992) Effects of the GABA uptake inhibitor tiagabine on inhibitory synaptic potentials in rat hippocampal slice cultures. J Neurophysiol 67(6):1698–1701. https://doi.org/10.1152/jn.1992.67.6.1698

Toodayan N (2018) James Parkinson's essay on the shaking palsy, 1817–2017. Med J Aust 208 (9):384–386. https://doi.org/10.5694/mja17.01085. PMID: 29764348

Tyagi RK, Bisht R, Pant J, Kumar P, Majeed AB, Prakash A (2015) Possible role of GABA-B receptor modulation in MPTP induced Parkinson's disease in rats. Exp Toxicol Pathol 67 (2):211–217. https://doi.org/10.1016/j.etp.2014.12.001

Tysnes OB, Storstein A (2017) Epidemiology of Parkinson's disease. J Neural Transm 124 (8):901–905. https://doi.org/10.1007/s00702-017-1686-y

Van Paesschen W, Revesz T, Duncan JS, King MD, Connelly A (1997) Quantitative neuropathology and quantitative magnetic resonance imaging of the hippocampus in temporal lobe epilepsy. Ann Neurol 42(5):756–766. https://doi.org/10.1002/ana.410420512

Vigot R, Batini C (1997) GABA(B) receptor activation of Purkinje cells in cerebellar slices. Neurosci Res 29(2):151–160. https://doi.org/10.1016/s0168-0102(97)00087-4

White JH, Wise A, Main MJ, Green A, Fraser NJ, Disney GH, Barnes AA, Emson P, Foord SM, Marshall FH (1998) Heterodimerization is required for the formation of a functional GABA(B) receptor. Nature 396(6712):679–682. https://doi.org/10.1038/25354

Williams S, Lacaille JC (1992) GABAB receptor-mediated inhibitory postsynaptic potentials evoked by electrical stimulation and by glutamate stimulation of interneurons in stratum lacunosum-moleculare in hippocampal CA1 pyramidal cells in vitro. Synapse 11(3):249–258. https://doi.org/10.1002/syn.890110309

Winblad B, Amouyel P, Andrieu S, Ballard C, Brayne C, Brodaty H, Cedazo-Minguez A, Dubois B, Edvardsson D, Feldman H, Fratiglioni L, Frisoni GB, Gauthier S, Georges J, Graff C, Iqbal K, Jessen F, Johansson G, Jönsson L, Kivipelto M, Knapp M, Mangialasche F, Melis R, Nordberg A, Rikkert MO, Qiu C, Sakmar TP, Scheltens P, Schneider LS, Sperling R, Tjernberg LO, Waldemar G, Wimo A, Zetterberg H (2016) Defeating Alzheimer's disease and other dementias: a priority for European science and society. Lancet Neurol 15(5):455–532. https://doi.org/10.1016/S1474-4422(16)00062-4

Wojcik WJ, Neff NH (1984) Gamma-aminobutyric acid B receptors are negatively coupled to adenylate cyclase in brain, and in the cerebellum these receptors may be associated with granule cells. Mol Pharmacol 25(1):24–28

Wojcik WJ, Travagli RA, Costa E, Bertolino M (1990) Baclofen inhibits with high affinity an L-type-like voltage-dependent calcium channel in cerebellar granule cell cultures. Neuropharmacology 29(10):969–972. https://doi.org/10.1016/0028-3908(90)90150-p

Wu C, Leung LS (1997) Partial hippocampal kindling decreases efficacy of presynaptic GABAB autoreceptors in CA1. J Neurosci 17(23):9261–9269. https://doi.org/10.1523/JNEUROSCI.17-23-09261.1997

Xu J, Wojcik WJ (1986) Gamma aminobutyric acid B receptor-mediated inhibition of adenylate cyclase in cultured cerebellar granule cells: blockade by islet-activating protein. J Pharmacol Exp Ther 239(2):568–573. PMID: 2430096

Yu Z, Cheng G, Hu B (1997) Mechanism of colchicine impairment on learning and memory, and protective effect of CGP36742 in mice. Brain Res 750(1–2):53–58. https://doi.org/10.1016/s0006-8993(96)01158-4

Zentner J, Hufnagel A, Wolf HK, Ostertun B, Behrens E, Campos MG, Solymosi L, Elger CE, Wiestler OD, Schramm J (1995) Surgical treatment of temporal lobe epilepsy: clinical, radiological, and histopathological findings in 178 patients. J Neurol Neurosurg Psychiatry 58 (6):666–673. https://doi.org/10.1136/jnnp.58.6.666

$GABA_B$ Receptors and Cognitive Processing in Health and Disease

Styliani Vlachou

Contents

Abstract $GABA_B$ receptors are implicated in numerous central nervous system-based behaviours and mechanisms, including cognitive processing in preclinical animal models. Homeostatic changes in the expression and function of these receptors across brain structures have been found to affect cognitive processing. Numerous preclinical studies have focused on the role of $GABA_B$ receptors in learning, memory and cognition per se with some interesting, although sometimes contradictory, findings. The majority of the existing clinical literature focuses on alterations in

---

S. Vlachou (✉)
Behavioural Neuroscience Laboratory, Neuropsychopharmacology Division, School of Psychology, Dublin City University, Dublin, Ireland
e-mail: stella.vlachou@dcu.ie

Curr Topics Behav Neurosci (2022) 52: 291–330
https://doi.org/10.1007/7854_2021_231
Published Online: 12 August 2021

$GABA_B$ receptor function in conditions and disorders whose main symptomatology includes deficits in cognitive processing. The aim of this chapter is to delineate the role of $GABA_B$ receptors in cognitive processes in health and disease of animal models and human clinical populations. More specifically, this review aims to present literature on the role of $GABA_B$ receptors in animal models with cognitive deficits, especially those of learning and memory. Further, it aims to capture the progress and advances of research studies on the effects of $GABA_B$ receptor compounds in neurodevelopmental and neurodegenerative conditions with cognitive dysfunctions. The neurodevelopmental conditions covered include autism spectrum disorders, fragile X syndrome and Down's syndrome and the neurodegenerative conditions discussed are Alzheimer's disease, epilepsy and autoimmune anti-$GABA_B$ encephalitis. Although some findings are contradictory, results indicate a possible therapeutic role of $GABA_B$ receptor compounds for the treatment of cognitive dysfunction and learning/memory impairments for some of these conditions, especially in neurodegeneration. Moreover, future research efforts should aim to develop selective $GABA_B$ receptor compounds with minimal, if any, side effects.

Keywords Animal models · Cognition · $GABA_B$ receptors · Humans · Learning · Memory · Neurodegeneration · Neurodevelopmental disorders · Pharmacotherapy · Treatment

1 Introduction

Metabotropic $GABA_B$ receptors ($GABA_BRs$) are implicated in numerous central nervous system (CNS)-based behaviours and mechanisms, such as the cognitive processes of learning and memory (Bowery 2006; Heaney and Kinney 2016; Serrats et al. 2017), in preclinical animal models. Homeostatic changes in the levels and functions of these receptors at molecular, cellular or neurochemical levels have been found to affect those cognitive processes. Preclinical studies using a variety of animal models have focused on the role of $GABA_BRs$ in learning, memory and cognition per se. Results from these studies, although sometimes contradictory, are valuable and promising.

Additionally, several clinical conditions and disorders with variable symptomatology that are associated with cognitive deficits also show alterations in $GABA_B$ receptor ($GABA_BR$) function (Heaney and Kinney 2016). These conditions include neurodevelopmental disorders such as Down's syndrome (DS), fragile X syndrome (FXS) and autism spectrum disorders (ASD), neurodegenerative disorders, such as Alzheimer's disease (AD) and epilepsy, and other neuropsychiatric disorders, such as schizophrenia. This correlation has spurred a number of studies that have focused on investigating the role of $GABA_BRs$ in cognitive processing in these clinical conditions. The development and advances in imaging techniques have allowed for a greater understanding of metabotropic GABAergic signalling pathways and their involvement in cognition (Murrell et al. 2020). An overall extensive account of the role of $GABA_BRs$ in neurodegeneration, with a focus on epilepsy and AD, is presented elsewhere in this book volume (Princivalle 2021). In this chapter the focus will be on cognitive processes identified in animal models of health and disease and in human populations whose cognitive symptomatology is affected or altered by $GABA_BR$ mechanisms.

In a broad sense, $GABA_BR$ agonists have been shown to impair learning and memory processes, while $GABA_BR$ antagonists improve them. These effects largely depend on either the regulation of postsynaptic excitability or on the presynaptic inhibition of neurotransmitter release. These effects also depend on the pre- and/or post-synaptic pathways affected, methodological components of the experiment (such as the tasks implemented, administration method and schedule, and the doses used) and the duration of the effects. It is important to understand the role of $GABA_BR$ agonists, antagonists and positive allosteric modulators (PAMs) on cognition in the general population, as well as in clinical populations, and the mechanisms of action underlying their effects. Thus, the aim of this review chapter is to examine these processes in detail and capture all existing literature in this area, in an effort to shed some light on the potential use of $GABA_BR$ compounds for the treatment of cognitive impairments in neurodevelopmental, neurodegenerative and/or neuropsychiatric disorders.

2 $GABA_B$ Receptors Mechanisms of Action

A detailed presentation of the $GABA_BR$ structure, function, locations and mechanisms of action is presented in this book volume chapters (Fritzius et al. 2020; Rose and Wickman 2020). Briefly, slow sustained neuronal inhibition is mediated by $GABA_BRs$ (Brenowitz et al. 1998), which are heterodimeric G-protein-coupled receptors (GPCRs) constructed from $GABA_BR1a$ or $GABA_BR1b$ and $GABA_BR2$ subunits present in neurons and glial cells throughout the CNS. These subunits need to co-exist for the receptors to be functional. Through the activation of Gi/o proteins, $GABA_BRs$ limit cAMP accumulation, decrease neurotransmitter release and induce neuronal hyperpolarization.

Taking into account their synaptic loci, $GABA_BR$ activation acts in two ways, both ways reducing glutamatergic signalling at excitatory synapses: (1) it reduces presynaptic GABA and glutamate release through inhibition of presynaptic Ca^+ channels in both inhibitory and excitatory synapses and (2) it causes hyperpolarization of postsynaptic neurons by activation of G protein-activated inwardly rectifying potassium (K^+) channels (GIRKs) (Pin and Bettler 2016). Further, in order for the $GABA_BRs$ to be activated, non-synaptic or volume transmission needs to take place, which requires high levels of GABA release. This can be achieved via several mechanisms, including simultaneous discharge of GABAergic interneurons (Holley et al. 2019), very intense discharges in the thalamus, or by the activation of neuroglia interneurons in the cortex (Sanchez-Vives et al. 2020).

3 Effects of $GABA_B$ Receptors on Cognitive Processes: Learning and Memory

Functional activation of $GABA_B$Rs inhibits learning and memory processes, though conflicting results have been reported. As cognitive processes and the effects of $GABA_B$Rs on them are assessed in the context of a vast number of conditions, this section will initially focus on aspects relating to synaptic plasticity and long-term potentiation (LTP) as they relate to cognition, and will then be divided into behavioural procedures used in animal studies for the assessment of cognitive processes in healthy and/or unhealthy populations. The effects of $GABA_B$Rs on cognition will be examined through a variety of procedures, such as: passive and active avoidance tests, Morris Water Maze (MWM), Radial Arm Maze (RAM), working memory tasks, novel object recognition (NOR) and (dis)location (NOL) tasks and the Barnes maze.

3.1 *$GABA_B$ Receptors, LTP and Synaptic Plasticity*

LTP is a long-lasting increase in synaptic effectiveness after repeated high-frequency stimulation, identified first in the hippocampus in 1973 by Bliss and Lomo (Bliss and Lømo 1973; Lømo 2003, 2018). LTP is a synaptic plasticity mechanism that is important for associative learning and creating memories and well-defined. Moreover, being well-defined, LTP is easy to measure and can be used as a readout for cognitive deficits.

Evidence from numerous studies shows that the GABAergic system is involved in LTP and other synaptic plasticity mechanisms in different areas of the brain, including the hippocampus and the perirhinal cortex. Importantly, oscillatory and synchronous activity is needed to induce synaptic plasticity and GABA seems to play an essential role in controlling these oscillations (Kohl and Paulsen 2010). Specific frequencies, such as gamma and theta, have been linked with memory formation in animals and humans [e.g., (Axmacher et al. 2006; Jutras and Buffalo 2010)], and GABA-initiated synchronous inhibitory postsynaptic potentials (IPSPs) in the hippocampus control these frequencies [e.g., (Gong et al. 2009; Mann and Mody 2010)].

As an example of these processes, there appear to be functional interactions between the GABAergic and cannabinoid (CB) systems in the hippocampus and other brain areas. After LTP induction in the dentate gyrus (DG) of rats, either the cannabinoid 1 receptor (CB1R) antagonist AM251 or the $GABA_B$R antagonist CGP55845 was administered. While AM251 increased LTP, CGP55845 decreased it, while their co-administration had differential effects on the population spike (PS) amplitude and field excitatory postsynaptic potential (fEPSP). Data from this study suggested that $GABA_B$R antagonists modulate cannabinoid outputs that decrease synaptic plasticity, while during co-administration, CB1R antagonists can alter the release of GABA ultimately resulting in enhancement of LTP induction (Nazari et al. 2016a, b). In a corresponding study by the same research group (Nazari et al. 2016a), where either AM251 or baclofen was administered, results showed that administration of either of the two compounds, AM251 or baclofen, increased both the PS amplitude and the fEPSP slope, while co-administration of AM251 and baclofen induced greater increases in PS amplitude and the fEPSP slope, further strengthening LTP induction through GABA$_B$Rs. These findings suggest that CB1R activation in the hippocampus seems to affect synaptic function of GABAergic interneurons located in the DG.

3.2 Active and Passive Avoidance Paradigms

In the active avoidance paradigm, animals (most commonly rodents) learn to avoid an aversive stimulus by initiating a specific locomotor response. In this task, animals are placed in a two-compartment shuttle box and learn the association between a conditioned stimulus (CS), such as light or tone, and an unconditioned stimulus (US), such as footshock. In this active avoidance paradigm, a conditioned avoidance response is defined as when the animal moves to the opposite compartment during the CS presentation (typically within 10 s) so as to avoid the shock. If the animal does not move to the other compartment, footshock is delivered. However, this footshock can be escaped by moving to the opposite compartment. Thus, performance is measured by the latency to avoid (i.e., enter the other chamber before the termination of the CS and the onset of the footshock) or escape (i.e., enter the other chamber after the onset of footshock) the US. There are two primary versions of this task: the one-way and the two-way. The first version always includes a defined safe chamber (i.e., a specific chamber where the animal will never receive the US). The second version does not define a specific chamber associated with the US. Instead, the CS is utilized to signal the onset of the US. The active avoidance paradigm is a cognitive task requiring spatial learning and cognitive coordination. It is useful for assessing associative learning (operant conditioning), short- and long-term memory, and it also provides procedures for testing acquisition, consolidation and retention processes.

The most extensively studied GABA$_B$R-active compound in this paradigm has been the GABA$_B$R agonist baclofen, with doses of 0.25 to 20 mg/kg. In the active avoidance task, as with other tasks described in this chapter, baclofen has shown contradictory findings, with both improved and impaired performance of mice and rats. Studies in CD1 mice (Farr et al. 2000), Sprague-Dawley (Fogel et al. 2010) and Long Evans rats (Stuchlik and Vales 2009) have shown an impairment in different versions of the active avoidance task, a study using Wistar rats has shown no effect (Kuziemka-Lęska et al. 1999) and other studies have shown memory improvement after administration of baclofen (Georgiev et al. 1988; Sharma and Kulkarni 1993), an effect that is consistently shown with GABA$_B$R antagonists.

In one of these studies, baclofen dose-dependently disrupted spatial learning and locomotion in the place avoidance task. Spatial learning and memory (i.e., spatial behaviour) refer to the ability of an organism to acquire and maintain, respectively, a mental representation of their environment. Spatial behaviour allows an animal to understand not only their location in relation to their surrounding environment, but also the location and dimension of objects and/or stimuli in the environment and how they are related to them and to each other. Importantly, assessment of the effects of $GABA_BR$ compounds on locomotor activity is a very important measure of possible therapeutic effects (e.g., spasticity, hyperactivity induced by psychostimulants) or locomotor side effects, as sometimes seen with $GABA_BR$ agonists, such as baclofen (e.g., motor impairment, muscle relaxation) in clinical settings and preclinical studies, that can significantly influence aspects of performance in a variety of animal models.

In the place avoidance task, animals are trained to move over a continuously rotating arena, on which an imperceptible to-be-avoided sector is defined, remaining stable with respect to the experimental room (Stuchlik and Vales 2009). In this study, baclofen administered at 3.5 mg/kg 30 min before the task selectively disrupted spatial behaviour in Long Evans rats, while doses of 4 and 6 mg/kg proved to decrease both avoidance efficiency and locomotor activity. Another study in which baclofen was administered 30 min before the task showed the opposite effect. In that study, which used LAKA mice, baclofen was administered at 0.25 and 0.5 mg/kg, and the 0.5 mg/kg dose was effective in improving one-way acquisition in mice (Sharma and Kulkarni 1993). In studies using Wistar rats, baclofen administered either 30 min pre-training (0.75 mg/kg, IP) or post-training (2, 5, 10, 20 mg/kg, IP) either had no effect (Kuziemka-Lęska et al. 1999) on the two-way acquisition or improved the two-way consolidation (Georgiev et al. 1988). Moreover, intra-pedunculopontine nucleus administration of baclofen 17 h post-training impaired two-way consolidation in Sprague-Dawley rats whose REM sleep was disrupted (Fogel et al. 2010). It is worth noting that the pedunculopontine nucleus has been implicated in the generation of REM sleep. Further, as memory consolidation may take place during increased post-learning REM sleep, any disruption of REM sleep can have an impact on the animals' subsequent performance in this task (Fogel et al. 2010). The variability of effects across doses already seen seems to be dependent on not only the exact procedures used, but also the route of administration and the species and strains used. The time of the administration does not seem to differ among studies already mentioned.

A recent study examined the role of dorsal hippocampal (CA1) $GABA_BRs$ on harmaline-induced memory consolidation deficits in mice using the step-down inhibitory avoidance task (Nasehi et al. 2017). Results from this study showed that post-training intra-CA1 injections of the $GABA_BR$ antagonist phaclofen, one of the first $GABA_BR$ antagonists described and the phosphonic acid derivative of baclofen, had no effect, while baclofen (0.1 μg/mouse) impaired animals' performance in this task, suggesting a modulation of storage of information. In further experiments, post-training infusion of harmaline (2 and 5 mg/kg, IP) decreased memory consolidation, while administration of phaclofen at a sub-effective dose (0.001 μg/mouse, intra-CA1) successfully blocked the deficits on memory consolidation induced by the highest doses of harmaline (2 and 5 mg/kg, IP). Further, a low dose of baclofen (0.001 μg/mouse, intra-CA1) potentiated impairment of memory consolidation induced by harmaline (2 mg/kg, IP). These findings suggest that the CA1 $GABA_BRs$ are involved in memory consolidation and that harmaline interacts with the CA1 $GABA_BRs$ to modulate this memory process (Nasehi et al. 2017).

Considerably more studies using the active avoidance task have used $GABA_BR$ antagonists rather than agonists (Getova et al. 1997; Getova and Bowery 1998; Farr et al. 2000; Getova and Dimitrova 2007). The vast majority of these studies have shown improvement in performance in the active avoidance paradigm. In these studies (Getova et al. 1997; Getova and Bowery 1998; Getova and Dimitrova 2007), Getova and colleagues used the $GABA_BR$ antagonists CGP 36742, CGP 55845, CGP56433, CGP 61344, CGP 62349, and CGP 71982, CGP 63360a, CGP76290a, and CGP 76291a showing different affinity for the $GABA_BR$ and different potency [for a detailed presentation of $GABA_BR$ antagonists, please see Nieto et al. (2021)]. All studies were conducted using the two-way version of the task and tested for both acquisition and retention in Wistar rats. With the exception of CGP36742, which impaired performance in the task, and CGP 76291a, which did not have an effect on acquisition, all other compounds improved performance. Interestingly, in the case of CGP36742, when the one-way version of the task was used, and the $GABA_BR$ antagonist was administered repeatedly for two weeks, prior to the task, it then also improved performance (Yu et al. 1997). Saclofen, another $GABA_BR$ antagonist, also improved performance in this task (Farr et al. 2000). Altogether, these data support the involvement of $GABA_BR$'s in memory processes.

In the passive avoidance paradigm, which is classically used to assess short- or long-term memory in animals (typically rats or mice), animals must react/behave contrary to their innate tendency to avoid light. The apparatus used in this paradigm is composed of two compartments, a black poorly illuminated compartment and a white illuminated compartment. During the acquisition/conditioning phase, the animal is placed in the white compartment. When the animal innately moves to the black compartment, it receives a mild footshock. Thus, during the initial phase the animal learns that moving to the dark compartment has negative consequences. During the test phase the animal is again placed in the white compartment and the passive avoidance response is examined. As opposed to the previous avoidance task discussed, the avoidance of the dark compartment requires the animal to remain in the white compartment and, therefore, the absence of movement (i.e., passive avoidance response).

The passive avoidance task is useful for assessing memory performance, which is positively correlated with the latency to escape from the white compartment; the better the recollection, the greater the latency to move. It is also useful for studying other mechanisms involved in cognition. A number of brain regions have been implicated in mechanisms active during the passive avoidance task [e.g., (Bradley et al. 2004)]; these include the hippocampus, medial prefrontal cortex (mPFC) and basal forebrain for acquisition (Sharma and Kulkarni 1993; Zarrindast et al. 1998; Kuziemka-Lęska et al. 1999; Car and Wiśniewska 2006; Car and Michaluk 2012;

Farahmandfar et al. 2017; Liu et al. 2017; Ebrahimi-Ghiri et al. 2018), and the hippocampus, amygdala, anterior cingulate cortex (ACC), mPFC and nucleus basalis magnocellularis (NBM) for consolidation (Zarrindast et al. 2002; Zarrindast et al. 2008b; Kunisawa et al. 2017).

Most of the studies on active and passive avoidance memory retention tests have used baclofen. Taking into account the species, the rodent strain, the routes of administration and the doses used, the majority of these studies have shown memory impairments, with some showing memory improvement and some showing no effect. Baclofen doses used would range between 0.125 and 20 mg/kg through systemic routes of administration, with variable directions of effect, while central administrations, in the amygdala or intracerebroventricularly (ICV), have shown impairments.

In one of the earliest studies, baclofen at doses of 2, 5 and 10 mg/kg injected intraperitoneally (IP) immediately after training improved retention in both active and passive avoidance tasks in male Wistar rats, while 20 mg/kg had no effect on active avoidance performance (Georgiev et al. 1988). Baclofen also enhanced memory in ICRC mice in the passive avoidance task, an effect that was blocked by the $GABA_BR$ antagonist CGP 35348 at a dose of 10 mg/kg (Saha et al. 1993). Further, both pre-training and post-training administration of CGP 35348 (75, 150 and 300 mg/kg, IP) significantly reduced the amnesic effect induced by scopolamine (1 mg/kg, IP) (Bianchi and Panerai 1993).

In another relevant study assessing cholinergic and GABAergic interactions on learning and memory (Sharma and Kulkarni 1993), (+/−)baclofen (0.25, 0.5 and 1 mg/kg) and (−)baclofen (0.25 and 0.5 mg/kg) also induced memory enhancement (for a discussion on the chirality of baclofen and the differences between the baclofen enantiomers, (R)-(−)-baclofen and (S)-(+)-baclofen, please see Nieto et al. (2021), whereas the $GABA_BR$ antagonist CGP 35348 did not show any effect per se, but reversed the (+/−)baclofen-induced delay in latency, without affecting the retention enhancing action of (+/−)baclofen. When sub-effective doses of GABA (50 mg/kg) and (+/−)baclofen (0.25 mg/kg) were co-administered, they induced a significant improvement in both acquisition and retention. In contrast, a different study assessing the effects of the cholinergic cognitive enhancer NS-105 on memory in the passive avoidance task in Wistar rats showed that NS-105 reversed cognitive impairment induced by baclofen (8 mg/kg, IP) (Ogasawara et al. 1999). Moreover, intra-hippocampal injection of baclofen (0.25, 0.5, 1 and 2 μg) reduced memory retention in rats, while intra-hippocampal injection of lower doses of the $GABA_BR$ antagonist CGP35348 (2.5, 5, 10 μg) did not affect and higher doses (25 and 50 μg) decreased memory retention. The same doses of CGP35348 (2.5, 5 and 10 μg), which did not have an effect per se, reduced the effect of baclofen (Zarrindast et al. 2002). Memory impairment with baclofen has been the primary result of other studies by the same research group in NMRI mice (Zarrindast et al. 2004). Interestingly, baclofen (1, 1.5 and 2 μg) prior to injection of morphine (20 mg/kg per day × 3 days) decreased the reversal of morphine-induced amnesia in morphine sensitized-mice (Zarrindast et al. 2008b). It appears that, at least in some of these studies, post-training administration of different doses of baclofen impairs memory retention while in some others even acquisition is impaired by administration of baclofen (Car and Wiśniewska 2006). It also appears that animals exhibiting different behavioural stereotypes such as aggressiveness or submissiveness are differentially affected by administration of 1 mg/kg of baclofen. In the passive avoidance test, it was shown that baclofen produced amnesia in aggressive mice and had a stronger amnesic effect in submissive mice (Dubrovina and Loskutova 2007). Moreover, a separate study has shown that $GABA_BR$ activation with baclofen accelerated extinction of fear memory in depressed-like mice (Dubrovina and Zinov'ev 2008).

Contrary to the above findings, numerous studies have shown no effect of baclofen in either of the passive avoidance measures (i.e., retrieval, consolidation, retention) (Swartzwelder et al. 1987; Castellano et al. 1989; Car and Wiśniewski 1998; Zarrindast et al. 1998, 2008a; Kuziemka-Łęska et al. 1999; Car et al. 2000; Cryan et al. 2004; Car and Wiśniewska 2006). The lack of effect of baclofen in the passive avoidance task measures in these studies may be due to differences in experimental condition methodology, drug doses and routes of administration or different rodent strains used.

Not many positive allosteric modulators (PAMs) for the $GABA_BRs$ have been developed, let alone tested for cognitive processes. In one of these few studies, the $GABA_BR$ PAM GS39783 did not show an effect in the passive avoidance task neither in CD1 mice (1, 3, 10, 30, 100 mg/kg) nor in Sprague-Dawley rats (25, 50, 100 mg/kg) at all of the doses tested (Cryan et al. 2004).

In studies utilizing $GABA_BR$ antagonists, administration of either of three $GABA_BR$ antagonists showed differential effects on the passive avoidance task, when the gamma-hydroxybutyrolactone (GHBL) animal model was used. The GHBL is an experimental model of absence seizures in rats and mice. Absence seizures in humans involve brief and sudden lapses of consciousness, as well as mild jerks of facial muscles electrophysiologically depicted as synchronous spike and wave discharges in the EEG (Snead 1991). In a similar manner, the GHBL animal model rapidly and consistently produces a combination of EEG and behavioural changes which resemble the human condition. CGP71982 improved both learning and memory retrieval, whereas CGP55845A and CGP62349 had no effect (Getova and Bowery 2001). Interestingly, administration of the $GABA_BR$ antagonist CGP36742 (SGS742) improved the learning capacity of mice, rats and rhesus monkeys, in a passive avoidance test, a partner recognition test and a conditional spatial colour task, respectively (Mondadori et al. 1993; Froestl et al. 2004). Moreover, treatment with either CGP35348 or CGP36742 significantly blocked acquisition impairment induced by baclofen, consolidation impairment induced by baclofen and NaNO2, and retrieval impairment induced by baclofen and 30% alcohol (Yu et al. 1996). Phaclofen had no effect in the passive avoidance task (Zarrindast et al. 1998; Dubrovina and Zinov'ev 2008).

Finally, some of the studies utilizing the passive avoidance task have examined mutant mice that lack specific $GABA_BR$ subtypes to further elucidate their role in learning and memory processes. In one of these studies, Jacobson and colleagues found that lack of either the $GABA_BR1a$ or $GABA_BR1b$ subunits had no effect on the passive avoidance task (Jacobson et al. 2007), while other studies have shown memory impairments in $GABA_BR1$ KO (Schuler et al. 2001), at a time when the splice variants of the heterodimeric receptor had not been characterized, or $GABA_BR2$ KO mice (Gassmann et al. 2004) in the same task. Interestingly, in the Jacobson study, latency to enter the dark side of a two-compartment trough-shaped apparatus was greater at retention testing than during training, although there was no effect of genotype on either training or retention latency (Jacobson et al. 2007). This indicated that both mutant strains could remember the shock they received in the dark compartment. The difference between the results of the Jacobson study and that of Schuler and colleagues (Schuler et al. 2001), which showed profound memory deficits in the retention phase of this test, can be attributed to the lack of both functional isoforms of the $GABA_BR1$ in the latter study. This further indicates that heterodimeric $GABA_BR1$ function is essential for the retention of passive shock-avoidance training, and that this may be accomplished with either of the $GABA_BR1a$ or $GABA_BR1b$ isoforms (Jacobson et al. 2007).

3.3 Morris Water Maze (MWM) and Radial Arm Maze (RAM)

Primarily designed to measure spatial learning and recall, the Morris Water Maze (MWM) is the most commonly used test to evaluate cognitive functions utilizing mnemonic mechanisms. By employing a variety of sophisticated protocols or procedures in the MWM, such as the visible or hidden platform test, the transfer test, and the relearning or repeated acquisition test, one can assess the acquisition and spatial location of relevant visual cues. These visual cues are further processed, consolidated, retained and retrieved, in order to successfully navigate, and thus locate, a hidden platform, with the goal of escaping the water and surviving (i.e., implementing memory functions as survival skills). The MWM has recently been used for the evaluation of ageing effects on memory, as well as those of drugs and experimental lesions in rodents, especially as these relate to animal models of neurodegenerative disorders with cognitive decline, such as Alzheimer's and Parkinson's disease. As with other tasks, findings from studies using the MWM are inconsistent. Studies using baclofen in mice or rats have shown impairment or no effect in the maze, while results from studies examining the effects of antagonists show effects in all directions, improvement, impairment or no effect.

More specifically, baclofen administered directly into the entorhinal cortex 15 min before training impaired learning and memory in the MWM in Sprague-Dawley rats (Deng et al. 2009). However, in the same strain of rats, baclofen administration 19 h before training systemically produced no effect (Li et al. 2014). Similar effects were identified in Wistar rats; intra-dorsal hippocampus administration of baclofen impaired performance (Arolfo et al. 1998), while systemic administration had no effect (Car and Michaluk 2012). In contrast, other studies in Wistar rats showed impairment in performance after systemic administration of baclofen 30 min before training (Nakagawa et al. 1995; Nakagawa and Takashima 1997). These contradictory findings may be due to the different time and doses of administration, taking into account that the same strain of rats was used in both studies, as well as systemic administration. Most recently, differently timed administrations of baclofen (right before pre-test or daily for 14 days, or repeatedly for a month) did not affect performance in the MWM at any of the doses tested in Wistar rats (Holajova and Franek 2018). Effect of $GABA_BR$ blockade by administration of the antagonist CGP35348 directly into the CA1 area of the hippocampus in Wistar rats largely depended on the timing of administration, with post-training and 15 min pre-test administration showing no effect, and 15 min pre-training administration showing an improvement in performance (Shahrzad and Nasser 2015).

In one study using numerous mouse strains to test the effect of CGP36742 on spatial learning and recall, administered systemically 40 min before training, C57BL/6 J and OF1 mice improved their performance compared to the control mice, while CD1 and DBA/2 mice demonstrated impaired task performance (Sunyer et al. 2007, 2009). Further, the performance of RGS7 KO mice indicated that RGS7, which regulates the interaction between $GABA_BRs$ and GIRK channels, is necessary to acquire this task (Ostrovskaya et al. 2014). Most recently, Phf24-null rats were developed by Serikawa and colleagues (Serikawa et al. 2019). Phf24 is known as Gαi-interacting protein (GINIP) and is associated with $GABA_BR$. Interestingly, Phf24 rats showed impairments in the MWM task. Another recent study focused on the effects of exercise and baclofen administration in striatum-lesioned rats and found that mild exercise and baclofen microinjection did not affect spatial learning or motor activity impairments, while the combination of them alleviated spatial learning and motor activity impairments in these rats (Modaberi et al. 2019).

The radial-arm maze (RAM) task is based on the natural survival tendency of food-deprived rodents to learn and remember different spatial locations for food in an eight-arm radial maze. A large number of protocols can be applied in this task which can be very useful in assessing neural bases for learning and memory. In the spatial working memory protocol, each arm of the maze is baited (i.e., food is placed in the arm) and repeat entries into an arm that has already been visited signify a working memory error. In the reference memory protocol, only some arms are baited at the beginning of the session. First entry into a non-baited arm signifies a reference memory error, and repeated entries to baited and non-baited arms are defined as working memory errors. As with some of the tasks previously described, this task can also be used with a large number of delays.

Most of the studies using the RAM to assess performance of animals per se, rather than symptomatology related to other conditions (e.g., memory processes in animal models of epilepsy), have used rats. Results from studies in rats using baclofen are inconsistent which may possibly be due to the different doses used. As an example, systemic administration of baclofen at a dose of 0.25 mg/kg (subcutaneously; SC) in female Sprague-Dawley rats 20 min before the test showed an improvement in performance, while a higher dose (1 mg/kg, SC) showed an impairment and all other doses in between had no significant effect (Levin et al. 2004). Higher doses

(1.25 and 2.5 m/kg, SC) had no effect in male Fisher 344 rats (Sidel et al. 1988). Moreover, intra-septal administration of baclofen in male Sprague-Dawley rats impaired consolidation in this task (Stackman and Walsh 1994). The septum has not been mentioned previously in this chapter. Thus, briefly, it is a subcortical structure in the forebrain that is found near the midline of the brain and its functions are not as well-known as those of other structures (e.g., amygdala, hippocampus, etc.). The septum in humans can be separated into two structures: the septum pellucidum and septum verum. The first seems to be a relay station between the hippocampus and the hypothalamus, and abnormalities in this structure are linked with various neurological conditions. The second is part of the limbic system; it receives afferent projections from the hippocampus, amygdala, and hypothalamus, and the ventral tegmental area (VTA). It also sends projections to the hippocampus, habenula, thalamus, VTA and hypothalamus and is linked with reward processes.

In one study using CD1 mice, either baclofen or the $GABA_BR$ antagonist CGP36742, both administered 15 min before the test, showed impairment and improvement of performance in the RAM, respectively (Carletti et al. 1993). Further, in two different studies, mice overexpressing $GABA_BR1a$ (Wu et al. 2007) or $GABA_BR1b$ (Stewart et al. 2009) showed an impairment in performance in this task, indicating the important role of $GABA_BRs$ in learning and memory.

With the exception of two studies where neither CGP35348 or CGP46381 showed an effect, most studies using CGP35348 (doses ranging from 12.5 to 300 mg/kg, IP) administered 20–30 min before training, or CGP36742 (150 mg/kg, IP) administered either 40 min before training or post-training, in male Long Evans rats, showed an improvement in performance in the RAM (in the 25 to 250 mg/kg range of doses for CGP35348) (Staubli et al. 1999; Helm et al. 2005; Chan et al. 2006).

In summary, these data suggest that $GABA_BR$ activation impairs performance in the MWM and RAM tasks in rodents, while the effects of $GABA_BR$ antagonists depend on the strain of rodents and doses of compounds used. Replication and further testing of these strains with both baclofen and $GABA_BR$ antagonists seems necessary so as to better understand the level of activation or de-activation of $GABA_BRs$ required for specific learning and memory processes to be successfully completed.

3.4 *Working Memory Tasks*

Working memory is defined as the capability to maintain the representation of a stimulus that is no longer present and perform mental operations on this representation so as to perform a specific task. A number of spontaneous alternation tasks can be used to study working memory in rodents. Some of the most commonly used ones are the T, Y and double-Y mazes. The T-maze in particular is a simple task commonly used for assessing spatial working memory in rats and mice, especially as a delayed alternation task. Similarly to the T-maze, the Y maze spontaneous alternation test is also commonly used for assessing spatial working memory in rats and mice, by measuring the willingness of rodents to spontaneously explore new environments. As seen with other similar tasks, rats and mice typically prefer to investigate a new arm of the maze rather than returning to one that was previously visited. Briefly, the Y-shaped maze has three white, opaque plastic arms at a 120° angle from each other. After introduction to the centre of the maze, the animal is allowed to freely explore the three arms. Over time and multiple arm entries, the animal naturally shows a tendency to enter a less recently visited arm. Performance is measured in terms of the percentage of alternated arm choices compared to total number of arm entries in order to calculate the percentage of alternations. The Y maze spontaneous alternation test is used to quantify cognitive performance in transgenic mice and identify novel potential therapeutic drugs for their effects on cognition. A higher percentage of alternations indicates better working memory (Miedel et al. 2017).

Another less commonly used set of tasks to assess spatial working memory are the delayed matching to position/non-matching to position tasks (DMTP/DNMTP). Either of these tasks, DMTP or DNMTP, measures short-term spatial memory, which is a component of the working memory construct (Dudchenko 2004; Teutsch and Kätzel 2019). When applied serially, they can also be used to measure reversal learning (Yhnell et al. 2016). In the DMTP task, a water-deprived animal interprets signals from two signal lights (left and right) and responds by pressing one of two levers (left and right). In the initial (sample) phase, one signal light is illuminated followed by a delay period where neither light is illuminated. Following the delay, during the choice phase, both lights are illuminated, and the animal has a limited amount of time to either respond at the lever in the same side to that which was signalled prior to the delay (DMTP) or opposite to that which was signalled prior to the delay (DNMTP), in the two versions of the task, respectively. The animal is given water reward for correct answers. These tests capture spatial working memory because the animal is required to maintain the representation of the stimulus, but it does not need to mentally manipulate the representation.

Experiments utilizing these tasks in rats have only taken place with Sprague-Dawley rats. In doses ranging from 0.03 to 3 nmol administered into the NBM, medial septum, mediodorsal thalamus, baclofen induced impairment, in a dose-dependent manner, although the exact effective doses differed depending on the timing of the administration (DeSousa et al. 1994; Romanides et al. 1999; Erickson et al. 2006). Contrary to these results, Escher and Mittleman showed that baclofen, in the lowest dose tested (2.5 mg/kg, IP), improves performance in female C57BL/6J and DBA/2 mice in the DMTP task when administered 10 min before the test (Escher and Mittleman 2004). The same performance improvement effect has been seen with phaclofen in this study (10, 30 mg/kg, IP), when administered 10 min before the test in both mouse strains.

Results of the effects of antagonists in these tasks have been inconsistent, possibly due to the route of administration, the doses, and the specific memory process captured through the timing of the administrations (i.e., pre-training, post-training, pre-test). Thus, CGP55845 showed no effect in male Fisher 344 rats when it was administered in the mPFC while it showed an impairment in performance when administered 40 min before training (0.01 mg/kg, IP) (Bañuelos et al. 2014). It also showed no effect in B6/C3H mice (0.5 mg/kg, IP), when administered 2–3 h before training (Kleschevnikov et al. 2012a). In contrast, saclofen administered directly into the DG of Sprague-Dawley rats 15 min before training improved performance (Liu et al. 2014).

Importantly, both $GABA_BR1a$ and $GABA_BR1b$ isoforms are necessary for standard or improved task performance in these mazes, as shown through impaired performance in this task exhibited by $GABA_BR1a$ or $GABA_BR1b$ KO mice (Jacobson et al. 2007). In a more recent study, hippocampal protein levels of both $GABA_BR1$ isoforms were increased in mice trained in this task compared to untrained mice (Falsafi et al. 2015).

3.5 Novel-Object Recognition (NOR) and (Dis)Location (NOL) Tasks

The NOR and NOL tasks are based on the natural tendency of rodents to explore novelty. Both the NOR and NOL tasks measure the ability of animals to discriminate between familiar and novel objects in their environment. In the acquisition/training phase of the task, animals are allowed to freely explore two objects that are identical in an experimental arena. The test phases of these tasks will occur within the same day for the examination of short-term memory or in more than 24 h for the examination of long-term memory. In the test phase of the NOR, two objects, including one of the two previously placed (i.e., familiar) objects (i.e., the objects present during the acquisition/training phase) and a novel one, are presented to the animal. In the NOL, both of the original objects are presented in the test phase. However, one object is placed at a different location. In either of the tasks, time spent investigating the objects is recorded at each phase of testing. If the animal remembers an object, it will not spend as much time investigating the familiar object in the NOR, while in the NOL, the animal should spend less time investigating the object that has not been relocated. The choice to explore the novel object reflects the use of recognition memory processes, which in these two tasks activate both the prefrontal cortex and the hippocampus (Lee et al. 2005; Akirav and Maroun 2006).

Importantly, a number of studies have shown that presynaptic $GABA_BRs$ are essential for NOR and NOL performance (Vigot et al. 2006; Jacobson et al. 2007; Cullen et al. 2014). Or, alternatively, they have shown that if presynaptic $GABA_BRs$ are not present, this has deleterious effects in the NOR and NOL. Another study using RGS7 KO mice showed the same impairments in both males and females. In this study, RGS7 ablation increased GABABR-GIRK coupling sensitivity and slowed GABABR-GIRK deactivation rates, indicating that these mice possibly experience enhanced postsynaptic $GABA_BR$-mediated signalling (Ostrovskaya et al. 2014). In contrast, mice with enhanced postsynaptic $GABA_BR$ signalling did not change their performance in the NOR, but were impaired in the NOL, suggesting that NOL may be a more sensitive task for this increased signalling than NOR (Terunuma et al. 2014).

In a study by Pitsikas and colleagues (Pitsikas et al. 2003), the role of $GABA_BR$ on recognition memory in Sprague-Dawley rats was assessed by administration of baclofen and the $GABA_BR$ antagonist CGP 35348 using the NOR task. Results showed that baclofen (0.5, 2 and 4 mg/kg, IP) dose-dependently impaired performance in this task, with the highest dose tested showing a significant impairment, suggesting an effect on acquisition and retention of information. Administration of the $GABA_BR$ antagonist CGP 35348 (100 and 300 mg/kg, IP) counteracted the baclofen-induced performance deficits. These results indicated that $GABA_BRs$ are involved in recognition memory. More recent data in Sprague-Dawley rats in which higher doses of baclofen (12.5, 25 mg/kg, IP) were administered at pre-training (19 h before training) had the same impairment effect (Li et al. 2014). Other studies have shown no effects of either baclofen or the $GABA_BR$ antagonist CGP55845, with systemic administration, in Wistar rats or NMRI mice, respectively (Car and Wiśniewski 1998; Khanegheini et al. 2015) in the NOR. Moreover, the $GABA_B$ receptor antagonist CGP35348, administered directly in the DG (1, 10 and 100 μg/μ L) was recently identified as a possible therapeutic agent against the progression of acute Aβ toxicity-induced memory impairment through its effects in the NOR in adult male rats (Almasi et al. 2018).

Most recently, the $GABA_BR$ PAMs GS39783 and CGP7930 and mGluR5 antagonist CDPPB were assessed in the NOR in an animal model of schizophrenia (i.e., the MK-801 model) in Wistar rats (Wierońska et al. 2015). MK-801 (i.e., dizocilpine) is a non-competitive antagonist with a high affinity for the N-Methyl-D-aspartate (NMDA) receptor, and a potent anticonvulsant. It binds on NMDA receptors in various brain sites, particularly in the hippocampus. It is a neuroprotective agent in animal models of stroke, trauma and Parkinsonism. However, it can also induce psychotic behaviour and neuronal degeneration. Directly relevant to this content, because of the above-mentioned effects, MK-801 is used as a cognitive impairment/dysfunction associated with schizophrenia (CIAS) animal model (Brown et al. 2014). Both CGP7930 and GS39783, administered IP 30 min before MK-801 administration, dose-dependently reversed MK-801-induced deficits. The effect of CGP7930 was observed at the dose of 1 mg/kg, while lower doses (0.1 and 0.5 mg/kg) and higher doses (2 and 5 mg/kg) were not effective. Further, in the co-administration experiment, CDPPB (1 mg/kg), and GS39783 (0.1 mg/kg) were administered at sub-effective doses, 30 min before acute MK-801 administration. The co-administration of these compounds induced clear antipsychotic-like effect in the NOR task. Interactions of the GABAergic system with mGluRs in the same animal model of cognitive symptoms in schizophrenia (i.e., the MK-801-induced deficits model) were further assessed by the same research group, this time using the mGluR4-selective orthosteric agonist, LSP4–2022 (Woźniak et al. 2016) with no interaction effect in the NOR. Similar promising results have been identified in the interactions between the muscarinic and GABAergic systems in the NOR as they relate to cognitive impairments in animal models of schizophrenia (Cieślik et al. 2019).

Additionally, in an effort to investigate whether the GABAergic system is involved in the beneficial effects of betaine in cognitive processes through the amelioration of water-immersion restraint stress (WIRS)-induced memory impairments, adult male mice were co-administered betaine and $GABA_BR$ agonists or antagonists after WIRS and were assessed for memory functions using the NOR 3–6 days after WIRS (Kunisawa et al. 2017). The co-administration of the $GABA_A$ receptor antagonist bicuculline (1 mg/kg) or the $GABA_BR$ antagonist phaclofen (10 mg/kg) 1 h after WIRS suppressed the memory-improving effects induced by betaine. Additionally, administration of the $GABA_A$ receptor agonist muscimol (1 mg/kg) or baclofen (10 mg/kg) 1 h after WIRS attenuated memory impairments. These interesting findings indicate that the beneficial effects of betaine may be partly mediated by the GABAergic system. Moreover, intra-hippocampal injection of Cortistatin-14 (CST-14), a neuropeptide related to somatostatin, impaired recognition memory consolidation in mice through activation of $GABA_BRs$, among other receptor systems, an effect that was shown through the reversal of CST-14 impairment by saclofen in the NOR (Jiang et al. 2017).

Specifically for the NOL, except for some of the data presented through other studies above, administration of baclofen or $GABA_BR$ antagonist phaclofen in NMRI mice had the same direction of effects, i.e., impairment in performance (Khanegheini et al. 2015), while CGP55845 in B6/C3H mice, similarly to the result in the NOR, had no effect (Kleschevnikov et al. 2012a).

3.6 Other Cognitive Tasks

Some other tests assessing the role of $GABA_BRs$ in cognitive processes in rodents have also been used, although the literature currently is not extensive. One of these tests is the Barnes maze, a popular test for assessing spatial learning and memory in rats and mice. In this test each animal is placed on top of the Barnes maze cyclic platform, a brightly lit environment consisting of a specific number of holes around its periphery. While on the open platform, rodents naturally seek a dark enclosed space, which is provided in the form of a dark box (goal box) under one of the round holes around the periphery of the platform. Except for examining the learning curve of the animal over consecutive trials, the searching strategy of the animal can also be important in this task. Although some animals randomly search for the correct hole, others may use a certain pattern behaviour and systematically check each hole. When animals move directly to the correct hole irrespective of their starting position, this indicates that spatial memory has formed. The amount of time required for the animal to locate the goal box using visuo-spatial cues surrounding the maze periphery is measured.

In one of the studies using the Barnes maze, Li and colleagues tested the effects of baclofen as well as the $GABA_BR$ PAM BHF177 in C57BL/6 J mice by administering them either before the training or before the test. Baclofen and BHF177 had no selective effects on spatial learning and memory in the Barnes maze, except at doses that were sedative or pro-convulsant (Li et al. 2013). In addition, in a most recent study (Sahraei et al. 2019), the role of systemic (1, 5 and 10 mg/kg, IP) and intra-nucleus accumbens (NAcc) (1, 5 and 10 μg/rat) injections of baclofen on spatial memory impairments in stress-exposed rats were assessed. Results showed that both the systemic and intra-NAcc administration of baclofen dose-dependently reduced escape latency and total distance and increased velocity in the treatment groups in the training trials. During the test, the rats that had received 5 mg/kg of baclofen had the highest target frequency, but there were no significant differences in velocity, duration or distance to the target between the groups. These findings suggested that baclofen can dose-dependently improve spatial memory, and GABA$_B$Rs in the NAcc play an important role in spatial memory.

Very few studies can also be identified using the conditional space colour task and the social recognition task. In the first case, primates learn to locate a food reward based on the colour and location of beakers. The complexity of the task can be gradually increased, with three conditions implemented and training in between. In the second case, animals are required to recognize a familiar animal by spending less time interacting with it, compared to a novel animal in the same environment, when both are presented at the same time. These tasks seem to require activation of the prefrontal cortex and hippocampus, and the amygdala and hippocampus, respectively. Studies by one research group have shown that either in Rhesus monkeys (0.5 mg/kg) or in Sprague-Dawley rats (0.003 to 300 mg/kg), systemic administration of CGP36742 improved performance in these tasks (Mondadori et al. 1993; Mondadori et al. 1996), indicating that inhibition of GABA$_B$R activation can improve learning and memory in complex tasks and recall in social memory.

4 GABA$_B$R Involvement in Cognitive Performance of Neurodevelopmental Conditions

4.1 GABA$_B$Rs During Embryo-Foetal Development

GABA$_B$Rs have been implicated in glial cell development in the peripheral nervous system (PNS), although the exact function of GABA signalling is not known. Recent studies have shown that GABA and GABA$_B$Rs are expressed in premyelinating and nonmyelinating Schwann cells throughout development and post injury, while GABA through GABA$_B$Rs does not seem to be involved in Schwann cell proliferation (Corell et al. 2015). Further, embryonic GABA$_B$R blockade alters cell migration, adult hypothalamic structure, and anxiety- and depression-like behaviours in mice in a sex-specific manner (Stratton et al. 2014), while GABA also regulates corticotropin releasing hormone levels in the paraventricular nucleus of the hypothalamus in newborn mice (Stratton et al. 2011). Another earlier study showed that

$GABA_BRs$ were present in the ventromedial nucleus of the hypothalamus at all ages examined, from embryonic day 13 to postnatal day 6 (Davis et al. 2002). A very interesting study in rats at about that time, using in situ hybridization and RNase protection assays (RPA) to investigate the early foetal expression of $GABA_BR1$ and $GABA_BR2$ mRNAs on the development of the rat CNS (Kim et al. 2003), showed that there was early and strong $GABA_BR1$ mRNA expression in the spinal cord, medullar and cerebral cortex neuroepithelium of discrete brain regions on gestational day (GD) 11.5. On GD 12.5, $GABA_BR1$ mRNAs were also found in the hippocampal formation, cerebral cortex, intermediate and posterior neuroepithelium, and the pontine neuroepithelium of the whole brain. Further, RPA results showed $GABA_BR1$ mRNA was intensely expressed on GD 11.5 and GD 12.5, when it was first detected in the ganglia, thalamus and cerebellum. Altogether, this data suggested that $GABA_BR1$ might have a role in the early foetal brain and spinal cord during pre- and post-synaptogenesis, neuronal maturation, proliferation and migration in the early development of the rat CNS. Data from the studies presented above indicate that $GABA_BRs$ play an important role in normal neurodevelopment and that, conversely, if $GABA_BR$ expression or function is for some reason (negatively) influenced during different phases of development, this can result in various symptoms and neurodevelopmental conditions, affecting not only development, but also cognition.

Changes in excitation and inhibition mechanisms are associated with neurodevelopmental disorders whose primary symptomatology includes cognitive and intellectual disabilities (Kramvis et al. 2020). Some of these conditions are ASD, FXS and DS, as well as Rett's and Tourette's syndrome, and neurofibromatosis (Deidda et al. 2014; Yamasue et al. 2019). Conversely, reduced GABAergic transmission, resulting from fewer GABA receptors may upset the excitatory/inhibitory balance which could result in both seizure-related disorders and intellectual impairments common to ASD (Fatemi et al. 2014). Delayed synaptic maturation, abnormal synaptic structure and/or function and alterations in intracellular signalling pathways have been linked to the pathogenesis of these conditions (Hampson et al. 2011). However, while DS and FXS are known to result from a specific genetic mutation, the causes of the majority of cases of ASD are unknown.

4.2 Autism Spectrum Disorder (ASD)

Growing evidence suggests a possible role for GABA in the neuropathophysiology of ASD (Blatt and Fatemi 2011; Enticott et al. 2013). Several lines of evidence mainly from animal studies, but also from clinical trials, point to an imbalance between neuronal excitation and inhibition in at least a subgroup of individuals with ASD or corresponding animal models (Gandal et al. 2012; Silverman et al. 2015; Frye et al. 2016; Sinclair et al. 2017; Veenstra-Vanderweele et al. 2017).

At the clinical level, continuously growing evidence suggests that the function of the GABAergic system is abnormally low in ASD (Fatemi et al. 2010). This indicates that administration of $GABA_BR$ agonists or positive modulators may be useful for the treatment of ASD symptomatology. In an earlier study (Oblak et al. 2010), $GABA_BR$ density in individuals with ASD and controls was quantified in the ACC and posterior cingulate cortex, areas important for socio-emotional and cognitive processing, and the fusiform gyrus, an area important for identification of faces and facial expressions. There were significant reductions in $GABA_BR$ density in all three regions examined, suggesting that alterations in this key inhibitory receptor may contribute to the functional deficits in individuals with ASD.

In a 10-week randomized-controlled study aimed at evaluating the potential of baclofen to enhance the effect of risperidone in children with ASD with moderate-to-severe irritability symptoms (Mahdavinasab et al. 2019), researchers used the Aberrant Behaviour Checklist-Community Edition (ABC-C) for the outcome measures on each of the follow-up visits. Results showed significant improvement for all the ABC subscales used, including that of inappropriate speech, of relevance to this chapter. Importantly, combined administration of baclofen and risperidone exerted a greater effect on improvement of hyperactivity symptoms at both midpoint and endpoint when compared with treatment with placebo plus risperidone, suggesting the possibility of using baclofen as an additional treatment to risperidone for further improvement of variable symptomatology in ASD.

Exploratory clinical trials conducted with STX209 (also known as arbaclofen or R-(−)-baclofen; the R-(−)-enantiomer of baclofen) found improvement on several outcome measures, including the ABC-Irritability (the primary endpoint) and the Lethargy/Social Withdrawal subscales, the Social Responsiveness Scale, the CY-BOCS-PDD and clinical global impression scales (Erickson et al. 2014; Veenstra-Vanderweele et al. 2017). Results from these studies indicated that arbaclofen may have the potential to improve symptoms in some children with ASD.

At the preclinical level, further exploration of the interactions between GABA and glutamate networks in relation to ASD showed that a particular type I transmembrane protein with preferential expression in the mammalian CNS, which is identified as PIANP, is involved in the control of behavioural traits in mammals and interacts with $GABA_BR1$ (Winkler et al. 2019). In a specific strain of global PIANP KO mice, researchers identified decreased size and altered cellular compositions of the DG as well as the cerebellum and decreased the number of cerebellar Purkinje cells. At a functional level, loss of PIANP led to impaired presynaptic $GABA_BR$-mediated inhibition of glutamate release and altered gene expression in the cortex, hippocampus, amygdala and hypothalamus including downregulation of Erdr1, a gene linked to autism-like behaviour. At a behavioural level, PIANP deficiency led to context-dependent enhanced anxiety and spatial learning deficits, an altered stress response, severely impaired social interaction, and enhanced repetitive behaviour, all representative characteristics of ASD phenotypes (Winkler et al. 2019). This study suggested PIANP as a potential new candidate gene involved in ASD, cerebellar and hippocampal pathology, and $GABA_BR$-mediated signalling.

In accordance with these recent findings, earlier results of the role of $GABA_BRs$ in N-methyl-D-aspartate-receptor (NMDAR) hypofunction were examined. NMDAR signalling and, in particular, hypofunction is associated with intellectual disability and disorders, such as ASD and schizophrenia, with phenotypes of social, cognitive and gamma (30–80 Hz) oscillatory abnormalities. Administration of the $GABA_BR$ agonist baclofen in NMDA-NR1(neo−/−) mice, a strain that shows reduced expression of the necessary NR1 subunit to model disrupted developmental NMDAR function, improved excitatory/inhibitory signalling balance, gamma-signal-to-noise ratio, which predicts deficits in working memory, and broadly reversed behavioural deficits (Gandal et al. 2012). These findings highlight a potential use of $GABA_BR$ agonists for the treatment of common phenotypes in different disorders showing NMDAR hypofunction.

Moreover, a very interesting recent study by Manz and colleagues (Manz et al. 2019) further elucidated the role of the interactions between the GABAergic and glutamatergic systems in reward circuits as they affect goal-directed behaviour, by identifying a new mechanism of feed-forward inhibition within the nucleus accumbens. This is highly relevant to not only autism, but any other disorder affected by reward circuit 'malfunctions', and motivational difficulties, such as major depressive disorder and addiction. In this study (Manz et al. 2019), $GABA_BR$ function at glutamatergic synapses within parvalbumin (PV)-expressing interneurons-embedded microcircuits in the NAcc core of male mice was explored. Results showed that PV-interneurons within feed-forward microcircuits target $GABA_B$ heteroreceptors on glutamate terminals. It is worth noting that activation of presynaptically-expressed $GABA_B$ heteroreceptors decreases glutamatergic synaptic strength by engaging a signalling pathway that interferes with the mechanism of vesicular exocytotic release.

4.3 *Fragile X Syndrome (FXS)*

Fragile X syndrome (FXS) is a common syndrome within ASD and a monogenetic cause of intellectual disability, co-existing with autism spectrum features, as well as psychiatric and medical problems. FXS is caused by the lack of the fragile X mental retardation protein (FMRP), a translational regulator of specific mRNAs at the postsynaptic level (Tassone and Hagerman 2003; Zafarullah and Tassone 2019) in different brain areas including the cerebellum (Maurin et al. 2015; Zhang et al. 2015). The lack of FMRP leads to atypical synaptic plasticity, potentially triggered by a homeostatic disturbance between excitatory and inhibitory network functioning at the level of the synapse (Zupan and Toth 2008; Pacey et al. 2009; Zeidler et al. 2018).

In the last decade or so, numerous studies have looked at the potential involvement of $GABA_BRs$ in FXS and its symptomatology (Adusei et al. 2010; Yamasue et al. 2019) through the development of a corresponding animal model of the Fragile X Mental Retardation 1 (*Fmr1*) knock out (*Fmr1*-KO) mouse (Kazdoba et al. 2014) and through clinical trials with the $GABA_BR$ agonists baclofen and arbaclofen (Hopkins 2011). *Fmr1*-KO mice share several phenotypes with FXS patients including cognitive deficits – of particular focus to this chapter – altered spine morphology, hyperactivity, sensory hypersensitivity, repetitive behaviours and macroorchidism (Bakker and Oostra 2003). In regard to the *Fmr1*-KO mouse model, there are divergent opinions in regard to the usefulness of it as a translational preclinical disease model, either because of discrepancies within the *Fmr1*-KO mouse model itself or because of the outcome measures currently used or because of both factors combined making it an over-predictive model of clinical efficacy (Berry-Kravis et al. 2018).

When the expression and presence of $GABA_B$Rs and their subunits was recently examined (Kang et al. 2017), selective deficits in the $GABA_B$R1a subunit expression were identified in *Fmr1*-KO mice, but the levels of the respective mRNAs remained unaltered. Similar trends of $GABA_B$R1a expression were seen in the hippocampus of a subset of FXS patients. Further, corresponding with other studies indicating that $GABA_B$Rs have a strong pre- and postsynaptic inhibitory effect on neurotransmission, with $GABA_B$R1a subunit-containing receptors mediating presynaptic inhibition in particular, deficits in the ability of $GABA_B$Rs to suppress glutamate release were identified in *Fmr1*-KO mice, while their ability to suppress GABA release and induce postsynaptic hyperpolarization was unaltered. In the same study, administration of arbaclofen reversed the imbalance between excitatory and inhibitory neurotransmission in *Fmr1*-KO mice. Altogether, results from this study showed that selective deficits in the activity of presynaptic $GABA_B$Rs contribute to the pathophysiology of FXS.

As seen in the study presented above, in general, administration of $GABA_B$R agonists, such as baclofen or arbaclofen (Henderson et al. 2012), and PAMs, such as CGP7930 (Zhang et al. 2015) or GS39783 (Pacey et al. 2011), corrects exacerbated protein synthesis and multiple phenotypes in *Fmr1*-KO mice (Henderson et al. 2012; Silverman et al. 2015), although in some cases tolerance also develops (Pacey et al. 2011), a side effect commonly seen with $GABA_B$R agonists, but less commonly occurring with $GABA_B$R PAMs for many of their actions (Vlachou et al. 2011). These are the two most commonly used $GABA_B$ compounds used in the FXS preclinical studies. Similarly, acamprosate, which activates both $GABA_A$ and $GABA_B$Rs, also ameliorates several phenotypes in *Fmr1*-KO mice (Schaefer et al. 2017). The focus of this section though is on the effects of these compounds on cognitive processes in these animal models or in the clinical trials.

In one of the recent studies, the *Fmr1*-KO mice showed increased auditory-evoked high-frequency gamma (30–80 Hz) power, indicating sensory hypersensitivity, compared to the C57BL/6 control mice, as measured by electroencephalography, with no sex differences identified. The same study also showed, among other behaviours, decreased T-maze spontaneous alternation, indicating diminished working memory. The $GABA_B$R agonist racemic baclofen (i.e., the racemic mixture of baclofen, containing the S-(+)-enantiomer and the R-(−)-enantiomer of baclofen) normalized auditory-evoked neural oscillations and behavioural deficits in the *Fmr1*-KO mice (Sinclair et al. 2017). In the most recent study by Kramvis and colleagues (Kramvis et al. 2020), GABAergic signalling was investigated in the mPFC of *Fmr1*-KO mice during prepubescence and adolescence. Molecular and functional changes were detected, with the second being most prominent during early postnatal development and resulting in stronger inhibition, through increased synaptic inhibitory drive and amplitude, and reduction of inhibitory short-term synaptic depression. Further, there was an increased number of receptors opening during peak current in *Fmr1*-KO inhibitory synapses during prepubescence. Changes in amplitudes and plasticity returned to normal during adolescence, although the inhibitory drive was reduced in *Fmr1*-KO, while synaptic kinetics were prolonged. Importantly, $GABA_{B1}R$ expression levels were different in *Fmr1*-KOs compared to their WT littermate controls. These findings further indicated the involvement of synaptic GABAergic changes in the mPFC, an area strongly implicated in cognitive processing, in FXS pathology (Kramvis et al. 2020). Further confirmation of an imbalance between the excitatory and inhibitory synapses in *Fmr1*-KO mice, this time in the cortico-hippocampal feed-forward circuits formed by the temporoammonic (TA) pathway also comes from another study (Wahlstrom-Helgren and Klyachko 2015). The TA is a direct, monosynaptic pathway leading from layer III of the entorhinal cortex to the distal dendritic region of area CA1 of the hippocampus. This projection has been implicated in a number of functions including memory processing (i.e., encoding and retrieval) and spatial navigation, generation of oscillatory activity, and control of hippocampal excitability (Dvorak-Carbone 1999). The above introduced study showed that inhibitory, but not excitatory, synapse dysfunction underlies cortico-hippocampal feed-forward circuit abnormalities in these mice, an effect primarily mediated by presynaptic $GABA_B$ receptor signalling in the TA pathway. The same study also found that GABA release is reduced in TA-associated inhibitory synapses of *Fmr1*-KO mice and this effect is regulated by $GABA_BRs$ (Wahlstrom-Helgren and Klyachko 2015).

Some studies have suggested an overlap in pathophysiology of 16p11.2 microdeletion syndrome and FXS. Human chromosome 16p11.2 microdeletion is among the most common gene copy number variations (CNVs) known to confer risk for intellectual disability and ASD and affects an estimated 3 in 10,000 people. This syndrome is characterized by intellectual disability, impaired language, communication and socialization skills, and ASD. Taking into account improvement in FXS phenotypes observed following chronic treatment with arbaclofen, a recent study aimed to examine the effects of chronic oral arbaclofen administration in two generated mouse models of 16p11.2 microdeletion syndrome. Results showed that chronic activation of $GABA_BRs$ improved performance on a series of cognitive and social tasks known to be impaired in the two different 16p11.2 deletion mouse models, indicating that arbaclofen may be useful in treating some of the core symptoms of human 16p11.2 microdeletion syndrome (Stoppel et al. 2018).

4.4 Down's Syndrome (DS)

DS is a neurodevelopmental disorder caused by the triplication of Chromosome 21 (trisomy 21) and characterized by numerous neurodevelopmental alterations and intellectual disability (Ohira et al. 1997; Hattori et al. 2000). Individuals with DS, as well as DS animal models, such as the Ts65Dn (Ts) mouse model, exhibit impairments in several cognitive processes, learning and memory domains, including hippocampus-dependent declarative (in humans) or spatial (in rodents) memory and visual recognition memory, the latter of which is largely controlled by the perirhinal cortex (Roncacé et al. 2017). Numerous preclinical research strategies have been examined in the last decade or so, with an aim to identify innovative therapeutic approaches to DS symptomatology, one of which focuses on the modulation of $GABA_B$Rs and the mechanisms or the pathways they influence.

The Ts65Dn (Ts) mouse model of DS is sensitive to an infantile spasms phenotype (Cortez et al. 2009; Blichowski et al. 2015). Further, it contains the core genomic triplication of the DS critical region, which includes 3 copies of the Kcnj6 gene that encodes the $GABA_B$R-coupled GIRK subunit 2 (GIRK2) channel (Best et al. 2007; Cramer et al. 2010; Kleschevnikov et al. 2012a, b; Joshi et al. 2016). GIRK channels hyperpolarize neurons to inhibit synaptic transmission throughout the nervous system (Harashima et al. 2006a, b; Zhou et al. 2012). The $GABA_B$R-coupled GIRK2 channel is necessary for the $GABA_B$R agonist-induced infantile spasms phenotype in hippocampal neurons and those of the DG, frontal cortex and substantia nigra (Harashima et al. 2006a, b; Fernandez et al. 2007) of the Ts mouse and may represent a novel therapeutic target for the treatment of infantile spasms in DS (Blichowski et al. 2015). Importantly, LTP, a cellular model for learning and memory, is impaired in the CA1 hippocampal area and the DG of Ts and other DS animal models (Siarey et al. 1997; Kleschevnicov et al. 2004; Costa and Grybko 2005; Belichenko et al. 2007; Yu et al. 2010; Kleschevnicov et al. 2017).

Contrary to the $GABA_B$R agonist-induced spasms in DS and other syndromes, treatment with $GABA_B$R antagonists seems to improve cognitive performance, especially hippocampus-based cognitive functions, with findings that suggest further exploration of the $GABA_B$R antagonists' therapeutic potential in DS. In one of these studies using $GABA_B$R antagonists, CGP55845 restored memory of Ts mice in the novel place recognition, NOR, and contextual fear conditioning tasks, although it did not affect locomotion and performance in the T-maze. Further, CGP55845 increased hippocampal levels of brain-derived neurotrophic factor (BDNF) in Ts mice, while treatment of hippocampal slices with the $GABA_B$R antagonists CGP55845 or CGP52432 enhanced LTP in the DG of Ts mice (Kleschevnikov et al. 2012a).

Altogether, this data indicate that either $GABA_B$R agonists or antagonists seem to enhance cognitive processes in neurodevelopmental conditions and may be used therapeutically to treat cognitive deficits, as long as the right $GABA_B$R location or pathway can be targeted.

5 Neurodegenerative Disorders

A detailed account of the role of $GABA_B$Rs in neurodegeneration, with a special focus on epilepsy and AD, is presented in Princivalle (2021). In this section, a brief summary of relevant research in neurodegenerative disorders as related to cognitive processes will be presented.

5.1 *Alzheimer's Disease (AD)*

AD pathophysiology is largely known to be based on amyloid-β plagues and neurofibrillary tangles accumulation in the brain tissue, mainly affecting regions of the hippocampus and cortex, but also expanding to other areas of the brain as the disease progresses. The degeneration of these brain areas is responsible for the symptomatology of AD patients which features memory loss (especially short-term memory and expanding to long-term memory) as its main or primary symptom, and includes lack of inhibition and other cognitive functions controlled by the frontal cortex, but also numerous gradual changes in behaviour.

As also discussed extensively in regard to most of the data on the effects of $GABA_B$Rs in cognitive and other processes, it appears that an imbalance between excitatory and inhibitory synapses (Gigout et al. 2015) may be responsible for the changes identified in $GABA_B$R numbers and/or function or their linked proteins in AD patients or animal models, and these changes may be triggered by amyloid-β in the hippocampus in the case of AD (Nava-Mesa et al. 2013; Mayordomo-Cava et al. 2015; Pilipenko et al. 2018, 2019). Thus, maintaining balance among the GABAergic and other neurotransmitter systems can be considered a beneficial strategy to slow down AD progression (Pilipenko et al. 2019). It is also important to note that these GABAergic and other neurotransmitter changes at cellular and network levels correlate with density of neurofibrillary tangles in the brain tissue.

As indicated previously, most of the recent data point out the role of presynaptic $GABA_B$R1 in AD pathophysiology at cellular level. A most recent review of the role of presynaptic function in AD gives a detailed account of these processes (Barthet and Mulle 2020), including the ones that focus on the presynaptic $GABA_B$R1.

An etiological factor for AD is chronic cerebral hypoperfusion (CCH), a condition that constitutes one of the causes of vascular dementia (VaD). CCH causes cognitive impairment and contributes to Alzheimer's pathology. In an effort to further understand this condition, a mouse model of CCH has been developed by unilateral common carotid artery occlusion (UCCAO). This animal model shows significant short-term memory deficits and mild long-term spatial memory impairment, and selective neurodegeneration in the brain (Zhao et al. 2014), among other effects on protein-related processes. Importantly, $GABA_B$R activation ameliorates cognitive impairment via restoring the balance of hyperpolarization-activated cyclic-nucleotide-gated cation nonselective (HCN) channels type 1 and 2 (HCN1 and

HCN2, respectively) surface expression in the hippocampal CA1 area in rats with CCH. HCN channels and $GABA_B$Rs mutually co-regulate the function of neurons in many brain areas including the hippocampus [for details on the molecular mechanism of this interaction, please see (Li et al. 2014)].

Extensive preclinical studies have used either $GABA_B$R agonists or antagonists to test for potential cognitive enhancing effects in AD animal models (Kumar et al. 2017; Almasi et al. 2018; Sahoo et al. 2018). In one of these studies, a non-transgenic rat model of AD obtained by intracerebroventricular streptozotocin (ICV STZ) injection was used (Pilipenko et al. 2018). Single or double ICV STZ injection (s) chronically decrease cerebral glucose uptake and produce multiple effects that resemble molecular, pathological and behavioural features of Alzheimer's disease (AD) (Grieb 2016). The effects of muscimol and baclofen at very low doses (0.01–0.05 mg/kg) on spatial memory were assessed and the expression of cortical and hippocampal proteins related to neuroinflammation. Baclofen enhanced memory and anti-inflammatory effects, strongly indicating an important role of $GABA_B$Rs in AD (Pilipenko et al. 2018). Interestingly, findings from this study indicated that low doses of $GABA_B$R agonists could be effective because they involve other allosteric sites or cell signalling and regulatory pathways in these processes (Pilipenko et al. 2018). Moreover, the effects of $GABA_B$R activation on spatial memory and learning ability in the AD rats were measured by the MWM, and results showed that baclofen restored spatial memory and learning ability of AD rats and suppressed the neuronal apoptosis and hippocampal atrophy by activating the PI3K/Akt signalling pathway, once again indicating a therapeutic role for $GABA_B$R manipulations in AD (Sun et al. 2020).

Importantly, one of the most successful progressions of $GABA_B$R-active compounds into clinical trials and potentially an approval for the treatment of cognitive dysfunction (i.e., memory and attentional performance) in AD was that of the $GABA_B$R antagonist CGP36742 (SGS742). CGP36742 showed promising results in animal models [e.g., (Brouillette et al. 2007)] and progressed to the first Phase II clinical trial where it showed improvement in attentional performance (i.e., choice reaction time and visual information processing) in mild AD patients. Was, Unfortunately, a second Phase II clinical trial showed no statistically significant improvement, and thus, CGP36742 did not reach Phase III trials (Steulet et al. 1996; Froestl et al. 2004; Davies et al. 2005; Serrats et al. 2017).

However, efforts in this direction continue at preclinical level. In a most recent study, the role of the $GABA_B$R antagonist CGP35348 was examined on the DG $GABA_B$R inhibition and its effects on learning and memory impairments that had been induced in adult male Wistar rats by microinjection of β-amyloid (Aβ) (Almasi et al. 2018). Consistent with data from earlier studies on $GABA_B$R antagonists for AD, results from this study indicated that microinjections of CGP35348 directly into the hippocampus counteract the learning, memory and cognitive impairments induced by Aβ and suggested that CGP35348 could be a possible therapeutic compound against the progression of acute Aβ toxicity-induced memory impairment (Almasi et al. 2018).

Altogether, studies focusing on further understanding the effects of $GABA_BR$ compounds on cognitive impairments characterizing AD and other relevant conditions have been promising and one can only hope that, although the exact targets may vary, the efforts to develop effective $GABA_BR$ drug treatments will continue in this direction.

5.2 *Epilepsy*

Epilepsy is a disorder occurring in various forms, some of which would be more common, such as the temporal lobe (TLE) or absence epilepsy (AE), and some others would be rare types, such as the Dravet syndrome. The main characteristic symptom of epilepsy is seizures. These seem to stem from an imbalance between excitatory and inhibitory processes, or either of these types of synapses, inhibitory or excitatory, being faulty. Importantly, in some forms of epilepsy, such as the TLE, cognitive disturbances including amnesia commonly occur (Zapata et al. 2017). Moreover, chronic atypical absence seizures characterizing AE are a component of the Lennox-Gastaut syndrome, a disorder invariably associated with severe cognitive impairment in children.

In clinical populations, case reports of specific types of epilepsy symptomatology are very common. In one of these recent case reports (Zeman et al. 2016), researchers focused on a patient in whom long-term, therapeutic administration of baclofen into the cerebrospinal fluid (CSF) induced three distinct varieties of memory impairment: (1) repeated, short periods of severe global amnesia, (2) accelerated long-term forgetting, evident over intervals of days and (3) a loss of established autobiographical memories. The latter of these impairments persisted after discontinuation of baclofen administration. These memory impairments are reported in TLE, specifically in the subtype of transient epileptic amnesia. This case report suggested a role for $GABA_B$ signalling in the modulation of human memory functioning for different duration and implicates it in 'epileptic amnesia' (Zeman et al. 2016).

A number of animal models have been developed in the past few years to resemble the symptomatology of epilepsy. In most recent years, animal models of epilepsy are also developed to resemble rare types of epilepsy or specific symptomatology of them (Löscher 2011; Grone and Baraban 2015; Pitkänen et al. 2017). In these animal models, typically, $GABA_BR$ agonists exacerbate, while $GABA_BR$ antagonists suppress the seizures (Han et al. 2013), suggesting a potential therapeutic role of the antagonists for epilepsy and its cognitive dysfunctions (Enna 1997). Importantly, and as it relates to AE, it appears that hippocampal circuitry in atypical absence seizures is possibly responsible for the cognitive impairment in that disorder (Han et al. 2013). $GABA_BR$ antagonists can reverse both the seizures and the impairment in cognition in experimental atypical absence seizures (Getova et al. 1997), suggesting a therapeutic use of these antagonists in the Lennox-Gastaut syndrome, another type of rare epilepsy. The AY9944 model of chronic atypical absence seizures in rats reliably reproduces the electrographic, behavioural, pharmacological and cognitive features of AE. Using this model, Chan and colleagues tested the hypothesis that the cognitive impairment associated with this disorder involves a $GABA_BR$-mediated mechanism (Chan et al. 2006). Results showed that CGP35348 blocked atypical absence seizures, restored LTP to normal level and reversed the cognitive deficit in the AY9944-treated animals, suggesting a possible therapeutic role of CGP35348 or other antagonists in this epilepsy type. Importantly, the same research group later showed that overexpression of $GABA_BR1a$ in mice [R1a(+)] resulted in an atypical absence seizure phenotype characterized by 3- to 6-Hz slow spike-and-wave discharges (SSWDs), reduced synaptic plasticity and cognitive impairment, while the overexpressed $GABA_BR1b$ mice (R1b(+)) showed a less pronounced reduction in hippocampal LTP (Stewart et al. 2009; Wang et al. 2009), thus indicating that, although abnormal levels of both $GABA_BR1$ subunits are involved in AE, the abnormal function of $GABA_BR1b$ appears to have less of an involvement in AE symptomatology and cognitive impairments. Acute or chronic administration of vigabatrin, which increases GABA concentration by inhibiting GABA transaminase, increased activation of $GABA_BR$ in the frontal cortex and the reticular thalamic nucleus (Perescis et al. 2020), areas extensively implicated in cognitive functioning directly or indirectly.

5.3 Autoimmune Anti-$GABA_B$ Encephalitis and Cognitive Impairments

Autoimmune synaptic encephalitis, although having been described in patients diagnosed with epilepsy (Dubey et al. 2014), can occur as paraneoplastic neurological syndromes, which are dysfunctions of the nervous system occurring in some cancer patients, especially in patients diagnosed with tumours and small cell lung cancer (Golombeck et al. 2016; Qiao et al. 2017; Li et al. 2018; Maureille et al. 2019). One such newly described, rare, but treatable form, is the autoimmune anti-$GABA_B$ limbic encephalitis, which is associated with CSF elevated levels of $GABA_BR$ antibody titre. The main characteristics of this type of encephalitis are cognitive decline, memory impairments, progressive seizures and behavioural disorder for a number of days or weeks (Lancaster et al. 2010; Su et al. 2015; Hui et al. 2016). In some of these patients, abnormalities in the hippocampal region, parahippocampal gyrus, temporal and occipital lobe have been identified using magnetic resonance imaging (MRI) (Cui et al. 2018). As this is a newly identified condition, literature is limited and is mainly based on case reports (Chung et al. 2019; Yao et al. 2019; Qin et al. 2020), some of which suggest that it is most common in middle-aged and elderly men (Zhu et al. 2020).

6 Conclusions

It is clear from all presented research studies on the involvement of $GABA_BRs$ in cognitive processes that GABA as a neurotransmitter and specifically $GABA_BRs$ are involved in different components of cognition. Taking into account the spread of $GABA_BRs$ in the brain as well as their primary role of inhibition of processes, this does not come as a surprise. However, although extensive research has focused on the role of $GABA_BRs$ in health and disease (Bowery 2006), in most cases there still appears to be lack of clarity on the exact mechanisms involved, taking into account the balance between presynaptic and postsynaptic $GABA_BR$ activation and/or inhibition throughout the brain. Yet again, the vast majority of these studies are conducted in male animals, not allowing for identification of potential sex differences in the effects of the drugs tested.

Thus, with the exception of baclofen, which has been approved as a muscle relaxant and anticonvulsant treatment, and not as a cognitive enhancer, no other $GABA_BR$ compound is used clinically. For example, there are no $GABA_BR$ antagonists in the clinic for any population. With all cognitive effect discrepancies taken into account, efforts should continue to help us elucidate the exact role of $GABA_BR$ subtypes in the different conditions affecting cognition, for both males and females, and more compounds, possibly selectively targeting those subtypes, should be developed with an aim to reach the bedside. It is a long effort ahead, but, existing literature suggests, it is worth aiming at.

References

Adusei DC et al (2010) Early developmental alterations in GABAergic protein expression in fragile X knockout mice. Neuropharmacology 59(3):167–171. https://doi.org/10.1016/j.neuropharm.2010.05.002

Akirav I, Maroun M (2006) Ventromedial prefrontal cortex is obligatory for consolidation and reconsolidation of object recognition memory. Cereb Cortex 16(12):1759–1765

Almasi A et al (2018) Influence of hippocampal GABAB receptor inhibition on memory in rats with acute β-amyloid toxicity. Metab Brain Dis 33(6):1859–1867. https://doi.org/10.1007/s11011-018-0292-5

Arolfo MP, Zanudio MA, Ramirez OA (1998) Baclofen infused in rat hippocampal formation impairs spatial learning. Hippocampus 8(2):109–113. https://doi.org/10.1002/(SICI)1098-1063(1998)8:2<109::AID-HIPO2>3.0.CO;2-G

Axmacher N et al (2006) Memory formation by neuronal synchronization. Brain Res Rev:170–182. https://doi.org/10.1016/j.brainresrev.2006.01.007

Bakker CE, Oostra BA (2003) Understanding fragile X syndrome: insights from animal models. Cytogenet Genome Res:111–123. https://doi.org/10.1159/000072845

Bañuelos C et al (2014) Prefrontal cortical GABAergic dysfunction contributes to age-related working memory impairment. J Neurosci 34(10):3457–3466. https://doi.org/10.1523/JNEUROSCI.5192-13.2014

Barthet G, Mulle C (2020) Presynaptic failure in Alzheimer's disease. Prog Neurobiol. https://doi.org/10.1016/j.pneurobio.2020.101801

Belichenko PV et al (2007) Synaptic and cognitive abnormalities in mouse models of down syndrome: exploring genotype-phenotype relationships. J Comp Neurol 504(4):329–345. https://doi.org/10.1002/cne.21433
Berry-Kravis EM et al (2018) Drug development for neurodevelopmental disorders: lessons learned from fragile X syndrome. Nat Rev Drug Discov:280–298. https://doi.org/10.1038/nrd.2017.221
Best TK, Siarey RJ, Galdzicki Z (2007) Ts65Dn, a mouse model of down syndrome, exhibits increased GABAB-induced potassium current. J Neurophysiol 97(1):892–900. https://doi.org/10.1152/jn.00626.2006
Bianchi M, Panerai AE (1993) Reversal of scopolamine-induced amnesia by the GABAB receptor antagonist CGP 35348 in the mouse. Cogn Brain Res 1(2):135–136. https://doi.org/10.1016/0926-6410(93)90018-z
Blatt GJ, Fatemi SH (2011) Alterations in GABAergic biomarkers in the autism brain: research findings and clinical implications. Anat Rec (Hoboken) 294(10):1646–1652. https://doi.org/10.1002/ar.21252
Blichowski M et al (2015) The GIRK2 subunit is involved in IS-like seizures induced by GABAB receptor agonists. Epilepsia 56(7):1081–1087. https://doi.org/10.1111/epi.13034
Bliss TVP, Lømo T (1973) Long-lasting potentiation of synaptic transmission in the dentate area of the anaesthetized rabbit following stimulation of the perforant path. J Physiol 232(2):331–356. https://doi.org/10.1113/jphysiol.1973.sp010273
Bowery NG (2006) GABAB receptor: a site of therapeutic benefit. Curr Opin Pharmacol 6 (1):37–43. https://doi.org/10.1016/j.coph.2005.10.002
Bradley PM et al (2004) Post-synaptic GABAB receptors – possible controllers of coincidence detection? Behav Brain Res 155(1):27–35. https://doi.org/10.1016/j.bbr.2004.03.032
Brenowitz S, David J, Trussell L (1998) Enhancement of synaptic efficacy by presynaptic GABA (B) receptors. Neuron 20(1):135–141. https://doi.org/10.1016/S0896-6273(00)80441-9
Brouillette J et al (2007) Hippocampal gene expression profiling reveals the possible involvement of Homer1 and GABA(B) receptors in scopolamine-induced amnesia. J Neurochem 102 (6):1978–1989. https://doi.org/10.1111/j.1471-4159.2007.04666.x
Brown JW, Rueter LE, Zhang M (2014) Predictive validity of a MK-801-induced cognitive impairment model in mice: implications on the potential limitations and challenges of modeling cognitive impairment associated with schizophrenia preclinically. Prog Neuropsychopharmacol Biol Psychiatry 49:53–62. https://doi.org/10.1016/j.pnpbp.2013.11.008
Car H, Michaluk P (2012) Baclofen influences acquisition and MMP-2, MMP-9 levels in the hippocampus of rats after hypoxia. Pharmacol Rep 64(3):536–545. https://doi.org/10.1016/S1734-1140(12)70849-6
Car H, Wiśniewska RJ (2006) Effects of baclofen and L-AP4 in passive avoidance test in rats after hypoxia-induced amnesia. Pharmacol Rep 58(1):91–100. http://www.ncbi.nlm.nih.gov/pubmed/16531635. Accessed 29 July 2020
Car H, Wiśniewski K (1998) The effect of baclofen and AP-7 on selected behavior in rats. Pharmacol Biochem Behav 59(3):685–689. https://doi.org/10.1016/s0091-3057(97)00462-0
Car H, Nadlewska A, Wiśniewski K (2000) 3,5-DHPG influences behavioral effects of baclofen in rats. Pol J Pharmacol 52(4):247–254. http://www.ncbi.nlm.nih.gov/pubmed/11345480. Accessed 29 July 2020
Carletti R, Libri V, Bowery NG (1993) The GABA-B antagonist CGP36742 enhances spatial learning performance and antagonises baclofen-induced amnesia in mice. Br J Pharmacol 109:1P
Castellano C et al (1989) 'Post-training systemic and intra-amygdala administration of the GABA-B agonist baclofen impairs retention. Behav Neural Biol 52(2):170–179. https://doi.org/10.1016/S0163-1047(89)90285-9
Chan KFY et al (2006) GABAB receptor antagonism abolishes the learning impairments in rats with chronic atypical absence seizures. Eur J Pharmacol 541(1–2):64–72. https://doi.org/10.1016/j.ejphar.2006.04.012

Chung HY et al (2019) Autoimmune encephalitis with anti-IgLON5 and anti-GABAB-receptor antibodies: a case report. Medicine 98(20):e15706. https://doi.org/10.1097/MD.0000000000015706

Cieślik P et al (2019) Simultaneous activation of muscarinic and GABAB receptors as a bidirectional target for novel antipsychotics. Behav Brain Res 359:671–685. https://doi.org/10.1016/j.bbr.2018.09.019

Corell M et al (2015) GABA and its B-receptor are present at the node of Ranvier in a small population of sensory fibers, implicating a role in myelination. J Neurosci Res 93(2):285–295. https://doi.org/10.1002/jnr.23489

Cortez MA et al (2009) Infantile spasms and down syndrome: a new animal model. Pediatr Res 65 (5):499–503. https://doi.org/10.1203/PDR.0b013e31819d9076

Costa ACS, Grybko MJ (2005) Deficits in hippocampal CA1 LTP induced by TBS but not HFS in the Ts65Dn mouse: a model of down syndrome. Neurosci Lett 382(3):317–322. https://doi.org/10.1016/j.neulet.2005.03.031

Cramer NP et al (2010) GABAB-GIRK2-mediated signaling in Down syndrome. Adv Pharmacol 58(C):397–426. https://doi.org/10.1016/S1054-3589(10)58015-3

Cryan JF et al (2004) Behavioral characterization of the novel GABAB receptor-positive modulator GS39783 (N,N′-dicyclopentyl-2-methylsulfanyl-5-nitro- pyrimidine-4,6-diamine): anxiolytic-like activity without side effects associated with baclofen or benzodiazepines. J Pharmacol Exp Ther 310(3):952–963. https://doi.org/10.1124/jpet.104.066753

Cui J et al (2018) The gamma-aminobutyric acid-B receptor (GABAB) encephalitis: clinical manifestations and response to immunotherapy. Int J Neurosci 128(7):627–633. https://doi.org/10.1080/00207454.2017.1408618

Cullen PK et al (2014) GABA-mediated presynaptic inhibition is required for precision of long-term memory. Learn Mem 21(4):180–184. https://doi.org/10.1101/lm.032961.113

Davies S, Castañer J, Castañer RM (2005) SGS-742: treatment of Alzheimer's dementia treatment of attention deficit hyperactivity disorder GABAB receptor antagonist. Drugs Future 30 (3):248–253. https://doi.org/10.1358/dof.2005.030.03.889852

Davis AM, Henion TR, Tobet SA (2002) γ-aminobutyric acidB receptors and the development of the ventromedial nucleus of the hypothalamus. J Comp Neurol 449(3):270–280. https://doi.org/10.1002/cne.10293

Deidda G, Bozarth IF, Cancedda L (2014) Modulation of GABAergic transmission in development and neurodevelopmental disorders: investigating physiology and pathology to gain therapeutic perspectives. Front Cell Neurosci 8(May):119. https://doi.org/10.3389/fncel.2014.00119

Deng PY et al (2009) GABAB receptor activation inhibits neuronal excitability and spatial learning in the entorhinal cortex by activating TREK-2 K+ channels. Neuron 63(2):230–243. https://doi.org/10.1016/j.neuron.2009.06.022

DeSousa NJ et al (1994) Stimulation of GABAB receptors in thebasal forebrain selectively impairs working memory of rats in the double Y-maze. Brain Res 641(1):29–38. https://doi.org/10.1016/0006-8993(94)91811-2

Dubey D et al (2014) Effectiveness of multimodality treatment for autoimmune limbic epilepsy. Epileptic Disord:494–499. https://doi.org/10.1684/epd.2014.0703

Dubrovina NI, Loskutova LV (2007) GABA(B) receptors and amnesia in mice withalternative stereotypes of behavior. Eksp Klin Farmakol 70(3):10–12. http://www.ncbi.nlm.nih.gov/pubmed/17650625. Accessed 29 July 2020

Dubrovina NI, Zinov'ev DR (2008) Contribution of GABA receptors to extinction of memory traces in normal conditions and in a depression-like state. Neurosci Behav Physiol 38 (8):775–779. https://doi.org/10.1007/s11055-008-9045-y

Dudchenko PA (2004) An overview of the tasks used to test working memory in rodents. Neurosci Biobehav Rev:699–709. https://doi.org/10.1016/j.neubiorev.2004.09.002

Dvorak-Carbone H (1999) Contribution of the temporoammonic pathway to hippocampal processing. California Institute of Technology, Pasadena. https://doi.org/10.7907/AYRT-5Y12

Ebrahimi-Ghiri M et al (2018) Role of CA1 GABAA and GABAB receptors on learning deficit induced by D-AP5 in passive avoidance step-through task. Brain Res 1678:164–173. https://doi.org/10.1016/j.brainres.2017.10.004
Enna SJ (1997) GABA(B) receptor agonists and antagonists: pharmacological properties and therapeutic possibilities. Expert Opin Investig Drugs:1319–1325. https://doi.org/10.1517/13543784.6.10.1319
Enticott PG et al (2013) GABAergic activity in autism spectrum disorders: an investigation of cortical inhibition via transcranial magnetic stimulation. Neuropharmacology 68:202–209. https://doi.org/10.1016/j.neuropharm.2012.06.017
Erickson EJ, Watts KD, Parent MB (2006) Septal co-infusions of glucose with a GABAB agonist impair memory. Neurobiol Learn Mem 85(1):66–70. https://doi.org/10.1016/j.nlm.2005.08.008
Erickson CA et al (2014) STX209 (Arbaclofen) for autism spectrum disorders: an 8-week open-label study. J Autism Dev Disord 44(4):958–964. https://doi.org/10.1007/s10803-013-1963-z
Escher T, Mittleman G (2004) Effects of ethanol and GABAB drugs on working memory in C57BL/6J and DBA/2J mice. Psychopharmacology (Berl) 176(2):166–174. https://doi.org/10.1007/s00213-004-1875-x
Falsafi SK et al (2015) Mouse hippocampal GABAB1 but not GABAB2 subunit-containing receptor complex levels are paralleling retrieval in the multiple-T-maze. Front Behav Neurosci 9. https://doi.org/10.3389/fnbeh.2015.00276
Farahmandfar M et al (2017) Recovery from ketamine-induced amnesia by blockade of GABA-A receptor in the medial prefrontal cortex of mice. Neuroscience 344:48–55. https://doi.org/10.1016/j.neuroscience.2016.02.056
Farr SA et al (2000) Modulation of memory processing in the cingulate cortex of mice. Pharmacol Biochem Behav 65(3):363–368. https://doi.org/10.1016/S0091-3057(99)00226-9
Fatemi SH et al (2010) mRNA and protein levels for GABA Aα4, α5, β1 and GABABR1 receptors are altered in brains from subjects with autism. J Autism Dev Disord 40(6):743–750. https://doi.org/10.1007/s10803-009-0924-z
Fatemi SH et al (2014) Downregulation of GABAA receptor protein subunits α6, β2, δ, ε, γ2, θ, and ρ2 in superior frontal cortex of subjects with autism. J Autism Dev Disord 44(8):1833–1845. https://doi.org/10.1007/s10803-014-2078-x
Fernandez F et al (2007) Pharmacotherapy for cognitive impairment in a mouse model of down syndrome. Nat Neurosci 10(4):411–413. https://doi.org/10.1038/nn1860
Fogel SM, Smith CT, Beninger RJ (2010) Increased GABAergic activity in the region of the Pedunculopontine and deep mesencephalic reticular nuclei reduces REM sleep and impairs learning in rats. Behav Neurosci 124(1):79–86. https://doi.org/10.1037/a0018244
Fritzius T et al (2020) Structural basis of GABAB receptor regulation and signaling. Curr Top Behav Neurosci (in this volume). https://doi.org/10.1007/7854_2020_147
Froestl W et al (2004) SGS742: The first GABAB receptor antagonist in clinical trials. Biochem Pharmacol:1479–1487. https://doi.org/10.1016/j.bcp.2004.07.030
Frye RE et al (2016) Neuropathological mechanisms of seizures in autism spectrum disorder. Front Neurosci. https://doi.org/10.3389/fnins.2016.00192
Gandal MJ et al (2012) GABAB-mediated rescue of altered excitatory-inhibitory balance, gamma synchrony and behavioral deficits following constitutive NMDAR-hypofunction. Transl Psychiatry 2:e142. https://doi.org/10.1038/tp.2012.69
Gassmann M et al (2004) Redistribution of GABAB(1) protein and atypical GABAB responses in GABAB(2)-deficient mice. J Neurosci 24(27):6086–6097. https://doi.org/10.1523/JNEUROSCI.5635-03.2004
Georgiev VP, Yonkov DI, Kambourova TS (1988) Interactions between angiotensin II and baclofen in shuttle-box and passive avoidance performance. Neuropeptides 12(3):155–158. https://doi.org/10.1016/0143-4179(88)90047-9
Getova D, Bowery NG (1998) The modulatory effects of high affinity GABA(B) receptor antagonists in an active avoidance learning paradigm in rats. Psychopharmacology (Berl) 137(4):369–373. http://www.ncbi.nlm.nih.gov/pubmed/9676897

Getova DP, Bowery NG (2001) Effects of high-affinity GABAB receptor antagonists on active and passive avoidance responding in rodents with gamma-hydroxybutyrolactone-induced absence syndrome. Psychopharmacology (Berl) 157(1):89–95. https://doi.org/10.1007/s002130100766

Getova DP, Dimitrova DD (2007) Effects of GABA B receptor antagonists CGP63360, CGP76290A and CGP76291A on learning and memory processes in rodents. Central Eur J Med 2(3):280–293. https://doi.org/10.2478/s11536-007-0033-3

Getova D, Bowery NG, Spassov V (1997) Effects of GABA(B) receptor antagonists on learning and memory retention in a rat model of absence epilepsy. Eur J Pharmacol 320(1):9–13. https://doi.org/10.1016/S0014-2999(96)00877-1

Gigout S et al (2015) Different pharmacology of N-desmethylclozapine at human and rat M2 and M 4 mAChRs in neocortex. Naunyn Schmiedebergs Arch Pharmacol 388(5):487–496. https://doi.org/10.1007/s00210-014-1080-3

Golombeck KS et al (2016) Evidence of a pathogenic role for CD8+ T cells in anti-GABAB receptor limbic encephalitis. Neurol Neuroimmunol NeuroInflamm 3(3). https://doi.org/10.1212/NXI.0000000000000232

Gong N et al (2009) GABA transporter-1 activity modulates hippocampal theta oscillation and theta burst stimulation-induced long-term potentiation. J Neurosci 29(50):15836–15845. https://doi.org/10.1523/JNEUROSCI.4643-09.2009

Grieb P (2016) Intracerebroventricular Streptozotocin injections as a model of Alzheimer's disease: in search of a relevant mechanism. Mol Neurobiol:1741–1752. https://doi.org/10.1007/s12035-015-9132-3

Grone B, Baraban S (2015) Animal models of epilepsy: legacies and new directions. bioRxiv, pp 013136. https://doi.org/10.1101/013136

Hampson DR, Adusei DC, Pacey LKK (2011) The neurochemical basis for the treatment of autism spectrum disorders and fragile X syndrome. Biochem Pharmacol:1078–1086. https://doi.org/10.1016/j.bcp.2011.02.005

Han HA, Cortez MA, Snead OC (2013) GABAB receptor and absence epilepsy. In: Jasper's basic mechanisms of the epilepsies. Oxford University Press, Oxford, pp 242–256. https://doi.org/10.1093/med/9780199746545.003.0019

Harashima C, Jacobowitz DM et al (2006a) Abnormal expression of the G-protein-activated inwardly rectifying potassium channel 2 (GIRK2) in hippocampus, frontal cortex, and substantia nigra of Ts65Dn mouse: a model of down syndrome. J Comp Neurol 494 (5):815–833. https://doi.org/10.1002/cne.20844

Harashima C, Jacobowitz DM et al (2006b) Elevated expression of the G-protein-activated inwardly rectifying potassium channel 2 (GIRK2) in cerebellar unipolar brush cells of a down syndrome mouse model. Cell Mol Neurobiol 26(4–6):719–734. https://doi.org/10.1007/s10571-006-9066-4

Hattori M et al (2000) The DNA sequence of human chromosome 21. Nature 405(6784):311–319. https://doi.org/10.1038/35012518

Heaney CF, Kinney JW (2016) Role of GABAB receptors in learning and memory and neurological disorders. Neurosci Biobehav Rev:1–28. https://doi.org/10.1016/j.neubiorev.2016.01.007

Helm KA et al (2005) GABAB receptor antagonist SGS742 improves spatial memory and reduces protein binding to the cAMP response element (CRE) in the hippocampus. Neuropharmacology 48(7):956–964. https://doi.org/10.1016/j.neuropharm.2005.01.019

Henderson C et al (2012) Reversal of disease-related pathologies in the fragile X mouse model by selective activation of GABAB receptors with arbaclofen. Sci Transl Med 4(152). https://doi.org/10.1126/scitranslmed.3004218

Holajova M, Franek M (2018) Effect of short- and long-term administration of baclofen on spatial learning and memory in rats. Physiol Res 67(1):133–141. https://doi.org/10.33549/physiolres.933554

Holley SM et al (2019) Major contribution of somatostatin-expressing interneurons and cannabinoid receptors to increased GABA synaptic activity in the striatum of Huntington's disease mice. Front Synaptic Neurosci 11:14. https://doi.org/10.3389/fnsyn.2019.00014

Hopkins CR (2011) ACS chemical neuroscience molecule spotlight on STX209 (Arbaclofen). ACS Chem Nerosci:381. https://doi.org/10.1021/cn200061g
Hui ATH et al (2016) A case of refractory seizure with cognitive impairment due to anti-GABA encephalitis. Hong Kong Med J:509–511. https://doi.org/10.12809/hkmj154604
Jacobson LH et al (2007) Specific roles of GABAB(1) receptor isoforms in cognition. Behav Brain Res 181(1):158–162. https://doi.org/10.1016/j.bbr.2007.03.033
Jiang J et al (2017) Intrahippocampal injection of Cortistatin-14 impairs recognition memory consolidation in mice through activation of sst2, ghrelin and GABAA/B receptors. Brain Res 1666:38–47. https://doi.org/10.1016/j.brainres.2017.04.017
Joshi K et al (2016) Infantile spasms in down syndrome: rescue by knockdown of the GIRK2 channel. Ann Neurol 80(4):511–521. https://doi.org/10.1002/ana.24749
Jutras MJ, Buffalo EA (2010) Synchronous neural activity and memory formation. Curr Opin Neurobiol:150–155. https://doi.org/10.1016/j.conb.2010.02.006
Kang JY et al (2017) Deficits in the activity of presynaptic-aminobutyric acid type B receptors contribute to altered neuronal excitability in fragile X syndrome. J Biol Chem 292 (16):6621–6632. https://doi.org/10.1074/jbc.M116.772541
Kazdoba TM et al (2014) Modeling fragile X syndrome in the Fmr1 knockout mouse. Intractable Rare Dis Res 3(4):118–133. https://doi.org/10.5582/irdr.2014.01024
Khanegheini A, Nasehi M, Zarrindast MR (2015) The modulatory effect of CA1 GABAb receptors on ketamine-induced spatial and non-spatial novelty detection deficits with respect to Ca2+. Neuroscience 305:157–168. https://doi.org/10.1016/j.neuroscience.2015.07.083
Kim MO et al (2003) Early fetal expression of GABAB1 and GABAB2 receptor mRNAs on the development of the rat central nervous system. Dev Brain Res 143(1):47–55. https://doi.org/10.1016/S0165-3806(03)00099-3
Kleschevnicov AM et al (2004) Hippocampal long-term potentiation suppressed by increased inhibition in the Ts65Dn mouse, a genetic model of down syndrome. J Neurosci 24 (37):8153–8160. https://doi.org/10.1523/JNEUROSCI.1766-04.2004
Kleschevnikov AM, Belichenko PV, Faizi M et al (2012a) Deficits in cognition and synaptic plasticity in a mouse model of down syndrome ameliorated by GABAB receptor antagonists. J Neurosci 32(27):9217–9227. https://doi.org/10.1523/JNEUROSCI.1673-12.2012
Kleschevnikov AM, Belichenko PV, Gall J et al (2012b) Increased efficiency of the GABAA and GABAB receptor-mediated neurotransmission in the Ts65Dn mouse model of down syndrome. Neurobiol Dis 45(2):683–691. https://doi.org/10.1016/j.nbd.2011.10.009
Kleschevnikov AM et al (2017) Evidence that increased Kcnj6 gene dose is necessary for deficits in behavior and dentate gyrus synaptic plasticity in the Ts65Dn mouse model of down syndrome. Neurobiol Dis 103:1–10. https://doi.org/10.1016/j.nbd.2017.03.009
Kohl MM, Paulsen O (2010) The roles of GABA B receptors in cortical network activity. In: Advances in pharmacology. Academic Press, Cambridge, pp 205–229. https://doi.org/10.1016/S1054-3589(10)58009-8
Kramvis I et al (2020) Dysregulated prefrontal cortex inhibition in prepubescent and adolescent fragile X mouse model. Front Mol Neurosci 13. https://doi.org/10.3389/fnmol.2020.00088
Kumar K, Kaur H, Deshmukh R (2017) Neuroprotective role of GABAB receptor modulation against streptozotocin-induced behavioral and biochemical abnormalities in rats. Neuroscience 357:67–74. https://doi.org/10.1016/j.neuroscience.2017.05.054
Kunisawa K et al (2017) Betaine attenuates memory impairment after water-immersion restraint stress and is regulated by the GABAergic neuronal system in the hippocampus. Eur J Pharmacol 796:122–130. https://doi.org/10.1016/j.ejphar.2016.12.007
Kuziemka-Łęska M, Car H, Wiśniewski K (1999) Baclofen and AII 3-7 on learning and memory processes in rats chronically treated with ethanol. Pharmacol Biochem Behav 62(1):39–43. https://doi.org/10.1016/S0091-3057(98)00091-4
Lancaster E et al (2010) Antibodies to the GABAB receptor in limbic encephalitis with seizures: case series and characterisation of the antigen. Lancet Neurol 9(1):67–76. https://doi.org/10.1016/S1474-4422(09)70324-2

Lee I, Hunsaker MR, Kesner RP (2005) The role of hippocampal subregions in detecting spatial novelty. Behav Neurosci 119(1):145–153. https://doi.org/10.1037/0735-7044.119.1.145

Levin ED, Weber E, Icenogle L (2004) Baclofen interactions with nicotine in rats: effects on memory. Pharmacol Biochem Behav 79(2):343–348. https://doi.org/10.1016/j.pbb.2004.08.013

Li X et al (2013) Comparison of the effects of the GABAB receptor positive modulator BHF177 and the GABAB receptor agonist baclofen on anxiety-like behavior, learning, and memory in mice. Neuropharmacology 70:156–167. https://doi.org/10.1016/j.neuropharm.2013.01.018

Li C-J et al (2014) Activation of GABAB receptors ameliorates cognitive impairment via restoring the balance of HCN1/HCN2 surface expression in the hippocampal CA1 area in rats with chronic cerebral hypoperfusion. Mol Neurobiol 50(2):704–720. https://doi.org/10.1007/s12035-014-8736-3

Li H et al (2018) Coexistence of Lambert-Eaton myasthenic syndrome and autoimmune encephalitis with anti-CRMP5/CV2 and anti-GABAB receptor antibodies in small cell lung cancer: a case report. Medicine 97(19). https://doi.org/10.1097/MD.0000000000010696

Liu QY et al (2014) Effects of intrahippocampal GABAB receptor antagonist treatment on the behavioral long-term potentiation and Y-maze learning performance. Neurobiol Learn Mem 114:26–31. https://doi.org/10.1016/j.nlm.2014.04.005

Liu L, Ito W, Morozov A (2017) GABAb receptor mediates opposing adaptations of GABA release from two types of prefrontal interneurons after observational fear. Neuropsychopharmacology 42(6):1272–1283. https://doi.org/10.1038/npp.2016.273

Lømo T (2003) The discovery of long-term potentiation. Philos Trans R Soc Lond B Biol Sci 358 (1432):617–620. https://doi.org/10.1098/rstb.2002.1226

Lømo T (2018) Discovering long-term potentiation (LTP) – recollections and reflections on what came after. Acta Physiol 222(2). https://doi.org/10.1111/apha.12921

Löscher W (2011) Critical review of current animal models of seizures and epilepsy used in the discovery and development of new antiepileptic drugs. Seizure:359–368. https://doi.org/10.1016/j.seizure.2011.01.003

Mahdavinasab SM et al (2019) Baclofen as an adjuvant therapy for autism: a randomized, double-blind, placebo-controlled trial. Eur Child Adolesc Psychiatry 28(12):1619–1628. https://doi.org/10.1007/s00787-019-01333-5

Mann EO, Mody I (2010) Control of hippocampal gamma oscillation frequency by tonic inhibition and excitation of interneurons. Nat Neurosci 13(2):205–212. https://doi.org/10.1038/nn.2464

Manz KM et al (2019) Heterosynaptic GABAB receptor function within feedforward microcircuits gates glutamatergic transmission in the nucleus Accumbens Core. J Neurosci 39 (47):9277–9293. https://doi.org/10.1523/JNEUROSCI.1395-19.2019

Maureille A et al (2019) Isolated seizures are a common early feature of paraneoplastic anti-GABA B receptor encephalitis. J Neurol 266(1):195–206. https://doi.org/10.1007/s00415-018-9132-0

Maurin T et al (2015) The FMRP/ GRK4 mRNA interaction uncovers a new mode of binding of the fragile X mental retardation protein in cerebellum. Nucleic Acids Res 43(17):8540–8550. https://doi.org/10.1093/nar/gkv801

Mayordomo-Cava J et al (2015) Amyloid-β(25-35) modulates the expression of GirK and KCNQ channel genes in the Hippocampus. Plos One 10(7):e0134385. https://doi.org/10.1371/journal.pone.0134385

Miedel CJ et al (2017) Assessment of spontaneous alternation, novel object recognition and limb clasping in transgenic mouse models of amyloid-β and tau neuropathology. J Vis Exp 2017 (123):55523. https://doi.org/10.3791/55523

Modaberi S et al (2019) Combination effects of forced mild exercise and GABA B receptor agonist on spatial learning, memory, and motor activity in striatum lesion rats. J Mot Behav 51 (4):438–450. https://doi.org/10.1080/00222895.2018.1505711

Mondadori C, Jaekel J, Preiswerk G (1993) CGP 36742: the first orally active GABAB blocker improves the cognitive performance of mice, rats, and rhesus monkeys. Behav Neural Biol 60 (1):62–68. https://doi.org/10.1016/0163-1047(93)90729-2

Mondadori C, Moebius HJ, Zingg M (1996) CGP 36 742, an orally active GABAB receptor antagonist, facilitates memory in a social recognition test in rats. Behav Brain Res 77 (1–2):227–229. https://doi.org/10.1016/0166-4328(95)00226-X

Murrell E et al (2020) Classics in neuroimaging: development of positron emission tomography tracers for imaging the GABAergic pathway. ACS Chem Nerosci 11(14). https://doi.org/10.1021/acschemneuro.0c00343

Nakagawa Y, Takashima T (1997) The GABA(B) receptor antagonist CGP36742 attenuates the baclofen- and scopolamine-induced deficit in Morris water maze task in rats. Brain Res 766 (1–2):101–106. http://www.ncbi.nlm.nih.gov/pubmed/9359592

Nakagawa Y et al (1995) Involvement of cholinergic systems in the deficit of place learning in Morris water maze task induced by baclofen in rats. Brain Res 683(2):209–214. https://doi.org/10.1016/0006-8993(95)00302-7

Nasehi M et al (2017) Possible involvement of the CA1 GABAergic system on harmaline induced memory consolidation deficit. Brain Res Bull 130:101–106. https://doi.org/10.1016/j.brainresbull.2017.01.011

Nava-Mesa MO et al (2013) Amyloid-β induces synaptic dysfunction through G protein-gated inwardly rectifying potassium channels in the fimbria-CA3 hippocampal synapse. Front Cell Neurosci 7:117. https://doi.org/10.3389/fncel.2013.00117

Nazari M, Komaki A, Salehi I et al (2016a) Interactive effects of AM251 and baclofen on synaptic plasticity in the rat dentate gyrus. Brain Res 1651:53–60. https://doi.org/10.1016/j.brainres.2016.09.029

Nazari M, Komaki A, Karamian R et al (2016b) The interactive role of CB(1) and GABA (B) receptors in hippocampal synaptic plasticity in rats. Brain Res Bull 120:123–130. https://doi.org/10.1016/j.brainresbull.2015.11.013

Nieto A, Bailey T, Kaczanowska K, McDonald P (2021) GABA B receptor chemistry and pharmacology: agonists, antagonists, and allosteric modulators. Curr Top Behav Neurosci, Curr Top Behav Neurosci (in this volume). https://doi.org/10.1007/7854_2021_232

Oblak AL, Gibbs TT, Blatt GJ (2010) Decreased GABAB receptors in the cingulate cortex and fusiform gyrus in Autism. J Neurochem 114(5):1414–1423. https://doi.org/10.1111/j.1471-4159.2010.06858.x

Ogasawara T et al (1999) Involvement of cholinergic and GABAergic systems in the reversal of memory disruption by NS-105, a cognition enhancer. Pharmacol Biochem Behav 64(1):41–52. https://doi.org/10.1016/S0091-3057(99)00108-2

Ohira M et al (1997) Gene identification in 1.6-Mb region of the down syndrome region on chromosome 21. Genome Res 7(1):47–58. https://doi.org/10.1101/gr.7.1.47

Ostrovskaya O et al (2014) RGS7/Gβ5/R7BP complex regulates synaptic plasticity and memory by modulating hippocampal GABABR-GIRK signaling. Elife 3:e02053. https://doi.org/10.7554/eLife.02053

Pacey LKK, Heximer SP, Hampson DR (2009) Increased GABAB receptor-mediated signaling reduces the susceptibility of fragile X knockout mice to audiogenic seizures. Mol Pharmacol 76 (1):18–24. https://doi.org/10.1124/mol.109.056127

Pacey LKK, Tharmalingam S, Hampson DR (2011) Subchronic administration and combination metabotropic glutamate and GABAB receptor drug therapy in fragile X syndrome. J Pharmacol Exp Ther 338(3):897–905. https://doi.org/10.1124/jpet.111.183327

Perescis MFJ, van Luijtelaar G, van Rijn CM (2020) Immediate versus late effects of vigabatrin on spike and wave discharges. Epilepsy Res 165. https://doi.org/10.1016/j.eplepsyres.2020.106379

Pilipenko V et al (2018) Very low doses of muscimol and baclofen ameliorate cognitive deficits and regulate protein expression in the brain of a rat model of streptozocin-induced Alzheimer's disease. Eur J Pharmacol 818:381–399. https://doi.org/10.1016/j.ejphar.2017.11.012

Pilipenko V et al (2019) GABA-containing compound gammapyrone protects against brain impairments in Alzheimer's disease model male rats and prevents mitochondrial dysfunction in cell culture. J Neurosci Res 97(6):708–726. https://doi.org/10.1002/jnr.24396

Pin J-P, Bettler B (2016) Organization and functions of mGlu and GABAB receptor complexes. Nature 540(7631):60–68. https://doi.org/10.1038/nature20566

Pitkänen A et al (eds) (2017) Models of seizures and epilepsy. 2nd edn. Academic Press, London. https://doi.org/10.1016/B978-0-12-804066-9.00084-5

Pitsikas N et al (2003) The GABAB receptor and recognition memory: possible modulation of its behavioral effects by the nitrergic system. Neuroscience 118(4):1121–1127. https://doi.org/10.1016/S0306-4522(03)00067-8

Princivalle AP (2021) GABAB receptors in neurodegeneration. Curr Top Behav Neurosci (in this volume). https://doi.org/10.1007/7854_2021_222

Qiao S et al (2017) Clinical, imaging, and follow-up observations of patients with anti-GABAB receptor encephalitis. Int J Neurosci 127(5):379–385. https://doi.org/10.1080/00207454.2016.1176922

Qin W et al (2020) Coexistence of anti-SOX1 and anti-GABAb receptor antibodies with autoimmune encephalitis in small cell lung cancer: a case report. Clin Interv Aging 15:171–175. https://doi.org/10.2147/CIA.S234660

Romanides AJ, Duffy P, Kalivas PW (1999) Glutamatergic and dopaminergic afferents to the prefrontal cortex regulate spatial working memory in rats. Neuroscience 92(1):97–106. https://doi.org/10.1016/S0306-4522(98)00747-7

Roncacé V et al (2017) Neuroanatomical alterations and synaptic plasticity impairment in the perirhinal cortex of the Ts65Dn mouse model of down syndrome. Neurobiol Dis 106:89–100. https://doi.org/10.1016/j.nbd.2017.06.017

Rose TR, Wickman K (2020) Mechanisms and regulation of neuronal GABAB receptor-dependent signaling. Curr Top Behav Neurosci (in this volume). https://doi.org/10.1007/7854_2020_129

Saha N et al (1993) Effects of post-training administration of (−)-baclofen and chlordiazepoxide on memory retention in ICRC Swiss mice: interactions with GABAA and GABAB receptor antagonists. Pharmacol Toxicol 72(3):159–162. https://doi.org/10.1111/j.1600-0773.1993.tb00309.x

Sahoo AK et al (2018) Features and outcomes of drugs for combination therapy as multi-targets strategy to combat Alzheimer's disease. J Ethnopharmacol 215:42–73. https://doi.org/10.1016/j.jep.2017.12.015

Sahraei H et al (2019) GABAB receptor activation ameliorates spatial memory impairments in stress-exposed rats. Neuropsychiatr Dis Treat 15:1497–1506. https://doi.org/10.2147/NDT.S205951

Sanchez-Vives MV et al (2020) GABAB receptor-modulation of thalamocortical dynamics and synaptic plasticity. Neuroscience. https://doi.org/10.1016/j.neuroscience.2020.03.011

Schaefer TL et al (2017) Acamprosate in a mouse model of fragile X syndrome: modulation of spontaneous cortical activity, ERK1/2 activation, locomotor behavior, and anxiety. J Neurodev Disord 9(1). https://doi.org/10.1186/s11689-017-9184-y

Schuler V et al (2001) Epilepsy, hyperalgesia, impaired memory, and loss of pre- and postsynaptic GABAB responses in mice lacking GABAB(1). Neuron 31(1):47–58. https://doi.org/10.1016/S0896-6273(01)00345-2

Serikawa T et al (2019) Increased seizure sensitivity, emotional defects and cognitive impairment in PHD finger protein 24 (Phf24)-null rats. Behav Brain Res 369:111922. https://doi.org/10.1016/j.bbr.2019.111922

Serrats J, Cunningham MO, Davies CH (2017) GABAB receptor modulation – to B or not to be B a pro-cognitive medicine? Curr Opin Pharmacol:125–132. https://doi.org/10.1016/j.coph.2017.09.012

Shahrzad P, Nasser N (2015) GABA-B receptor antagonist CGP35348 improves testosterone induced spatial acquisition impairment in adult male rat. J Behav Brain Sci 5(11):491–502. https://doi.org/10.4236/jbbs.2015.511047

Sharma AC, Kulkarni SK (1993) (+/−)Baclofen sensitive scopolamine-induced short-term memory deficits in mice. Indian J Exp Biol 31(4):348–352. http://www.ncbi.nlm.nih.gov/pubmed/8395465. Accessed 29 July 2020

Siarey RJ et al (1997) Altered long-term potentiation in the young and old Ts65Dn mouse, a model for down syndrome. Neuropharmacology 36(11–12):1549–1554. https://doi.org/10.1016/S0028-3908(97)00157-3

Sidel ES et al (1988) Potential interactions between GABAb and cholinergic systems: baclofenaugments scopolamine-induced performance deficits in the eight-arm radialmaze. 96:116–120

Silverman JL et al (2015) GABA B receptor agonist R-baclofen reverses social deficits and reduces repetitive behavior in two mouse models of Autism. Neuropsychopharmacology 40 (9):2228–2239. https://doi.org/10.1038/npp.2015.66

Sinclair D et al (2017) Disorders of the nervous system GABA-B agonist baclofen normalizes auditory-evoked neural oscillations and behavioral deficits in the Fmr1 knockout mouse model of fragile X syndrome. eNeuro 4(1). https://doi.org/10.1523/ENEURO.0380-16.2017

Snead OC (1991) The γ-hydroxybutyrate model of absence seizures: correlation of regional brain levels of γ-hydroxybutyric acid and γ-butyrolactone with spike wave discharges. Neuropharmacology 30(2):161–167. https://doi.org/10.1016/0028-3908(91)90199-L

Stackman RW, Walsh TJ (1994) Baclofen produces dose-related workingmemory impairments after intraseptal injection. Behav Neural Biol 61:181–185

Staubli UV, Scafidi J, Chun D (1999) GABAB receptor antagonism: facilitatoryeffects on memory parallel those on LTP induced by TBS but not HFS. J Neurosci 19:4609

Steulet AF et al (1996) Determination of rat brain and plasma levels of the orally active GABAB antagonist 3-amino-propyl-n-butyl-phosphinic acid (CGP 36742) by a new GC/MS method. Biochem Pharmacol 51(5):613–619. https://doi.org/10.1016/S0006-2952(95)02183-3

Stewart LS et al (2009) Severity of atypical absence phenotype in GABAB transgenic mice is subunit specific. Epilepsy Behav 14(4):577–581. https://doi.org/10.1016/j.yebeh.2009.01.019

Stoppel LJ et al (2018) R-baclofen reverses cognitive deficits and improves social interactions in two lines of 16p11.2 deletion mice. Neuropsychopharmacology 43(3):513–524. https://doi.org/10.1038/npp.2017.236

Stratton MS, Searcy BT, Tobet SA (2011) GABA regulates corticotropin releasing hormone levels in the paraventricular nucleus of the hypothalamus in newborn mice. Physiol Behav 104 (2):327–333. https://doi.org/10.1016/j.physbeh.2011.01.003

Stratton MS et al (2014) Embryonic GABAB receptor blockade alters cell migration, adult hypothalamic structure, and anxiety- and depression-like behaviors sex specifically in mice. PLoS One 9(8). https://doi.org/10.1371/journal.pone.0106015

Stuchlik A, Vales K (2009) Baclofen dose-dependently disrupts learning in a place avoidance task requiring cognitive coordination. Physiol Behav 97(3–4):507–511. https://doi.org/10.1016/j.physbeh.2009.03.024

Su M, Xu D, Tian R (2015) 18F-FDG PET/CT and MRI findings in a patient with anti-GABAB receptor encephalitis. Clin Nucl Med 40(6):515–517. https://doi.org/10.1097/RLU.0000000000000676

Sun Z, Sun L, Tu L (2020) GABAB receptor-mediated PI3K/Akt signaling pathway alleviates oxidative stress and neuronal cell injury in a rat model of Alzheimer's disease. J Alzheimers Dis:1–14. https://doi.org/10.3233/JAD-191032

Sunyer B et al (2007) Strain-dependent effects of SGS742 in the mouse. Behav Brain Res 181 (1):64–75. https://doi.org/10.1016/j.bbr.2007.03.025

Sunyer B et al (2009) Hippocampal levels of phosphorylated protein kinase a (phosphor-S96) are linked to spatial memory enhancement by SGS742. Hippocampus 19(1):90–98. https://doi.org/10.1002/hipo.20484

Swartzwelder HS et al (1987) Baclofen disrupts passive avoidance retention in rats. Psychopharmacology (Berl) 92(3):398–401. https://doi.org/10.1007/BF00210851

Tassone F, Hagerman PJ (2003) Expression of the FMR1 gene. Cytogenet Genome Res:124–128. https://doi.org/10.1159/000072846

Terunuma M et al (2014) Postsynaptic GABAB receptor activity regulates excitatory neuronal architecture and spatial memory. J Neurosci 34(3):804–816. https://doi.org/10.1523/JNEUROSCI.3320-13.2013

Teutsch J, Kätzel D (2019) Operant assessment of DMTP spatial working memory in mice. Front Behav Neurosci 13:193. https://doi.org/10.3389/fnbeh.2019.00193

Veenstra-Vanderweele J et al (2017) Arbaclofen in children and adolescents with autism Spectrum disorder: a randomized, controlled, phase 2 trial. Neuropsychopharmacology 42(7):1390–1398. https://doi.org/10.1038/npp.2016.237

Vigot R et al (2006) Differential compartmentalization and distinct functions of GABAB receptor variants. Neuron 50(4):589–601. https://doi.org/10.1016/j.neuron.2006.04.014

Vlachou S et al (2011) Repeated administration of the GABAB receptor positive modulator BHF177 decreased nicotine self-administration, and acute administration decreased cue-induced reinstatement of nicotine seeking in rats. Psychopharmacology (Berl) 215 (1):117–128. https://doi.org/10.1007/s00213-010-2119-x

Wahlstrom-Helgren S, Klyachko VA (2015) GABAB receptor-mediated feed-forward circuit dysfunction in the mouse model of fragile X syndrome. J Physiol 593(22):5009–5024. https://doi.org/10.1113/JP271190

Wang X et al (2009) The circuitry of atypical absence seizures in GABABR1a transgenic mice. Pharmacol Biochem Behav 94(1):124–130. https://doi.org/10.1016/j.pbb.2009.07.017

Wierońska JM et al (2015) mGlu5-GABAB interplay in animal models of positive, negative and cognitive symptoms of schizophrenia. Neurochem Int 88:97–109. https://doi.org/10.1016/j.neuint.2015.03.010

Winkler M et al (2019) Pianp deficiency links GABAB receptor signaling and hippocampal and cerebellar neuronal cell composition to autism-like behavior. Mol Psychiatry. https://doi.org/10.1038/s41380-019-0519-9

Woźniak M et al (2016) Involvement of GABA B receptor signaling in antipsychotic-like action of the novel Orthosteric agonist of the mGlu 4 receptor, LSP4-2022. Curr Neuropharmacol 14 (5):413–426. https://doi.org/10.2174/1570159x13666150516000630

Wu Y et al (2007) Transgenic mice over-expressing GABABR1a receptors acquire an atypical absence epilepsy-like phenotype. Neurobiol Dis 26(2):439–451. https://doi.org/10.1016/j.nbd.2007.01.013

Yamasue H, Aran A, Berry-Kravis E (2019) Emerging pharmacological therapies in fragile X syndrome and Autism. Curr Opin Neurol:635–640. https://doi.org/10.1097/WCO.0000000000000703

Yao L et al (2019) Clinical features and long-term outcomes of seizures associated with autoimmune encephalitis: a follow-up study in East China. J Clin Neurosci 68:73–79. https://doi.org/10.1016/j.jocn.2019.07.049

Yhnell E, Dunnett SB, Brooks SP (2016) The utilisation of operant delayed matching and non-matching to position for probing cognitive flexibility and working memory in mouse models of Huntington's disease. J Neurosci Methods 265:72–80. https://doi.org/10.1016/j.jneumeth.2015.08.022

Yu ZF, Hu BR, Cheng GJ (1996) Improvement of learning and memory functions by GABAB receptor antagonists in mice. Yao Xue Xue Bao 31(11):801–805. http://www.ncbi.nlm.nih.gov/pubmed/9863249. Accessed 29 July 2020

Yu ZF, Cheng GJ, Hu BR (1997) Mechanism of colchicine impairment of learning and memory, and protective effect of CGP36742 in mice. Brain Res 750(1–2):53–58. https://doi.org/10.1016/S0006-8993(96)01158-4

Yu T et al (2010) Effects of individual segmental trisomies of human chromosome 21 syntenic regions on hippocampal long-term potentiation and cognitive behaviors in mice. Brain Res 1366:162–171. https://doi.org/10.1016/j.brainres.2010.09.107

Zafarullah M, Tassone F (2019) Molecular biomarkers in fragile x syndrome. Brain Sci 9(5). https://doi.org/10.3390/brainsci9050096

Zapata J et al (2017) Epilepsy and intellectual disability linked protein Shrm4 interaction with GABA B Rs shapes inhibitory neurotransmission. Nat Commun 8:14536. https://doi.org/10.1038/ncomms14536

Zarrindast MR et al (1998) Effects of GABAergic drugs on physostigmine-induced improvement in memory acquisition of passive avoidance learning in mice. Gen Pharmacol 31(1):81–86. https://doi.org/10.1016/S0306-3623(97)00441-2

Zarrindast MR et al (2002) Effects of intrahippocampal injection of GABAergic drugs on memory retention of passive avoidance learning in rats. J Psychopharmacol 16(4):313–319. https://doi.org/10.1177/026988110201600405

Zarrindast MR et al (2004) Influence of β-adrenoceptor agonists and antagonists on baclofen-induced memory impairment in mice. Behav Pharmacol 15(4):293–297. https://doi.org/10.1097/01.fbp.0000137211.95623.07

Zarrindast M-R et al (2008a) Effect of intra-amygdala injection of nicotine and GABA receptor agents on anxiety-like behaviour in rats. Pharmacology 82(4):276–284. https://doi.org/10.1159/000161129

Zarrindast MR, Hoghooghi V, Rezayof A (2008b) Inhibition of morphine-induced amnesia in morphine-sensitized mice: involvement of dorsal hippocampal GABAergic receptors. Neuropharmacology 54(3):569–576. https://doi.org/10.1016/j.neuropharm.2007.11.004

Zeidler S et al (2018) Paradoxical effect of baclofen on social behavior in the fragile X syndrome mouse model. Brain Behav 8(6). https://doi.org/10.1002/brb3.991

Zeman A et al (2016) The GABAB receptor agonist, baclofen, contributes to three distinct varieties of amnesia in the human brain – a detailed case report. Cortex 74:9–19. https://doi.org/10.1016/j.cortex.2015.10.005

Zhang W et al (2015) GABA B receptor upregulates fragile X mental retardation protein expression in neurons. Sci Rep 5. https://doi.org/10.1038/srep10468

Zhao Y et al (2014) Chronic cerebral hypoperfusion causes decrease of O-GlcNAcylation, hyperphosphorylation of tau and behavioral deficits in mice. Front Aging Neurosci 6:10. https://doi.org/10.3389/fnagi.2014.00010

Zhou H et al (2012) GIRK channel modulation by assembly with allosterically regulated RGS proteins. Proc Natl Acad Sci U S A 109(49):19977–19982. https://doi.org/10.1073/pnas.1214337109

Zhu F et al (2020) Clinical characteristics of anti-GABA-B receptor encephalitis. Front Neurol:11. https://doi.org/10.3389/fneur.2020.00403

Zupan B, Toth M (2008) Inactivation of the maternal fragile X gene results in sensitization of GABAB receptor function in the offspring. J Pharmacol Exp Ther 327(3):820–826. https://doi.org/10.1124/jpet.108.143990

The manufacturer's authorised representative in the EU is Springer Nature Customer Service Centre GmbH, Europaplatz 3, 69115 Heidelberg, Germany. If you have any concerns regarding our products, please contact ProductSafety@springernature.com

Printed and bound by CPI Group (UK) Ltd, Croydon, CR0 4YY
28/09/2026
02224090-0017